STEPHEN HAWKING

HIS LIFE AND WORK

The Story and Science of One of the Most Extraordinary, Celebrated and Courageous Figures of Our Time

KITTY FERGUSON

BANTAM BOOKS

LONDON • TORONTO • SYDNEY • AUCKLAND • JOHANNESBURG

TRANSWORLD PUBLISHERS
61–63 Uxbridge Road, London W5 5SA
A Random House Group Company
www.transworldbooks.co.uk

STEPHEN HAWKING
A BANTAM BOOK: 9780857500748

First published in Great Britain
in 2011 by Bantam Press
an imprint of Transworld Publishers
Bantam edition published 2012

Illustrations on pp. 37, 96, 97, 100, 101, 102, 122, 200, 201, 202, 204, 206, 208, 212, 213, 252 by Patrick Mulrey

A CIP catalogue record for this book
is available from the British Library.

Addresses for Random House Group Ltd companies outside the UK
can be found at: www.randomhouse.co.uk
The Random House Group Ltd Reg. No. 954009

The Random House Group Limited supports The Forest Stewardship Council (FSC®), the leading international forest-certification organization. Our books carrying the FSC label are printed on FSC®-certified paper. FSC is the only forest-certification scheme endorsed by the leading environmental organizations, including Greenpeace. Our paper-procurement policy can be found at www.randomhouse.co.uk/environment

Typeset in 11.5/13.5pt Bembo by Falcon Oast Graphic Art Ltd.
Printed and bound by CPI Group (UK) Ltd, Croydon, CR0 4YY.

2 4 6 8 10 9 7 5 3 1

Kitty Fer[illegible]
his f[illegible]
Camb[illegible]
care[illegible] [illegible]nductor and [illegible]nd
lect[illegible] scientists for readers and au[illegible] with
little or no scientific background. Her seven books have appeared to critical acclaim all over the world, in twenty-seven languages. In 2000 she worked with Hawking, helping edit his book *The Universe in a Nutshell*.

Kitty grew up in San Antonio, Texas, moved to New York City at the age of nineteen to study at the Juilliard School of Music, and lived for forty-eight years in New York City and Chester, New Jersey. She and her husband now divide their time between Cambridge and South Carolina. They have three grown children and two grandchildren.

Visit Kitty's website at www.kitty-ferguson.com

Acclaim for *Stephen Hawking; His Life and Work*:

'As both a global icon and an innovative theoretical physicist, Stephen Hawking is well served by science writer Kitty Ferguson's fascinating biography. Ferguson explains in accessible terms the major themes that Hawking has explored in his career, and creates a portrait of the private man by drawing on her close personal contact with him and his family . . . Ferguson's sympathetic and informed take on an individual who has enriched human knowledge against the odds is an excellent summing-up . . . of his unique and creative contribution to both science and humanity' George Ellis, *Nature*

'Proximity gives Ferguson privileged access to Hawking's private world, and her scientific fluency enables her to weave a biography that moves seamlessly between his intellectual and personal development. Ferguson is adept at making his complex theories accessible to non-experts. Beyond explaining his scientific achievements, she brings us closer to the man, [and] ultimately, the most astonishing aspect of Ferguson's Hawking is not his otherness, but the extraordinary things that one determined individual can accomplish once he realizes that "life is worth living"'
The Montreal Gazette

'K[illegible]guson g[illegible] a thorough account of Hawking the man, as well as of his [illegible]traordinary body of work in cosmology. The human elements are beautifully told, and there's the very real feeling not just that Ferguson knows Hawking well and has done for a long time, but that she genuinely cares for him and his family. The overriding sense is that Kitty Ferguson has endeavoured to understand the great scientist as a human being, and has directed her warm but by no means gushing biography at similarly interested lay readers who also harbour an interest in the fundamental questions that have gone some way to being answered by Stephen Hawking' *Entertainment Focus*

'Ferguson replaces the iconic but static image of cosmologist Hawking with flesh and blood in this vivid portrait . . . [She] captures the very full life and work of one of the most vibrant minds of our time' *Publishers Weekly*, starred review

'Ferguson shines at explaining Hawking's theories, the jovially competitive academic world in which they are hammered out and her subject's distinctive and evolving intellectual style . . . An irresistible story. His puckish humor and exuberance for life and for ideas are infectious even at a remove. And the ideas themselves could not ask for a better elucidator, not even, dare I say it, Hawking himself. Giving the rest of us a glimmer of the wonders swirling inside his head is no small feat, and may be the truest portrait of all' *Salon*

'Kitty Ferguson's *Stephen Hawking: His Life and Work* is a really fascinating book, brilliantly written, a lively page-turner. Lovely explanations of the science' Richard Bacon, BBC Radio 5

'Ferguson gives a lucid account of his important work on black holes and stays quite intelligible even when discussing his ideas about the overall shape of space-time' *The Telegraph*

'Ferguson's biography of this remarkable man is based on personal interviews and on sound understanding of the physics behind Hawking's ideas. She presents it all with clarity, and diagrams rather than formulas. This is a delightful book' *Hot Spot*

To my granddaughters Grace and Alice

Stephen Hawking, a line drawing (ink and emulsion) done in 2010 by Cambridge artist Oliver Wallington.

Acknowledgements

I WISH TO THANK STEPHEN HAWKING FOR HIS TIME AND patience in helping me understand his theories, and for putting up with some terribly naïve questions from me.

I am grateful to my agents, Brie Burkeman and Rita Rosenkranz, and my editors, Sally Gaminara at Transworld Publishers and Luba Ostashevsky at Palgrave Macmillan.

I am also grateful to the following for their assistance in many ways, including reading and checking over portions of this book and conversing with me about the subjects discussed in it. Some of the people listed below have had no direct involvement. Some are no longer alive. But the extent to which they have helped me, through the years, understand Stephen Hawking and his work and the cience related to it, means it would be unconscionable not to thank them here.

Sidney Coleman, Judith Croasdell, Paul Davies, Bryce DeWitt, Yale Ferguson, Matthew Fremont, Joan Godwin, Andrei Linde, Sue Masey, Don Page, Malcolm Perry, Brian Pippard, Joanna Sanferrare, Leonard Susskind, Neil Turok, Herman and Tina Vetter, John A. Wheeler and Anna Zytkow.

Nonetheless, any shortcomings in this book are my full responsibility.

Kitty Ferguson

Also by Kitty Ferguson

BLACK HOLES IN SPACETIME

THE FIRE IN THE EQUATIONS: SCIENCE, RELIGION & THE SEARCH FOR GOD

PRISONS OF LIGHT: BLACK HOLES

MEASURING THE UNIVERSE: THE HISTORICAL QUEST TO QUANTIFY SPACE

STEPHEN HAWKING: QUEST FOR A THEORY OF EVERYTHING (*1991 and 2001*)

THE NOBLEMAN AND HIS HOUSEDOG: TYCHO AND KEPLER – THE UNLIKELY PARTNERSHIP THAT FOREVER CHANGED OUR UNDERSTANDING OF THE HEAVENS

THE MUSIC OF PYTHAGORAS (2008)

Contents

PART I
1942–1975

1

'The quest for a Theory of Everything'

1980

IN THE CENTRE OF CAMBRIDGE, ENGLAND, THERE ARE A handful of narrow lanes that seem hardly touched by the twentieth or twenty-first centuries. The houses and buildings represent a mixture of eras, but a step around the corner from the wider thoroughfares into any of these little byways is a step back in time, into a passage leading between old college walls or a village street with a medieval church and churchyard or a malt house. Traffic noises from equally old but busier roads nearby are barely audible. There is near silence, birdsong, voices, footsteps. Scholars and townspeople have walked here for centuries.

When I wrote my first book about Stephen Hawking in 1990, I began the story in one of those little streets,

Free School Lane. It runs off Bene't Street, beside the church of St Bene't's with its eleventh-century bell tower. Around the corner, in the lane, flowers and branches still droop through the iron palings of the churchyard, as they did twenty years ago. Bicycles tethered there belie the antique feel of the place, but a little way along on the right is a wall of black, rough stones with narrow slit windows belonging to the fourteenth-century Old Court of Corpus Christi College, the oldest court in Cambridge. Turn your back to that wall and you will see, high up beside a gothic-style gateway, a plaque that reads, THE CAVENDISH LABORATORY. This gateway and the passage beyond are a portal to a more recent era, oddly tucked away in the medieval street.

There is no hint of the friary that stood on this site in the twelfth century or the gardens that were later planted on its ruins. Instead, bleak, factory-like buildings, almost oppressive enough to be a prison, tower over grey asphalt pavement. The situation improves further into the complex, and in the two decades since I first wrote about it some newer buildings have gone up, but the glass walls of these well-designed modern structures are still condemned to reflect little besides the grimness of their older neighbours.

For a century, until the University of Cambridge built the 'New' Cavendish Labs in 1974, this complex housed one of the most important centres of physics research in the world. Here, 'J. J.' Thomson discovered the electron, Ernest Rutherford probed the structure of the atom – and the list goes on and on. When I attended lectures here in the 1990s (for not everything moved to the New Cavendish in 1974), enormous chalk-boards were still in use, hauled noisily up and down with crank-driven chain-pulley systems to make room for the endless strings of equations in a physics lecture.

The Cockcroft Lecture Room, part of this same site, is a much more up-to-date lecture room. Here, on 29 April 1980, scientists, guests and university dignitaries gathered in steep tiers of seats, facing a two-storey wall of chalk-board and slide screen – still well before the advent of PowerPoint. The occasion was the inaugural lecture of a new Lucasian Professor of Mathematics, 38-year-old mathematician and physicist Stephen William Hawking. He had been named to this illustrious chair the previous autumn.

The title announced for his lecture was a question: 'Is the End in Sight for Theoretical Physics?' Hawking startled his listeners by announcing that he thought it was. He invited them to join him in a sensational escape through time and space on a quest to find the Holy Grail of science: the theory that explains the universe and everything that happens in it – what some were calling the Theory of Everything.

Watching Stephen Hawking, silent in a wheelchair while one of his students read his lecture for the audience, no one unacquainted with him would have thought he was a promising choice to lead such an adventure. Theoretical physics was for him the great escape from a prison more grim than any suggested by the Old Cavendish Labs. Beginning when he was a graduate student in his early twenties, he had lived with encroach-ing disability and the promise of an early death. Hawking has amyotrophic lateral sclerosis, known in America as Lou Gehrig's disease after the New York Yankees' first baseman, who died of it.* The progress of the disease in Hawking's case had been slow, but by the time he became Lucasian Professor he could no longer walk, write, feed himself, or raise his head if it tipped forward. His speech was slurred and almost unintelligible except to those who knew him best. For the Lucasian lecture, he had painstakingly dictated his text earlier, so that it could be read by the student. But Hawking certainly was and is no invalid. He is an active mathematician and physicist, whom some were even then calling the most brilliant since Einstein. The Lucasian Professorship is an extremely prestigious position in the University of Cambridge, dating from 1663. The second holder of the chair was Sir Isaac Newton.

It was typical of Hawking's iconoclasm to begin this distinguished professorship by predicting the end of his own field. He said he thought there was a good chance the so-called Theory of Everything would be found before the close of the twentieth century, leaving little for theoretical physicists like himself to do.

Since that lecture, many people have come to think of Stephen Hawking as the standard-bearer of the quest for that theory. However, the candidate he named for Theory of Everything was not one of his own theories but 'N=8 supergravity', a theory which many physicists at that time hoped might unify all the particles and the forces of nature. Hawking is quick to point out that his work is only one part of a much larger picture, involving physicists all over the world, and also part of a very old

* There has been recent evidence that Gehrig may not have had amyotrophic lateral sclerosis, but another disease similar to it.

quest. The longing to understand the universe must surely be as ancient as human consciousness. Ever since human beings first began to look at the night skies as well as at the enormous variety of nature around them, and considered their own existence, they've been trying to explain all this with myths, religion, and, later, mathematics and science. We may not be much nearer to understanding the complete picture than our remotest ancestors, but most of us like to think, as does Stephen Hawking, that we are.

Hawking's life story and his science continue to be full of paradoxes. Things are often not what they seem. Pieces that should fit together refuse to do so. Beginnings may be endings; cruel circumstances can lead to happiness, although fame and success may not; two brilliant and highly successful scientific theories taken together yield nonsense; empty space isn't empty; black holes aren't black; the effort to unite everything in a simple explanation reveals, instead, a fragmented picture; and a man whose appearance inspires shock and pity takes us joyfully to where the boundaries of time and space ought to be – but are not.

Anywhere we look in our universe, we find that reality is astoundingly complex and elusive, sometimes alien, not always easy to take, and often impossible to predict. Beyond our universe there may be an infinite number of others. The close of the twentieth century has come and gone, and no one has discovered the Theory of Everything. Where does that leave Stephen Hawking's prediction? Can any scientific theory truly explain it *all*?

2

'Our goal is nothing less than a complete description of the universe we live in'

THE IDEA THAT ALL THE AMAZING INTRICACY AND variety we experience in the world and the cosmos may come down to something remarkably simple is not new or far-fetched. The sage Pythagoras and his followers in southern Italy in the sixth century BC studied the relationships between lengths of strings on a lyre and the musical pitches these produced, and realized that hidden behind the confusion and complexity of nature there is pattern, order, rationality. In the two and a half millennia since, our forebears have continued to find – often, like the Pythagoreans, to their surprise and awe – that nature is less complicated than it first appears.

Imagine, if you can, that you are a super-intelligent alien who has absolutely no experience of our universe: is there a set of rules so complete that by studying them you could discover exactly what our universe is like? Suppose someone handed you that rule book. Could it possibly be a *short* book?

For decades, many physicists believed that the rule book is not lengthy and contains a set of fairly simple principles, perhaps even just one principle that lies behind everything that has happened, is happening, and ever will happen in our universe. In 1980, Stephen Hawking made the brash claim that we would hold the rule book in our hands by the end of the twentieth century.

My family used to own a museum facsimile of an ancient board game. Archaeologists digging in the ruins of the city of Ur in Mesopotamia had unearthed an exquisite inlaid board with a few small carved pieces. It was obviously an elaborate game, but no one knew its rules. The makers of the facsimile had tried to deduce them from the design of the board and pieces, but those like ourselves who bought the game were encouraged to make our own decisions and discoveries about how to play it.

You can think of the universe as something like that: a magnificent, elegant, mysterious game. Certainly there are rules, but the rule book didn't come with the game. The universe is no beautiful relic like the game found at Ur. Yes, it is old, but the game continues. We and everything we know about (and much we do not) are in the thick of the play. If there is a Theory of Everything, we and everything in the universe must be obeying its principles, even while we try to discover what they are.

You would expect the complete, unabridged rules for the universe to fill a vast library or supercomputer. There would be rules for how galaxies form and move, for how human bodies work and fail to work, for how humans relate to one another, for how subatomic particles interact, how water freezes, how plants grow, how dogs bark – intricate rules within rules within rules. How could anyone think this could be reduced to a few principles?

Richard Feynman, the American physicist and Nobel laureate, gave an excellent example of the way the reduction process happens. There was a time, he pointed out, when we had something we called motion and something else called heat and something else again called sound. 'But it was soon discovered,' wrote Feynman:

> after Sir Isaac Newton explained the laws of motion, that some of these apparently different things were aspects of the same thing. For example, the phenomena of sound could be completely understood as the motion of atoms in the air. So sound was no longer considered something in addition to motion. It was also discovered that heat phenomena are easily understandable from the laws of motion. In this way, great globs of physics theory were synthesized into a simplified theory.'[1]

Life among the Small Pieces

All matter as we normally think of it in the universe – you and I, air, ice, stars, gases, microbes, this book – is made up of minuscule building blocks called atoms. Atoms in turn are made up of smaller objects, called particles, and a lot of empty space.

The most familiar matter particles are the electrons that orbit the nuclei of atoms and the protons and neutrons that are clustered in the nuclei. Protons and neutrons are made up of even tinier particles of matter called 'quarks'. All matter particles belong to a class of particles called 'fermions', named for the great Italian physicist Enrico

Fermi. They have a system of messages that pass among them, causing them to act and change in various ways. A group of humans might have a message system consisting of four different services: telephone, fax, e-mail and 'snail mail'. Not all the humans would send and receive messages and influence one another by means of all four message services. You can think of the message system among the fermions as four such message services, called forces. There is another class of particles that carry these messages among the fermions, and sometimes among themselves as well: 'messenger' particles, more properly called 'bosons'. Apparently every particle in the universe is either a fermion or a boson.

One of the four fundamental forces of nature is gravity. One way of thinking about the gravitational force holding us to the Earth is as 'messages' carried by bosons called gravitons between the particles of the atoms in your body and the particles of the atoms in the Earth, influencing these particles to draw closer to one another. Gravity is the weakest of the forces, but, as we'll see later, it is a very long-range force and acts on everything in the universe. When it adds up, it can dominate all the other forces.

A second force, the electromagnetic force, is messages carried by bosons called photons among the protons in the nucleus of an atom, between the protons and the electrons nearby, and among electrons. The electromagnetic force causes electrons to orbit the nucleus. On the level of everyday experience, photons show up as light, heat, radio waves, microwaves and other waves, all known as electromagnetic radiation. The electromagnetic force is also long-range and much stronger than gravity, but it acts only on particles with an electric charge.

A third message service, the strong nuclear force, causes the nucleus of the atom to hold together.

A fourth, the weak nuclear force, causes radioactivity and plays a necessary role, in stars and in the early universe, in the formation of the elements.

The gravitational force, the electromagnetic force, the strong nuclear force, and the weak nuclear force . . . the activities of those four forces are responsible for all messages among all fermions in the universe and for all interactions among them. Without the four forces, every fermion (every particle of matter) would exist, if it existed at all, in isolation, with no means of contacting or influencing any other, oblivious to every other. To put it bluntly, whatever *doesn't* happen by means of one of the four forces doesn't happen. If that is true, a complete understanding of the forces would give us an understanding of the principles underlying everything that happens in the universe. Already we have a remarkably condensed rule book.

Much of the work of physicists in the twentieth century was aimed at learning more about how the four forces of nature operate and how they are related. In our human message system, we might discover that telephone, fax and e-mail are not really so separate after all, but can be thought of as the same thing showing up in three different ways. That discovery would 'unify' the three message services. In a similar way, physicists have sought, with some success, to unify the forces. They hope ultimately to find a theory which explains all four forces as one showing up in different ways – a theory that may even unite both fermions and bosons in a single family. They speak of such a theory as a unified theory.

A theory explaining the universe, the Theory of Everything, must go several steps further. Of particular interest to Stephen Hawking, it must answer the question, what was the universe like at the instant of beginning, before any time whatsoever had passed? Physicists phrase that question: what are the 'initial conditions' or the 'boundary conditions at the beginning of the universe'? Because this issue of boundary conditions has been and continues to be at the heart of Hawking's work, it behooves us to spend a little time with it.

The Boundary Challenge

Suppose you put together a layout for a model railway, then position several trains on the tracks and set the switches and throttles controlling the train speeds as you want them, all before turning on the power. You have set up boundary conditions. For this session with your train set, reality is going to begin with things in precisely this state and not in any other. Where each train will be five minutes after you turn on the power, whether any train will crash with another, depends heavily on these boundary conditions.

Imagine that when you have allowed the trains to run for ten minutes, without any interference, a friend enters the room. You switch off the power. Now you have a second set of boundary conditions: the precise position of everything in the layout at the second you switched it off. Suppose you challenge your friend to try to work out exactly where all the trains started out ten minutes earlier. There would be a host of questions besides the simple matter of where the trains are standing and how the throttles and switches are set. How quickly does each of the trains accelerate and slow down? Do certain parts of the tracks offer more resistance than others? How steep are the gradients? Is the power supply constant? Is it certain there has been nothing to interfere with the running of the train set – something no longer evident? The whole exercise would indeed be daunting. Your friend would be in something like the position of a modern physicist trying to work out how the universe began – what were the boundary conditions at the beginning of time.

Boundary conditions in science do not apply only to the history of the universe. They simply mean the lie of the land at a particular point in time, for instance the start of an experiment in a laboratory. However, unlike the situation with the train set or a lab experiment, when considering the universe, one is often not allowed to *set up* boundary conditions. One of Hawking's favourite questions is how many ways the universe could have begun and still ended up the way we observe it today, assuming that we have correct knowledge and understanding of the laws of physics and they have not changed. He is using 'the way we observe the universe today' as a boundary condition and also, in a more subtle sense, using the laws of physics and the assumption that they have not changed as boundary conditions. The answer he is after is the reply to the question, what were the boundary conditions at the beginning of the universe, or the 'initial conditions of the universe' – the exact layout at the word go, including the minimal laws that had to be in place at that moment in order to produce at a certain time in the future the universe as we know it today? It is in considering this question that he has produced some of his most interesting work and surprising answers.

A unified description of the particles and forces, and knowledge of the boundary conditions for the origin of the universe, would be a stupendous scientific achievement, but it would not be a Theory of Everything. In addition, such a theory must account for values that are 'arbitrary elements' in all present theories.

Language Lesson

Arbitrary elements include such 'constants of nature' as the mass and charge of the electron and the velocity of light. We observe what these are, but no theory explains or predicts them. Another example: physicists know the strength of the electromagnetic force and the weak nuclear force. The electroweak theory is a theory that unifies the two, but it cannot tell us how to calculate the difference in strength between the two forces. The difference in strength is an 'arbitrary element', not predicted by the theory. We know what it is from observation, and so we put it into a theory 'by hand'. This is considered a weakness in a theory.

When scientists use the word *predict*, they do not mean telling the future. The question 'Does this theory predict the speed of light?' isn't asking whether the theory tells us what that speed will be next Tuesday. It means, would this theory make it possible for us to work out the speed of light if it were impossible to observe what that speed is? As it happens, no present theory does predict the speed of light. It is an arbitrary element in all theories.

One of Hawking's concerns when he wrote *A Brief History of Time* was that there be a clear understanding of what is meant by a *theory*. A theory is not Truth with a capital *T*, not a rule, not fact, not the final word. You might think of a theory as a toy boat. To find out whether it floats, you set it on the water. You test it. When it flounders, you pull it out of the water and make some changes, or you start again and build a different boat, benefiting from what you've learned from the failure.

Some theories are good boats. They float a long time. We may know there are a few leaks, but for all practical purposes they serve us well. Some serve us so well, and are so solidly supported by experiment and testing, that we begin to regard them as truth. Scientists, keeping in mind how complex and surprising our universe is, are extremely wary about calling them that. Although some theories do have a lot of experimental success to back them up and others are hardly more than a glimmer in a theorist's eyes – brilliantly designed boats that have never been tried on the water – it is risky to assume that *any* of them is absolute, fundamental scientific 'truth'.

It is important, however, not to dither around for ever, continuing to call into question well-established theories without having a good reason for doing so. For science to move ahead, it is necessary to decide whether some theories are dependable enough, and match observation sufficiently well, to allow us to use them as building blocks and proceed from there. Of course, some new thought or discovery might come along and threaten to sink the boat. We'll see an example of that later in this book.

In *A Brief History of Time* Stephen Hawking wrote that a scientific theory is 'just a model of the universe, or a restricted part of it, and a set of rules that relate quantities in the model to observations that we make. It exists only in our minds and does not have any other reality (whatever that may mean).'[2] The easiest way to understand this definition is to look at some examples.

There is a film clip showing Hawking teaching a class of graduate students, probably in the early 1980s, with the help of his graduate assistant. By this time Hawking's ability to speak had deteriorated so seriously that it was impossible for anyone who did not know him well to understand him. In the clip, his graduate assistant interprets Hawking's garbled speech to say, 'Now it just so happens that we have a model of the universe here', and places a large cardboard cylinder upright on the seminar table. Hawking frowns and mutters something that only the assistant can understand. The assistant apologetically picks up the cylinder and turns it over to stand on its other end. Hawking nods approval, to general laughter.

A 'model', of course, does not have to be something like a cardboard cylinder or a drawing that we can see and touch. It can be a mental picture or even a story. Mathematical equations or creation myths can be models.

Getting back to the cardboard cylinder, how does it resemble the universe? To make a full-fledged theory out of it, Hawking would have to explain how the model is related to what we actually see around us, to 'observations', or to what we might observe if we had better technology. However, just because someone sets a piece of cardboard on the table and tells how it is related to the actual universe does not mean anyone should accept this as *the* model of the universe. We are to consider it, not swallow it hook, line and sinker. It is an idea, existing 'only in our minds'. The cardboard cylinder may turn out to be a useful model. On the other hand, some evidence may turn up to prove that it is not. We shall have found that we are part of a slightly different game from the one the model suggested we were playing. Would that mean the theory was 'bad'? No, it may have been a very good theory, and everyone may have learned a great deal from considering it, testing it, and having to change it or discard it. The effort to shoot it down may have required innovative thinking and experiments that will lead to something more successful or pay off in other ways.

What is it then that makes a theory a good theory? Quoting Hawking again, it must 'accurately describe a large class of observations on the basis of a model that contains only a few arbitrary elements, and it must make definite predictions about the results of future observations'.[3]

For example, Isaac Newton's theory of gravity describes a very large class of observations. It predicts the behaviour of objects dropped or thrown on Earth, as well as planetary orbits.

It's important to remember, however, that a good theory does not have to arise entirely from observation. A good theory can be a wild theory, a great leap of imagination. 'The ability to make these intuitive leaps is really what characterizes a good theoretical physicist,' says Hawking.[4] However, a good theory should not be at odds with things already observed, unless it gives convincing reasons for seeming to be at odds. Superstring theory, one of the most exciting current theories, predicts more than three dimensions of space, a prediction that certainly seems inconsistent with observation. Theorists explain the discrepancy by suggesting the extra dimensions are curled up so small we are unable to recognize them.

We've already seen what Hawking means by his second

requirement, that a theory contain only a few arbitrary elements.

The final requirement, according to Hawking, is that it must suggest what to expect from future observations. It must challenge us to test it. It must tell us what we will observe if the theory is correct. It should also tell us what observations would prove that it is *not* correct. For example, Albert Einstein's theory of general relativity predicts that beams of light from distant stars bend a certain amount as they pass massive bodies like the sun. This prediction is testable. Tests have shown Einstein was correct.

Some theories, including most of Stephen Hawking's, are impossible to test with our present technology, perhaps even with any conceivable future technology. They *are* tested with mathematics. They must be mathematically consistent with what we do know and observe. But we cannot observe the universe in its earliest stages to find out directly whether his 'no-boundary proposal' (to be discussed later) is correct. Although some tests were proposed for proving or disproving 'wormholes', Hawking does not think they would succeed. But he has told us what he thinks we will find if we ever do have the technology, and he is convinced that his theories are consistent with what we have observed so far. In some cases he has risked making some very specific predictions about the results of experiments and observations that push at the boundaries of our present capabilities.

If nature is perfectly unified, then the boundary conditions at the beginning of the universe, the most fundamental particles and the forces that govern them, and the constants of nature, are interrelated in a unique and completely compatible way, which we might be able to recognize as inevitable, absolute and self-explanatory. To reach that level of understanding would indeed be to discover the Theory of Everything . . . of Absolutely Everything . . . even the answer, perhaps, to the question of *why* does the universe fit this description . . . to 'know the Mind of God', as Hawking termed it in *A Brief History of Time*, or 'the Grand Design', as he would phrase it less dramatically in a more recent book by that name.

Laying Down the Gauntlet

We are ready to list the challenges that faced any 'Theory of Everything' candidate when Hawking delivered his Lucasian Lecture in 1980. You'll learn in due course how some requirements in this list have changed subtly since then.

- It must give us a model that unifies the forces and particles.
- It must answer the question, what were the 'boundary conditions' of the universe, the conditions at the very instant of beginning, before any time whatsoever passed?
- It must be 'restrictive', allowing few options. It should, for instance, predict precisely how many types of particles there are. If it leaves options, it must somehow account for the fact that we have the universe we have and not a slightly different one.
- It should contain few arbitrary elements. We would rather not have to peek too often at the actual universe for answers. Paradoxically, the Theory of Everything itself may be an arbitrary element. Few scientists expect it to explain why there should exist either a theory or anything

at all for it to describe. It is not likely to answer Stephen Hawking's question: 'Why does the universe [or, for that matter, the Theory of Everything] go to all the bother of existing?'[5]

- It must predict a universe like the universe we observe or else explain convincingly why there are discrepancies. If it predicts that the speed of light is ten miles per hour, or disallows penguins or pulsars, we have a problem. A Theory of Everything must find a way to survive comparison with what we observe.
- It should be simple, although it must allow for enormous complexity. The physicist John Archibald Wheeler of Princeton wrote:

  > Behind it all
  > is surely an idea so simple,
  > so beautiful,
  > so compelling that when –
  > in a decade, a century,
  > or a millennium –
  > we grasp it,
  > we will all say to each other,
  > how could it have been otherwise?
  > How could we have been so stupid
  > for so long?[6]

  The most profound theories, such as Newton's theory of gravity and Einstein's relativity theories, are simple in the way Wheeler described.
- It must solve the enigma of combining Einstein's theory of general relativity (a theory that explains gravity) with quantum mechanics (the theory we use successfully when talking about the other three forces). This is a challenge that Stephen Hawking has taken up. We introduce the

problem here. You will understand it better after reading about the uncertainty principle of quantum mechanics in this chapter and about general relativity later.

Theory Meets Theory

Einstein's theory of general relativity is the theory of the large and the very large – stars, planets, galaxies, for instance. It does an excellent job of explaining how gravity works on that level.

Quantum mechanics is the theory of the very small. It describes the forces of nature as messages among fermions (matter particles). Quantum mechanics also contains something extremely frustrating, the uncertainty principle: we can never know precisely both the *position* of a particle and its *momentum* (how it is moving) at the same time. In spite of this problem, quantum mechanics does an excellent job of explaining things on the level of the very small.

One way to combine these two great twentieth-century theories into one unified theory would be to explain gravity, more successfully than has been possible so far, as an exchange of messenger particles, as we do with the other three forces. Another avenue is to rethink general relativity in the light of the uncertainty principle.

Explaining gravity as an exchange of messenger particles presents problems. When you think of the force holding you to the Earth as the exchange of gravitons (messenger particles of gravity) between the matter particles in your body and the matter particles that make up the Earth, you are describing the gravitational force in a quantum-mechanical way. But because all these gravitons are also exchanging gravitons among themselves, mathematically this is a messy business. We get infinities, mathematical nonsense.

Physical theories cannot really handle infinities. When they have appeared in other theories, theorists have resorted to something known as 'renormalization'. Richard Feynman used renormalization when he developed a theory to explain the electromagnetic force, but he was far from pleased about it. 'No matter how clever the word,' he wrote, 'it is what I would call a dippy process!'[7] It involves putting in other infinities and letting the infinities cancel each other out. It does sound dubious, but in many cases it seems to work in practice. The resulting theories agree with observation remarkably well.

Renormalization works in the case of electromagnetism, but it fails in the case of gravity. The infinities in the gravitational force are of a much nastier breed than those in the electromagnetic force. They refuse to go away. Supergravity, the theory Hawking spoke about in his Lucasian lecture, and superstring theory, in which the basic objects in the universe are not pointlike particles but tiny strings or loops of string, began to make promising inroads in the twentieth century; and later in this book we shall be looking at even more promising recent developments. But the problem is not completely solved.

On the other hand, what if we allow quantum mechanics to invade the study of the very large, the realm where gravity seems to reign supreme? What happens when we rethink what general relativity tells us about gravity in the light of what we know about the uncertainty principle, the principle that you can't measure accurately the position and the momentum of a particle at the same time? Hawking's work along these lines has had bizarre results: black holes aren't black, and the boundary conditions may be that there are no boundaries.

While we are listing paradoxes, here's another: empty space isn't empty. Later in this book we'll discuss how we arrive at that conclusion. For now be content to know that the uncertainty principle means that so-called empty space teems with particles and antiparticles. (The matter–antimatter used in science fiction is a familiar example.)

General relativity tells us that the presence of matter or energy makes spacetime curve, or warp. We've already mentioned one result of that curvature: the bending of light beams from distant stars as they pass a massive body like the sun.

Keep those two points in mind: (1) 'Empty' space is filled with particles and antiparticles, adding up to an enormous amount of energy. (2) The presence of this energy causes curvature of spacetime.

If both are true the entire universe ought to be curled up into a small ball. This hasn't happened. When general relativity and quantum mechanics work together, what they predict seems to be dead wrong. Both general relativity and quantum mechanics are exceptionally good theories, two of the outstanding intellectual achievements of the twentieth century. They serve us magnificently not only for theoretical purposes but in many practical ways. Nevertheless, put together they yield infinities and nonsense. The Theory of Everything must somehow resolve that nonsense.

Predicting the Details

Once again imagine that you are an alien who has never seen our universe. With the Theory of Everything you ought nevertheless to be able to predict everything about it . . . right? It's possible you can predict suns and planets and galaxies and black holes and quasars – but can you predict next year's Derby winner? How specific can you be? Not very.

The calculations necessary to study all the data in the universe are ludicrously far beyond the capacity of any imaginable computer. Hawking points out that although we can solve the equations for the movement of two bodies in Newton's theory of gravity, we can't solve them exactly for *three* bodies, not because Newton's theory doesn't work for three bodies but because the maths is too complicated. The real universe, needless to say, has more than three bodies in it.

Nor can we predict our health, although we understand the principles that underlie medicine, the principles of chemistry and biology, extremely well. The problem again is that there are too many billions upon billions of details in a real-life system, even when that system is just one human body.

With the Theory of Everything in our hands we'd still be a staggeringly long way from predicting everything. Even if the underlying principles are simple and well understood, the way they work out is enormously complicated. 'A minute to learn, the lifetime of the universe to master', to paraphrase an advertising slogan. 'Lifetime of the universe to master' is a gross understatement.*

Where does that leave us? What horse will win the Grand National next year *is* predictable with the Theory

* The advertising slogan for the game Othello is 'A minute to learn, a lifetime to master'.

of Everything, but no computer can hold all the data or do the maths to make the prediction. Is that correct?

There's a further problem. We must look again at the uncertainty principle of quantum mechanics.

The Fuzziness of the Very Small

At the level of the very small, the quantum level of the universe, the uncertainty principle also limits our ability to predict.

Think of all those odd, busy inhabitants of the quantum world, both fermions and bosons. They're an impressive zoo of particles. Among the fermions there are electrons, protons and neutrons. Each proton or neutron is, in turn, made up of three quarks, which are also fermions. Then we have the bosons: photons (messengers of the electromagnetic force), gravitons (the gravitational force), gluons (the strong force), and Ws and Zs (the weak force). It would be helpful to know where all these and many others are, where they are going, and how quickly they are getting there. Is it possible to find out?

The diagram of an atom in Figure 2.1 is the model proposed by Ernest Rutherford at the Cavendish Labs in Cambridge early in the twentieth century. It shows electrons orbiting the nucleus of the atom as planets orbit the sun. We now know that things never really look like this on the quantum level. The orbits of electrons cannot be plotted as though electrons were planets. We do better to picture them swarming in a cloud around the nucleus. Why the blur?

The uncertainty principle makes life at the quantum level a fuzzy, imprecise affair, not only for electrons but for all the particles. Regardless of how we go about trying

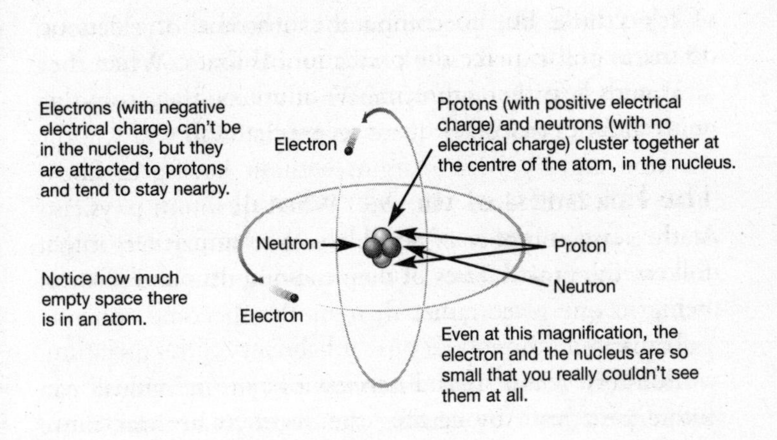

Figure 2.1. In the Rutherford model of a helium atom, the electrons orbit the nucleus the way planets orbit the sun. We now know that, because of the uncertainty principle of quantum mechanics, electron orbits are not really well-defined paths as shown in this model.

to observe what happens, it is impossible to find out precisely both the *momentum* and the *position* of a particle at the same time. The more accurately we measure how the particle is moving, the less accurately we know its position, and vice versa. It works like a seesaw: when the accuracy of one measurement goes up, the accuracy of the other must go down. We pin down one measurement only by allowing the other to become more uncertain.

The best way to describe the activity of a particle is to study all the possible ways it might be moving and then calculate how likely one way is as opposed to another. It becomes a matter of probabilities. A particle has *this* probability to be moving – *that* way – or it has *that* probability to be – *here*. Those probabilities are nevertheless very useful information.

It's a little like predicting the outcome of elections. Election poll experts work with probabilities. When they deal with large enough numbers of voters, they come up with statistics that allow them to predict who will win the election and by what margin, without having to know how each individual will vote. When quantum physicists study a large number of possible paths that particles might follow, the probabilities of their moving thus and so or of being in one place rather than another become concrete information.

Pollsters admit that interviewing an individual can influence a vote by causing the voter to become more aware of issues. Physicists have a similar dilemma. Probing the quantum level influences the answers they find.

Thus far the comparison between predicting elections and studying the quantum level seems a good one. Now it breaks down: on election day, each voter does cast a definite vote one way or another, secret perhaps but not uncertain. If pollsters placed hidden cameras in voting booths – and were not arrested – they could find out how each individual voted. It is not like that in quantum physics. Physicists have devised ingenious ways of sneaking up on particles, all to no avail. The world of elementary particles does not just *seem* uncertain because we haven't been clever enough to find a successful way to observe it. It *really* is uncertain. No wonder Hawking, in his Lucasian lecture, called quantum mechanics 'a theory of what we do not know and cannot predict'.[8]

Taking this limitation into account, physicists have redefined the goal of science: the Theory of Everything will be a set of laws that make it possible to predict events *up to the limit set by the uncertainty principle*, and that means in many cases satisfying ourselves with statistical probabilities, not specifics.

Hawking sums up our problem. In answer to the question of whether everything is predetermined either by the Theory of Everything or by God, he says yes, he thinks it is. 'But it might as well not be, because we can never know what is determined. If the theory has determined that we shall die by hanging, then we shall not drown. But you would have to be awfully sure that you were destined for the gallows to put to sea in a small boat during a storm.'[9] He regards the idea of free will as 'a very good approximate theory of human behaviour'.[10]

Is There Really a Theory of Everything?

Not all physicists believe there is a Theory of Everything, or, if there is, that it is possible for anyone to find it. Science may go on refining what we know by making discovery after discovery, opening boxes within boxes but never arriving at the ultimate box. Others argue that events are not entirely predictable but happen in a random fashion. Some believe God and human beings have far more freedom of give-and-take within this creation than a deterministic Theory of Everything would allow. They believe that as in the performance of a great piece of orchestral music, though the notes are written down, there may yet be enormous creativity in the playing of the notes that is not at all predetermined.

Whether a complete theory to explain the universe is within our reach or ever will be, there are those among us who want to make a try. Humans are intrepid beings with insatiable curiosity. Some, like Stephen Hawking, are particularly hard to discourage. One spokesman for those who are engaged in this science, Murray Gell-Mann, described the quest:

> It is the most persistent and greatest adventure in human history, this search to understand the universe, how it works and where it came from. It is difficult to imagine that a handful of residents of a small planet circling an insignificant star in a small galaxy have as their aim a complete understanding of the entire universe, a small speck of creation truly believing it is capable of comprehending the whole.[11]

3

'Equal to anything!'

WHEN STEPHEN HAWKING WAS TWELVE YEARS OLD, TWO of his schoolmates made a bet about his future. John McClenahan bet that Stephen 'would never come to anything'; Basil King, that he would 'turn out to be unusually capable'.[1] The stake was a bag of sweets.

Young S. W. Hawking was no prodigy. Some reports claim he was brilliant in a haphazard way, but Hawking remembers that he was just another ordinary English schoolboy, slow learning to read, his handwriting the despair of his teachers. He ranked no more than halfway up in his school class, though he now says, in his defence, 'It was a very bright class.'[2] Maybe someone might have predicted a career in science or engineering from the fact that Stephen was intensely interested in learning the secrets of how things such as clocks and radios work. He took them apart to find out, but he could seldom put them back together. Stephen was never well-coordinated physically, and he was not keen on sports or other physical activities. He was almost always the last to be chosen for any sports team. John McClenahan had good reason to think he would win the wager.

Basil King probably was just being a loyal friend or liked betting on long shots. Maybe he did see things about Stephen that teachers, parents and Stephen himself couldn't see. He hasn't claimed his bag of sweets, but it's time he did. Because Stephen Hawking, after such an unexceptional beginning, is now one of the intellectual giants of our modern world – and among its most heroic figures. How such transformations happen is a mystery that biographical details alone cannot explain. Hawking would have it that he is still 'just a child who has never grown up. I still keep asking these how and why questions. Occasionally I find an answer.'[3]

1942–1959

Stephen William Hawking was born during the Second World War, on 8 January 1942, in Oxford. It was a winter of discouragement and fear, not a happy time to be born. Hawking likes to recall that his birth was exactly three hundred years after the death of Galileo, who is called the father of modern science. But few people in January 1942 were thinking about Galileo.

Stephen's parents, Frank and Isobel Hawking, were not wealthy. Frank's very prosperous Yorkshire grandfather had over-extended himself buying farm land and then gone bankrupt in the great agricultural depression of the early twentieth century. His resilient wife, Frank's grandmother and Stephen's great-grandmother, saved the family from complete ruin by opening a school in their home.

Her ability and willingness to take this unusual step are evidence that reading and education must already have been a high priority in the family.

Isobel, Stephen's mother, was the second oldest of seven children. Her father was a family doctor in Glasgow. When Isobel was twelve, they moved to Devon.

It wasn't easy for either family to scrape together money to send a child to Oxford, but in both cases they did. Taking on a financial burden of this magnitude was especially unusual in the case of Isobel's parents, for few women went to university in the 1930s. Though Oxford had been admitting female students since 1878, it was only in 1920 that the university had begun granting degrees to women. Isobel's studies ranged over an unusually wide curriculum in a university where students tended to be much more specialized than in an American liberal arts college or university. She studied philosophy, politics and economics.[4]

Stephen's father Frank was a meticulous, determined young man who kept a journal every day from the age of fourteen and would continue it until the end of his life.[5] He was at Oxford earlier than Isobel, studying medical science with a speciality in tropical medicine. When the Second World War broke out he was in East Africa doing field research, and he intrepidly found his way overland to take ship for England and volunteer for military service. He was assigned instead to medical research.

Isobel held several jobs after graduation from Oxford, all of them beneath her ability and credentials as a university graduate. One was as an inspector of taxes. She so loathed that that she gave it up in disgust to become a secretary at a medical institute in Hampstead. There she met Frank Hawking. They were married in the early years of the war.

In January 1942 the Hawkings were living in Highgate, north London. In the London area hardly a night passed without air raids, and Frank and Isobel Hawking decided Isobel should go to Oxford to give birth to their baby in safety. Germany was not bombing Oxford or Cambridge, the two great English university towns, reputedly in return for a British promise not to bomb Heidelberg and Göttingen. In Oxford, the city familiar from her recent university days, Isobel spent the final week of her pregnancy first in a hotel and then, as the birth grew imminent and the hotel grew nervous, in hospital, but she was still able to go out for walks to fill her time. On one of those leisurely winter days, she happened into a bookshop and, with a book token, bought an astronomical atlas. She would later regard this as a rather prophetic purchase.[6]

Not long after Stephen's birth on 8 January his parents took him back to Highgate. Their home survived the war, although a V-2 rocket hit a few doors away when the Hawkings were absent, blowing out the back windows of their house and leaving glass shards sticking out of the opposite wall like little daggers.[7] It had been a good moment to be somewhere else.

After the war the family lived in Highgate until 1950. Stephen's sister Mary was born there in 1943 (when Stephen was less than two years old), and a second daughter, Philippa, arrived in 1946. The family would adopt another son, Edward, in 1955, when Stephen was a teenager. In Highgate Stephen attended the Byron House School, whose 'progressive methods' he would later blame for his not learning to read until after he left there.

When Dr Frank Hawking, beginning to be recognized as a brilliant leader in his field, became head of the Division of Parasitology at the National Institute for Medical Research, the family moved to St Albans.

Eccentric in St Albans

The Hawkings were a close family. Their home was full of good books and good music, often reverberating with the operas of Richard Wagner played at high volume on the record player. Frank and Isobel Hawking believed strongly in the value of education, a good bit of it occurring at home. Frank gave his children a grounding in, among other things, astronomy and surveying, and Isobel took them often to the museums in South Kensington, where each child had a favourite museum and none had the slightest interest in the others' favourites. She would leave Stephen in the Science Museum and Mary in the Natural History Museum, and then stay with Philippa – too young to be left alone – at the Victoria and Albert. After a while she would collect them all again.[8]

In St Albans the Hawkings were regarded as a highly intelligent, eccentric family. Their love of books extended to such compulsive reading habits that Stephen's friends found it odd and a little rude of his family to sit at the dining table, uncommunicative, their noses buried in their books. Reports that the family car was a used hearse are false. For many years the Hawkings drove around in a succession of used London taxis of the black, boxlike sort. This set them apart not only because of the nature of the vehicle, but also because after the war cars of any kind were not easily available. Only families who were fairly wealthy had them at all. Frank Hawking installed a table in the back of the taxi, between the usual bench seat and the fold-down seats, so that Stephen and his siblings could play cards and games. The car and the game table were put to especially good use getting to their usual holiday location, a painted gypsy caravan and an enormous army tent set up in a field at Osmington Mills, in Dorset. The Hawking campsite was only a hundred yards from the beach. It was a rocky beach, not sand, but it was an interesting part of the coast – smuggler territory in a past age.

In the post-war years it was not unusual for families to live frugally with few luxuries, unable to afford home repairs, and, out of generosity or financial constraint, house more than two generations under one roof. But the Hawkings, though their house in St Albans was larger than many British homes, carried frugality and disrepair to an extreme. In this three-storey, strangely put together red-brick dwelling, Frank kept bees in the cellar, and Stephen's Scottish grandmother lived in the attic, emerging regularly to play the piano spectacularly well for local folk dances. The house was in dire need of work when the Hawkings moved in, and it stayed that way. According to Stephen's adopted younger brother Edward, 'It was a very large, dark house . . . really rather spooky, rather like a nightmare.'[9] The leaded stained glass at the front door must originally have been beautiful but was missing pieces. The front hall was lit only by a single bulb and its fine authentic William Morris wall covering had darkened. A greenhouse behind the rotting porch lost panes whenever there was a wind. There was no central heating, the carpeting was sparse, broken windows were not replaced.

The books, packed two deep on shelves all over the house, added a modicum of insulation. Frank Hawking would brook no complaints. One had only to put on more clothes in winter, he insisted. Frank himself was often away on research trips to Africa during the coldest months. Stephen's sister Mary recalls thinking that fathers were 'like migratory birds. They were there for Christmas and then they vanished until the weather got warm.'[10] She thought that fathers of her friends who *didn't* disappear were 'a bit odd'.[11]

The house lent itself to imaginative escapades. Stephen and Mary competed in finding ways to get in, some of them so secret that Mary was never able to discover more than ten of the eleven that Stephen managed to use. As if one such house were not enough, Stephen had another imaginary one in an imaginary place he called Drane. It seemed he did not know where this was, only that it existed. His mother became a little frantic, so determined was he to take a bus to find it, but later, when they visited Kenwood House in Hampstead Heath, she heard him declare that this was it, the house he had seen in a dream.[12]

'Hawkingese' was the name Stephen's friends gave the Hawking 'family dialect'. Frank Hawking himself had a stutter and Stephen and his siblings spoke so rapidly at home that they also stumbled over their words and invented their own oral shorthand.[13] That did not prevent Stephen from being, according to his mother, 'always extremely conversational'. He was also 'very imaginative . . . loved music and acting in plays', also 'rather lazy' but 'a self-educator from the start . . . like a bit of blotting paper, soaking it all up'.[14] Part of the reason for his lack of distinction in school was that he could not be bothered with things he already knew or decided he had no need to know.

Stephen had a rather commanding nature in spite of being smaller than most of his classmates. He was well-organized and capable of getting other people organized. He was also known as something of a comedian. Getting knocked around by larger boys didn't bother him much, but he had his limits, and he could, when driven to it, turn rather fierce and daunting. His friend Simon Humphrey had a heftier build than Stephen, but Simon's mother recalled that it was Stephen, not Simon, who on one memorable occasion swung around with his fists clenched to confront the much larger bullies who were teasing them. 'That's the sort of thing he did – he was equal to anything.'[15]

The eight-year-old Stephen's first school in St Albans was the High School for Girls, curiously named since its students included young children well below 'high school' age, and its Michael House admitted boys. A seven-year-old named Jane Wilde, in a class somewhat younger than Stephen's, noticed the boy with 'floppy-golden-brown hair' as he sat 'by the wall in the next-door classroom',[16] but she didn't meet him. She would later become his wife.

Stephen attended that school for only a few months, until Frank needed to stay in Africa longer than usual and Isobel accepted an invitation to take the children for four months to Majorca, off the east coast of Spain. Balmy, beautiful Majorca, the home of Isobel's friend from her Oxford days, Beryl, and Beryl's husband, the poet Robert Graves, was an enchanting place to spend the winter. Education was not entirely neglected for there was a tutor for Stephen and the Graveses' son William.[17]

Back in St Albans after this idyllic hiatus, Stephen went for one year to Radlett, a private school, and then did well enough in his tests to qualify for a place at the more selective St Albans School, also a private school, in the shadow of the Cathedral. Though in his first year at St Albans he managed to rank no better than an astonishing third from the bottom of his class, his teachers were beginning to perceive that he was more intelligent than he was demonstrating in the classroom. His friends dubbed him 'Einstein', either because he seemed more intelligent than they or because they thought he was eccentric. Probably both. His friend Michael Church remembers that he had a sort of 'overarching arrogance . . . some overarching sense of what the world was about'.[18]

'Einstein' soon rose in ranking to about the middle of the class. He even won the Divinity prize one year. From Stephen's earliest childhood, his father had read him stories from the Bible. 'He was quite well versed in religious things,' Isobel later told an interviewer.[19] The family often enjoyed having theological debates, arguing quite happily for and against the existence of God.

Undeterred by a low class placing, ever since the age of eight or nine Stephen had been thinking more and more seriously about becoming a scientist. He was addicted to questioning how things worked and trying to find out. It seemed to him that in science he could find out the truth, not only about clocks and radios but also about everything else around him. His parents planned that at thirteen he would go to Westminster School. Frank Hawking thought his own advancement had suffered because of his parents' poverty and the fact that he had not attended a prestigious school. Others with less ability but higher social standing had got ahead of him, or so he felt. Stephen was to have something better.

The Hawkings could not afford Westminster unless Stephen won a scholarship. Unfortunately, he was prone at this age to recurring bouts of a low fever, diagnosed as glandular fever, that sometimes was serious enough to keep him home from school in bed. As bad luck would have it, he was ill at the time of the scholarship examination. Frank's hopes were dashed and Stephen continued at St Albans School, but he believes his education there was at least as good as the one he would have received at Westminster.

After the Hawkings adopted Edward in 1955, Stephen was no longer the only male sibling. Stephen accepted his new younger brother in good grace. He was, according to Stephen, 'probably good for us. He was a rather difficult child, but one couldn't help liking him.'[20]

Continuing at St Albans School rather than heading off to Westminster had one distinct advantage. It meant being able to continue growing up in a little band of close friends who shared with Stephen such interests as the hazardous manufacture of fireworks in the dilapidated greenhouse and inventing board games of astounding complexity, and who relished long discussions on a wide range of subjects. Their game 'Risk' involved railways, factories, manufacturing, and its own stock exchange, and took days of concentrated play to finish. A feudal game had dynasties and elaborate family trees. According to Michael Church, there was something that particularly intrigued Stephen about conjuring up these worlds and setting down the laws that governed them.[21] John McClenahan's father had a workshop where he allowed John and Stephen to construct model aeroplanes and boats, and Stephen later remarked that he liked 'to build working models that I could control . . . Since I began my Ph.D., this need has been met by my research into cosmology. If you understand how the universe operates, you control it in a way.'[22] In a sense, Hawking's grown-up models of the universe stand in relation to the 'real' universe in the same way his childhood model aeroplanes and boats stood in relation to real aeroplanes and boats. They give an agreeable, comforting feeling of control while, in actuality, representing no control at all.

Stephen was fifteen when he learned that the universe was expanding. This shook him. 'I was sure there must be some mistake,' he says. 'A static universe seemed so much more natural. It could have existed and could continue to exist for ever. But an expanding universe would change with time. If it continued to expand, it would become virtually empty.'[23] That was disturbing.

Like many other teenagers of their generation, Stephen and his friends became fascinated with extrasensory perception (ESP). They tried to dictate the throw of dice with their minds. However, Stephen's interest turned to disgust when he attended a lecture by someone who had investigated famous ESP studies at Duke University in the United States. The lecturer told his audience that whenever the experiments got results, the experimental techniques were faulty, and whenever the experimental techniques were not faulty, they got no results. Stephen concluded that ESP was a fraud. His scepticism about claims for psychic phenomena has not changed. To his way of thinking, people who believe such claims are stalled at the level where he was at the age of fifteen.

Ancestor of 'Cosmos'

Probably the best of all the little group's adventures and achievements – and one that captured the attention and admiration of the entire town of St Albans – was building a computer that they called LUCE (Logical Uniselector Computing Engine). Cobbled together out of recycled pieces of clocks and other mechanical and electrical items, including an old telephone switchboard, LUCE could perform simple mathematical functions. Unfortunately that teenage masterpiece no longer exists. Whatever remained of it was thrown away eventually when a new head of computing at St Albans went on a cleaning spree.[24]

The most advanced version of LUCE was the product of Stephen's and his friends' final years of school before university. They were having to make hard choices about the future. Frank Hawking encouraged his son to follow him into medicine. Stephen's sister Mary would do that, but Stephen found biology too imprecise to suit him. Biologists, he thought, observed and described things but didn't explain them on a fundamental level. Biology also involved detailed drawings, and he wasn't good at drawing. He wanted a subject in which he could look for exact answers and get to the root of things. If he'd known about molecular biology, his career might have been very different. At fourteen, particularly inspired by a teacher named Mr Tahta, he had decided that what he wanted to do was 'mathematics, more mathematics, and physics'.

Stephen's father insisted this was impractical. What jobs were there for mathematicians other than teaching? Moreover he wanted Stephen to attend his own college, University College, Oxford, and at 'Univ' one could not read mathematics. Stephen followed his father's advice and began boning up on chemistry, physics and only a little maths, in preparation for entrance to Oxford. He would apply to Univ to study mainly physics and chemistry.

In 1959, during Stephen's last year before leaving home for university, his mother Isobel and the three younger children accompanied Frank when he journeyed to India for an unusually lengthy research project. Stephen stayed in St Albans and lived for the year with the family of his friend Simon Humphrey. He continued to spend a great deal of time improving LUCE, though Dr Humphrey interrupted regularly to insist he write letters to his family – something Stephen on his own would have happily neglected. But the main task of that year had to be studying for scholarship examinations coming up in March. It was essential that Stephen perform extremely well in these examinations if there was to be even an outside chance of Oxford's accepting him.

Students who rank no higher than halfway up in their school class seldom get into Oxford unless someone pulls strings behind the scenes. Stephen's lacklustre performance in school gave Frank Hawking plenty of cause to think he had better begin pulling strings. Stephen's headmaster at St Albans also had his doubts about Stephen's chances of acceptance and a scholarship, and he suggested Stephen might wait another year. He was young to be applying to university. The two other boys planning to take the exams with him were a year older. However, both headmaster and father had underestimated Stephen's intelligence and knowledge, and his capacity to rise to a challenge. He achieved nearly perfect marks in the physics section of the entrance examinations. His interview at

Oxford with the Master of University College and the physics tutor, Dr Robert Berman, went so well there was no question but that he would be accepted to read physics and be given a scholarship. A triumphant Stephen joined his family in India for the end of their stay.

Not a Grey Man

In October 1959, aged seventeen, Hawking went up to Oxford to enter University College, his father's college. 'Univ' is in the heart of Oxford, on the High Street. Founded in 1249, it is the oldest of the many colleges that together make up the University. Stephen would study natural science, with an emphasis on physics. By this time he had come to consider mathematics not as a subject to be studied for itself but as a tool for doing physics and learning how the universe behaves. He would later regret that he had not exerted more effort mastering that tool.

Oxford's architecture, like Cambridge's, is a magnificent hodge-podge of every style since the Middle Ages. Its intellectual and social traditions predate even its buildings and, like those of any great university, are a mix of authentic intellectual brilliance, pretentious fakery, innocent tomfoolery and true decadence. For a young man interested in any of these, Stephen's new environment had much to offer. Nevertheless, for about a year and a half, he was lonely and bored. Many students in his year were considerably older than he, not only because he had sat his examinations early but because others had taken time off for national service. He was not inspired to relieve his boredom by exerting himself academically. He had discovered he could get by better than most by doing virtually no studying at all.

Contrary to their reputation, Oxford tutorials are often not one-to-one but two or three students with one tutor. A young man named Gordon Berry became Hawking's tutorial partner. They were two of only four physics students who entered Univ that Michaelmas (autumn) term of 1959. This small group of newcomers – Berry, Hawking, Richard Bryan and Derek Powney – spent most of their time together, somewhat isolated from the rest of the College.

It wasn't until he was halfway through his second year that Stephen began enjoying Oxford. When Robert Berman describes him, it's difficult to believe he's speaking of the same Stephen Hawking who seemed so ordinary a few years earlier and so bored the previous year. '[H]e did, I think, positively make an effort to sort of come down to [the other students'] level and you know, be one of the boys. If you didn't know about his physics and to some extent his mathematical ability, he wouldn't have told you . . . He was very popular.'[25] Others who remember Stephen in his second and third years at Oxford describe him as lively, buoyant and adaptable. He wore his hair long, was famous for his wit, and liked classical music and science fiction.

The attitude among most Oxford students in those days, Hawking remembers, was 'very antiwork': 'You were supposed either to be brilliant without effort, or to accept your limitations and get a fourth-class degree. To work hard to get a better class of degree was regarded as the mark of a grey man, the worst epithet in the Oxford vocabulary.' Stephen's freewheeling, independent spirit and casual attitude towards his studies fitted right in. In a typical incident one day in a tutorial, after reading a solution he had worked out, he crumpled up the paper disdainfully and propelled it across the room into the wastepaper basket.

The physics curriculum, at least for someone with Hawking's abilities, could be navigated successfully without rising above this blasé approach. Hawking described it as 'ridiculously easy. You could get through without going to any lectures, just by going to one or two tutorials a week. You didn't need to remember many facts, just a few equations.'[26] You could also, it seems, get through without spending very much time doing experiments in the laboratory. Gordon and he found ways to use shortcuts in taking data and fake parts of the experiments. 'We just didn't apply ourselves,' remembers Berry. 'And Steve was right down there in not applying himself.'[27]

Derek Powney tells the story of the four of them receiving an assignment having to do with electricity and magnetism. There were thirteen questions, and their tutor, Dr Berman, told them to finish as many as they could in the week before the next tutorial. At the end of the week Richard Bryan and Derek had managed to solve one and a half of the problems; Gordon only one. Stephen had not yet begun. On the day of the tutorial Stephen missed three morning lectures in order to work on the questions, and his friends thought he was about to get his come-uppance. His bleak announcement when he joined them at noon was that he had been able to solve only ten. At first they thought he was joking, until they realized he *had* done ten. Derek's comment was that this was the moment Stephen's friends recognized 'that it was not just that we weren't in the same street, we weren't on the same planet'.[28] 'Even in Oxford, we must all have been remarkably stupid by his standards.'[29]

His friends were not the only ones who sometimes found his intelligence impressive. Dr Berman and other dons were also beginning to recognize that Hawking had a brilliant mind, 'completely different from his contemporaries'. 'Undergraduate physics was simply not a challenge for him. He did very little work, really, because anything that was do-able he could do. It was only necessary for him to know something could be done, and he could do it without looking to see how other people did it. Whether he had any books I don't know, but he didn't have very many, and he didn't take notes.'[30] 'I'm not conceited enough to think that I ever taught him anything.'[31] Another tutor called him the kind of student who liked finding mistakes in the textbooks better than working out the problems.

The Oxford physics course was scheduled in a way that made it easy not to see much urgent need for work. It was a three-year course with no exams until the end of the third year. Hawking calculates he spent on the average about one hour per day studying: about one thousand hours in three years. 'I'm not proud of this lack of work,' he says. 'I'm just describing my attitude at the time, which I shared with most of my fellow students: an attitude of complete boredom and feeling that nothing was worth making an effort for. One result of my illness has been to change all that: when you are faced with the possibility of an early death, it makes you realize that life is worth living, and that there are lots of things you want to do.'

One major explanation why Stephen's spirits improved dramatically in the middle of his second year was that he and Gordon Berry joined the college Boat Club. Neither of them was a hefty hunk of the sort who make the best rowers. But both were light, wiry, intelligent and quick, with strong, commanding voices, and these are the attributes that college boat clubs look for when recruiting a coxswain (cox) – the person who sits looking forward, facing the line of four or eight rowers, and steers the boat with handles attached to the rudder. The position of cox is definitely a position of control, something that Hawking has said appealed to him with model boats, aeroplanes and universes – a man of slight build commanding eight muscle-men.

Stephen exerted himself far more on the river, rowing and coxing for Univ, than he did at his studies. One sure way to be part of the 'in' crowd at Oxford was to be a member of your college rowing team. If intense boredom and a feeling that nothing was worth making an effort for were the prevailing attitudes elsewhere, all that changed on the river. Rowers, coxes and coaches regularly assembled at the boathouse at dawn, even when there was a crust of ice on the river, to perform arduous calisthenics and lift the racing shell into the water. The merciless practice went on in all weather, up and down the river, coaches bicycling along the towpath exhorting their crews. On race days emotions ran high and crowds of rowdy well-wishers sprinted along the banks of the river to keep up with their college boats. There were foggy race days when boats appeared and vanished like ghosts, and drenching race days when water filled the bottom of the boat. Boat club dinners in formal dress in the college hall lasted late and ended in battles of wine-soaked linen napkins.

All of it added up to a stupendous feeling of physical

well-being, camaraderie, all-stops-out effort, and of living college life to the hilt. Stephen became a popular member of the boating crowd. At the level of intercollege competition he did well. He'd never before been good at a sport, and this was an exhilarating change. The College Boatsman of that era, Norman Dix, remembered him as an 'adventurous type; you never knew quite what he was going to do'.[32] Broken oars and damaged boats were not uncommon as Stephen steered tight corners and attempted to take advantage of narrow manoeuvring opportunities that other coxes avoided.

At the end of the third year, however, examinations suddenly loomed larger than any boat race. Hawking almost floundered. He'd settled on theoretical physics as his speciality. That meant a choice between two areas for graduate work: cosmology, the study of the very large; or elementary particles, the study of the very small. Hawking chose cosmology. 'It just seemed that cosmology was more exciting, because it really did seem to involve the big question: Where did the universe come from?'[33] Fred Hoyle, the most distinguished British astronomer of his time, was at Cambridge. Stephen had become particularly enthusiastic about the idea of working with Hoyle when he took a summer course with one of Hoyle's most outstanding graduate students, Jayant Narlikar. Stephen applied to do Ph.D. research at Cambridge and was accepted with the condition that he get a First from Oxford.

One thousand hours of study was meagre preparation for getting a First. However, an Oxford examination offers a choice from many questions and problems. Stephen was confident he could get through successfully by doing problems in theoretical physics and avoiding any questions that required knowledge of facts. As the examination day approached, his confidence faltered. He decided, as a fail-safe, to take the Civil Service exams and apply for a job with the Ministry of Works.

The night before his Oxford examinations Stephen was too nervous to sleep. The examination went poorly. He was to take the Civil Service exams the next morning, but he overslept and missed them. Now everything hung on his Oxford results.

As Stephen and his friends waited on tenterhooks for their results to be posted, only Gordon was confident he had done well in his examinations – well enough for a First, he believed. Gordon was wrong. He and Derek received Seconds, Richard a disappointing Third. Stephen ended up disastrously on the borderline between a First and a Second.

Faced with a borderline result, the examiners summoned Hawking for a personal interview, a 'viva'. They questioned him about his plans. In spite of the tenseness of the situation, with his future hanging in the balance, Stephen managed to come up with the kind of remark for which he was famous among his friends: 'If I get a First, I shall go to Cambridge. If I receive a Second, I will remain at Oxford. So I expect that you will give me a First.' He got his First. Dr Berman said of the examiners: 'They were intelligent enough to realize they were talking to someone far cleverer than most of themselves.'[34]

That triumph notwithstanding, all was not well. Hawking's adventures as a cox, his popularity, and his angst about his exams had pushed into the background a problem that he had first begun to notice that year and that refused to go away. 'I seemed to be getting more clumsy, and I fell over once or twice for no apparent reason,'[35] he remembers. The problem had even invaded his halcyon existence on the river when he began to have difficulty sculling (rowing a one-man boat). During his final Oxford term, he tumbled down the stairs and landed on his head. His friends spent several hours helping him overcome a temporary loss of short- and long-term memory, insisted he go to a doctor to make sure no serious damage had been done, and encouraged him to take a Mensa intelligence test to prove to them and to himself that his mind was not affected. All seemed well, but they found it difficult to believe that his fall had been a simple accident.

There was indeed something amiss, though not as a result of his tumble . . . and not with his mind. That summer, on a trip he and a friend took to Persia (now Iran), he became seriously ill, probably from a tourist stomach problem or a reaction to the vaccines required for the trip.[36] It was a harrowing journey in other ways, more harrowing for his family back home than for Stephen. They lost touch with him for three weeks, during which time there was a serious earthquake in the area where he was travelling. Stephen, as it turned out, had been so ill and riding on such a bumpy bus that he didn't notice the earthquake at all. He finally got back home, depleted and unwell. Later there would be speculation about whether a non-sterile smallpox vaccination prior to the trip had caused his illness in Persia and also his ALS, but the latter had, in fact, begun earlier. Nevertheless, because of his illness in Persia and the increasingly troubling symptoms he was experiencing, Stephen arrived at Cambridge a more unsettled and weaker twenty-year-old than he had been at Oxford the previous spring. He moved into Trinity Hall for the Michaelmas term in the autumn of 1962.

During the summer before Stephen left for Cambridge, Jane Wilde saw him while she was out walking with her friends in St Albans. He was a 'young man with an awkward gait, his head down, his face shielded from the world under an unruly mass of straight brown hair . . . immersed in his own thoughts, looking neither right nor left . . . lolloping along in the opposite direction'.[37] Jane's friend Diana King, sister of Stephen's friend Basil King, astonished her friends by telling them that she had gone out with him. 'He's strange but very clever. He took me to the theatre once. He goes on Ban the Bomb marches.'[38]

4

'The realization that I had an incurable disease, that was likely to kill me in a few years, was a bit of a shock'

HAWKING'S FIRST YEAR AT CAMBRIDGE WAS LARGELY A disaster. Fred Hoyle already had his full quota of graduate students and Stephen was assigned Denis Sciama instead. Sciama was a lesser name in physics than Hoyle – in fact, Stephen had never heard of him – but others knew him to be a fine mentor who cared deeply about his students. He also was far more available in Cambridge than Hoyle, who was an international figure and spent much of his time at observatories in other parts of the world. Sciama and Hoyle both favoured the 'Steady State' theory of the universe. Hoyle, with Hermann Bondi and Tom Gold, had fathered that theory.

The Steady State theory recognized the expansion of the universe but, unlike the Big Bang theory, did not require the universe to have a beginning in time. The proposal was that as the universe expands, with galaxies moving apart from one another, new matter appears to fill the increasingly large gaps among them, eventually forming new stars and galaxies. At any moment in its history and future, the universe looks pretty much the same as it does at any other time. The Steady State was destined to lose the contest with the Big Bang theory, but for a while it seemed a brilliantly viable competitor.

For someone with Hawking's slipshod mathematics background, general relativity was rough going, and he soon sorely regretted allowing his father to steer him away from maths at Oxford. Sciama suggested that he might concentrate on astrophysics, but Stephen had firmly set his sights on general relativity and cosmology. Keeping his head barely above water, he undertook a quick, arduous self-education. At King's College London, Hermann Bondi had started a course in general relativity. Stephen, along with other Cambridge graduate students, travelled there regularly for lectures.

Relativity and cosmology were risky choices even for those sufficiently well prepared in mathematics. The scientific community regarded cosmology with some suspicion and disfavour. As Hawking would later recall, 'Cosmology used to be considered a pseudoscience and the preserve of physicists who might have done useful work in their earlier years, but who had gone mystic in their dotage.'[1] It was highly speculative, with insufficient observational data to curb or shape speculation.[2] Sciama himself, just two years before Hawking met him, wrote that cosmology was 'a highly controversial subject, which contains little or no agreed body of doctrine'.[3]

Hawking was aware of these difficulties, but the challenge of working at the frontiers and venturing into unexplored territory was irresistible. Cosmology and general relativity were 'neglected fields that were ripe for development at that time. Unlike elementary particles, there was a well-defined theory, Einstein's general theory of relativity, thought to be impossibly difficult. People were so pleased to find any solution to [Einstein's] field equations; they didn't ask what physical significance, if any, it had.'[4]

Einstein's general theory of relativity was indeed, as Hawking pointed out, a well-defined theory in which gravity is explained as the warping of spacetime, but Sciama had been right about cosmology. The battle still raged about which theory could correctly describe the history of the universe – the Big Bang theory or the Steady State theory. Had the universe had a beginning or not? In the twenty-first century, it seems incredible that when Hawking came up to Cambridge as a graduate student in 1962, that contest had not yet by any means been settled.

Hawking's failure to get Hoyle as his supervisor and his mathematical deficiencies were setbacks, but no more than typical for a first-year graduate student. While he struggled to catch up on general relativity and find a way through the mathematical maze needed to understand it, a far more unusual and merciless problem was overtaking him in the autumn of 1962, threatening to make all this effort meaningless. The clumsiness he had noticed during his third year at Oxford kept getting worse. That first autumn in Cambridge, he had trouble tying his shoes and sometimes had a problem talking. His speech became slurred, enough so that those who met him for the first time assumed he had a slight speech impediment.

When he went home to St Albans for Christmas after his first term in Cambridge, Stephen's physical problems were too obvious to conceal from his parents. Frank Hawking took his son to the family doctor. That doctor referred them to a specialist. They made an appointment for after the holidays.

Shortly after his twenty-first birthday in January, Hawking found himself not heading back to Cambridge for the Lent term but in St Bartholomew's Hospital in London for tests. Perhaps it made the situation a little less daunting that his sister Mary, preparing to follow their father into medicine, was in training at 'Bart's'. Stephen refused the private hospital room his parents had wanted for him, because of his 'socialist principles'. In hospital, specialists took a muscle sample from his arm, stuck electrodes into him, and injected radio-opaque fluid into his spine and watched it going up and down with X-rays while they tilted the bed on which he lay. After two weeks they released him, telling him vaguely that what he had wasn't a 'typical case' and that it wasn't multiple sclerosis. The doctors suggested he go back to Cambridge and get on with his work. 'I gathered,' Hawking remembers, 'that they expected it to continue to get worse, and that there was nothing they could do, except give me vitamins. I could see that they didn't expect them to have much effect. I didn't feel like asking for more details, because they were obviously bad.'

Isobel Hawking did not immediately learn how seriously ill her son was, until, out ice skating with her, he fell and couldn't get up. Finally off the ice, she bundled him into a café and pressured him to talk about the physical difficulties he was having and what the doctors were saying. She insisted on conferring with his doctor herself, and was given the same devastating news.[5]

Hawking had contracted a rare disease for which there is no known cure, amyotrophic lateral sclerosis (ALS), known in Britain as motor neurone disease, in America as Lou Gehrig's disease. It causes a gradual disintegration of the nerve cells in the spinal cord and brain that regulate voluntary muscle activity. The first symptoms are usually weakness and twitching of the hands, and perhaps slurred speech and difficulty in swallowing. As nerve cells disintegrate, the muscles they control atrophy. Eventually this happens to every voluntary muscle of the body. Movement becomes impossible. Speech and all other means of communication are lost. Though Hawking is not the only patient to have survived for several decades, death almost always occurs within two or three years as a result of pneumonia or suffocation when the respiratory muscles fail. The disease does not affect involuntary muscles of the heart, muscles of waste elimination or the sexual organs, and the brain remains completely lucid to the end. To some this seems an advantage, to others a horror. Patients in the final stages of the disease are often given morphine, not for pain – there is none – but for panic and depression.

For Hawking everything had changed. With typical understatement, he describes his reaction: 'The realization that I had an incurable disease, that was likely to kill me in a few years, was a bit of a shock. How could something like that happen to me? Why should I be cut off like this? However, while I had been in hospital, I had seen a boy I vaguely knew die of leukaemia in the bed opposite me. It had not been a pretty sight. Clearly there were people who were worse off than I. At least, my condition didn't make me feel sick. Whenever I feel inclined to be sorry for myself, I remember that boy.'

Nevertheless, at first Hawking went into a deep depression. He didn't know what he ought to do, what was going to happen to him, how quickly he would get worse, or what it would be like. His doctors had told him to continue his Ph.D. research, but that had already been going poorly. This fact was almost as depressing to him as his illness. It seemed pointless to try to continue working towards a doctorate he wouldn't live to receive, nothing but a foolish device for keeping his mind preoccupied while his body was dying. He holed up miserably in his college rooms at Trinity Hall, but he insists: 'Reports in magazine articles that I drank heavily are an exaggeration. I felt somewhat of a tragic character. I took to listening to Wagner.

'My dreams at that time were rather disturbed,' he remembers. 'Before my condition had been diagnosed, I had been very bored with life. There had not seemed to be anything worth doing. But shortly after I came out of hospital, I dreamt that I was going to be executed. I suddenly realized that there were a lot of worthwhile things I could do, if I were reprieved. Another dream that I had several times was that I would sacrifice my life to save others. After all, if I were going to die anyway, it might as well do some good.'

Frank Hawking took advantage of all the connections his stature in the medical profession made available to him. He contacted experts in every possibly related disease, but all was in vain. Hawking's doctors hoped his condition would stabilize, but the disease progressed rapidly. They soon informed him that he did indeed have only about two years to live. At that point his father appealed to Denis Sciama to help Stephen finish his dissertation early. Sciama, knowing Hawking's potential and unwilling to let him compromise even if he was dying, turned the request down.

Two years passed. The progression of the disease slowed. 'I didn't die. In fact, although there was a cloud hanging over my future, I found to my surprise that I was enjoying life in the present more than before.' He had to use a cane, but his condition wasn't all that bad. Total disability and death, though still a not-too-distant certainty, were postponed. Sciama suggested that since he was going to live a while longer, he ought to finish his thesis. Hawking had his reprieve, a precarious and temporary one, but life was precious and full of worthwhile things.

In January 1963, just before Hawking entered the hospital for tests, Basil King and his sister Diana had hosted a New Year's party in St Albans. There, Hawking had met Diana's friend Jane Wilde,[6] who was just finishing at St Albans High School and had been accepted to study languages the next autumn at Westfield College in the University of London. Jane later described Stephen as she caught sight of him at that party – 'slight of frame, leaning against the wall in a corner with his back to the light, gesticulating with long thin fingers as he spoke – his hair falling across his face over his glasses – and wearing a dusty black-velvet jacket and a red-velvet bow tie.'[7] Embroidering somewhat on the story of his viva at Oxford (the oral exam that finally won him his First), he regaled her and a friend of his from Oxford with the story that he had tempted the examiners to give him a First and let him go to Cambridge by giving them the opportunity to send him, like a Trojan horse, into the rival university.[8] To Jane this dishevelled graduate student seemed terribly intelligent, eccentric and rather arrogant. But he was interesting, and she liked his self-mocking wit. He said he was studying cosmology. She didn't know what that meant.

Stephen and Jane exchanged names and addresses at the party, and a few days later Jane received an invitation to a birthday celebration – his twenty-first – on 8 January. The party was Jane's first experience of the eccentric Hawking home at 14 Hillside Road, St Albans. Though she recognized most of the family as familiar faces in St Albans, she felt painfully unsophisticated among them and their friends, and she spent most of the evening in a corner near the fire, trying to stay warm in the icy cold house, holding Stephen's younger brother Edward on her lap. The evening cannot have been a total success for Stephen either. His physical problems were becoming impossible to conceal. He had trouble pouring the drinks.

About a month later, Jane overheard Diana King and a friend discuss the news that Stephen had been diagnosed with 'some terrible, paralysing incurable disease . . . a bit like multiple sclerosis, but it's not multiple sclerosis and they reckon he's probably only got a couple of years to live'.[9] Diana's brother Basil had been to visit him in the hospital.

It came as a surprise when Jane encountered Stephen a week later on the railway platform in St Albans, looking much as he had before but more conventionally dressed and with a neater haircut. They were both waiting for the train to London. On the journey, they sat together and talked. When Jane mentioned that she had been sorry to hear about his hospital stay, Stephen wrinkled his nose and said nothing.[10] She dropped the subject. He asked her whether she would like to go to the theatre with him some weekend when he was home from Cambridge. She said she would.

Their first date was for dinner and the theatre in London. The evening turned out to be so expensive that when they had boarded the bus back towards the railway station, Stephen realized that he had run out of funds. There were no ATMs in those days. After treating this young lady to a truly lavish first date, he had to ask her whether she could pay the bus fare. Rummaging around in her handbag, Jane discovered that her purse was not there, and so began their first adventure together.

Scuttling off the bus before anyone could ask for their fare, Jane and Stephen returned to the darkened, shut Old Vic theatre and found a way in by the stage door. Jane's purse was under the seat where it had fallen and all seemed to be turning out well when the lights went out completely. Stephen took her hand, and they groped their way back on to the stage and across it in total darkness, and out of the stage door again, Jane following Stephen's sure lead 'with silent admiration'.[11]

Stephen was definitely not the pizza-and-a-movie type of man, for his next invitation was to the Trinity Hall May Ball. Dinner and the theatre in London and a Cambridge May Ball were certainly among the most seriously splendid dates a girl might hope for.

When Stephen arrived in June* to take Jane to

* May Balls in Cambridge usually happen in June.

Cambridge, the deterioration of his physical condition shocked her. She wondered whether this 'slight, frail, limping figure who appeared to use the steering wheel to hoist himself up to see over the dashboard'[12] was capable of driving the car to Cambridge. The hazards of the journey, however, stemmed not from Stephen's disability but from the recklessness and speed of his driving. They arrived with Jane vowing to herself to take the train home rather than repeat this experience.[13]

Though Trinity Hall is small compared with other Cambridge colleges like Trinity and St John's, its May Ball proved to be the magical experience a May Ball at its best should be. The lawns and flower beds, falling away towards the river and the meadows of the Backs beyond, were romantically lit, and everyone in formal attire looked remarkably better than usual. There was music to suit all tastes, in different parts of the college. A string quartet in an elegantly panelled room. A cabaret in the Hall. A jazz band. A Jamaican steel band. Champagne was served from a bathtub, and there was a lavish buffet. The festivities continued until dawn and breakfast, and the next day included a punt on the river. Jane was at first mystified but then impressed by the ability of Hawking's friends to argue quite nastily with him one minute on some intellectual subject and then, the next, treat him with extraordinary gentleness and care for his weakened physical condition. When it was all over, Stephen, to Jane's chagrin, would not hear of her taking the train back rather than driving with him. She arrived home so flustered and disgusted with his driving that she got out of the car, left him at the kerb and stalked into the house. At her mother's insistence she went back to invite him in for tea. In spite of the extravagant date invitations, this was no romance yet, though Hawking thought she was 'a very nice girl',[14] and it was at about this time that Derek Powney became puzzled about his old friend's sudden interest in John Donne's elegies, some of the most beautiful and explicit love poems ever written.[15]

After seeing Stephen on a few more occasions with his family and hers, Jane set off for a summer in Spain, a requirement for her language degree from Westfield College. When she returned, Stephen had departed for Cambridge again, and Jane soon left St Albans herself to live in London and begin her studies. It wasn't until November that she heard from him. He was coming to London for a dental appointment and invited her to go with him to the Wallace Collection (a famous display of art, furniture, porcelain, arms and armour), to dinner and to Wagner's opera *The Flying Dutchman*. On this date, Stephen stumbled and fell in the middle of Lower Regent Street. Jane dragged him to his feet. She noticed that as his walking became increasingly unsteady, his opinions became stronger and more defiant. On this occasion, not long after the assassination of US President Kennedy, he expressed disapproval of Kennedy's handling of the Cuban Missile Crisis.[16]

That winter, Stephen came frequently to London for seminars and dental appointments and seemed to have a steady supply of opera tickets. Jane also travelled often to Cambridge to see him at the weekends. By this time she was definitely 'in love with Stephen, with his wicked sense of humour. The light in his eyes was magnetic',[17] but she refused to have a short-term affair with him. Short term was unfortunately all he could foresee, and their weekends were not happy. Jane returned to London many times in tears.

One problem was that, voluble as Stephen could be on most matters, he was not willing to discuss his illness or share his feelings about it. Though this troubled Jane at the time, she didn't put pressure on him. It was only later that she realized they had set a precedent of non-communication that would serve them ill in the future.[18] One day in late winter she met him after he had an appointment with his Harley Street consultant. When she asked him how it had gone, 'he grimaced' and told her the doctor had told him 'not to bother to come back, because there's nothing he can do'.[19] End of conversation.

Jane's first year at Westfield was a period of spiritual questioning. It would not have been difficult to be won over to agnosticism or even atheism by this charismatic, intellectually brilliant young man, beside whom she still felt a little like an awkward teenager. But Jane stuck with the faith in God ingrained in her from childhood by her mother and also to the belief that good can come out of any disaster. She concluded that she would have to 'maintain sufficient faith for the two of us if any good were to come of our sad plight'.[20] Stephen, though he never shared her faith, admired her energy and her optimism and gradually began to find them contagious.

Not everything was an upward curve. In spite of their closeness in the winter, when Jane spent a term in Spain that spring of 1964, her letters to Stephen went unanswered. During a short interval in St Albans before she departed again to spend the summer touring Europe with her family, Jane found Stephen depressed and cynical, playing Wagner at high volume, not bothering, for a change, to hide his sense of futility and frustration, and seemingly determined to do all he could to alienate her. As she would later tell an interviewer, 'he was really in quite a pathetic state. I think he'd lost the will to live. He was very confused.'[21] They were apart for most of the summer. Stephen went with his sister Philippa to Bayreuth for Wagner's *Ring* cycle and, from there, on a journey behind the Iron Curtain to Prague.

Near the end of her family's European travels, Jane found a postcard from Stephen waiting for her at their hotel in Venice. It was marvel enough that there should be one at all, but it was cheerful and informative. The picture was of the castle-fortress that towers over Salzburg, in Austria, and Stephen's message exclaimed about the Salzburg Festival, Bayreuth and Prague. Clutching that postcard, Jane explored Venice in a romantic haze, hardly able to contain her eagerness to get back to England and Stephen.

When Jane got home to St Albans, she found Stephen in much better spirits than he had been earlier in the summer, in spite of having knocked out his front teeth in a fall while travelling on a train in Germany – a sad shame after all that dental work that had brought him to London. His physical condition appeared to have stabilized. He was daring to look ahead.

On a wet Cambridge autumn evening at the beginning of the Michaelmas term, Stephen proposed marriage and Jane agreed. 'I wanted to find some purpose to my existence,' she says, 'and I suppose I found it in the idea of looking after him. But we were in love, we got married, there didn't seem much choice in the matter. I just decided what I was going to do, and I did it.'[22] They had come to realize, 'that together we could make something worthwhile of our lives'.[23]

For Stephen the engagement made 'all the difference'. 'The engagement changed my life. It gave me something to live for. It made me determined to live. Without the help that Jane has given I would not have been able to carry on, nor have had the will to do so.'

Jane's father gave his consent to the marriage on the condition that she complete her college education and that unreasonable demands not be placed on her. Frank Hawking suggested that because of his son's brief life expectancy they have children as soon as possible. As a medical man, he assured her that Stephen's condition was not hereditary.[24]

One obstacle to their marriage had to be dealt with immediately. Westfield College did not permit its undergraduates to marry. An exception was made on the grounds that Jane's betrothed husband might not live until their wedding date if it were postponed. Jane was, however, required to move out of the college into private accommodation in London. There she would spend her week days, returning to Cambridge and Stephen at weekends.[25] Stephen also had to move out of college housing and find new lodgings.

Hawking's natural buoyancy returned. He found an ingenious way of phoning London for only the cost of a local Cambridge call, and in long telephone conversations 'illness assumed the proportions of a minor background irritant as we talked about job prospects, housing, wedding arrangements, and our first trip to the United States . . . due to start just ten days after the wedding'.[26] Hawking was at last making progress with his studies. He decided to count himself supremely lucky that his illness would never touch his mind, no matter how it might paralyse his body. Work in theoretical physics was going to take place almost entirely in his mind. It was one of the few careers he might have chosen in which physical disability wouldn't be a serious handicap.

This attitude sounds courageous, but it embarrasses Stephen Hawking to hear himself described that way. It would have been courageous and required tremendous willpower, he thinks, to have chosen such a difficult course deliberately, but that wasn't how it happened. He simply did the only thing possible. As he puts it, 'One has to be grown up enough to realize that life is not fair. You just have to do the best you can in the situation you are in.'[27] It was true in 1964, and still is today, that, as far as he is concerned, the less made of his physical problems the better. If this book were to talk about his scientific work and fail entirely to mention that doing such work possibly represents more of an achievement for him than it would for most people, that would suit him fine. One of the most important things you can learn about him is how unimportant his disability is. It isn't accurate to call him a sick man. Health involves much more than physical condition, and in this broader sense for most of his life he's been one of the healthiest persons around. That message comes through loud and clear in his writing and in most of the things written about him, and it is even more apparent when you're with him. That's the Hawking image, and though we should take seriously his warning 'You shouldn't believe everything you read', it isn't a fake image.

Meanwhile, no marriage was possible until he had a job, and no job was possible without a Ph.D. He began looking for an idea with which to complete his thesis.

Challenging the Future

Though Hawking's life had been in turmoil since the diagnosis in the winter of 1963, neither his deteriorating physical condition nor his growing preoccupation with Jane Wilde had eclipsed his interest in cosmology. His office in the Department of Applied Maths and Theoretical Physics was next door to Jayant Narlikar, whom Hawking had met on a summer course before coming up to Cambridge. Narlikar was one of Hoyle's students and working with Hoyle on possible modifications to general relativity that might reconcile the Steady State model with recent observations that called it into question. This challenge piqued Hawking's curiosity.

In June 1964, prior to publication of Hoyle's work with Narlikar, Hoyle gave a lecture about it at the Royal Society. Hawking travelled to London to attend. When the floor was opened for questions, Hawking rose to his feet with the help of his stick and challenged one of Hoyle's results. An astonished Hoyle asked Hawking how he could possibly judge whether the result was right or wrong. Hawking replied that he had 'worked it out'. Unaware that Hawking and Narlikar had discussed the results many times and that Hawking had done calculations of his own, Hoyle and the audience assumed that this unknown research student had 'worked it out' in his head right there at the lecture. The audience was impressed; Hoyle was infuriated. Surprisingly, Hawking seems not to have lost the friendship of Narlikar. In any case, his reputation for brilliance and brashness had begun, and so had his interest in calculations and speculation having to do with the expanding universe.

Hawking learned about a theory of the British mathematician and physicist Roger Penrose concerning what happens when a star has no nuclear fuel left to burn and collapses under the force of its own gravity. Penrose, building on earlier work by such physicists as Subrahmanyan Chandrasekhar and John Wheeler, claimed that even if the collapse isn't perfectly smooth and symmetrical, the star will nevertheless be crushed to a tiny point of infinite density and infinite curvature of spacetime, a singularity at the heart of a black hole.

Hawking took off from there by reversing the direction of time, imagining a point of infinite density and infinite curvature of spacetime – a singularity – exploding outwards and expanding. Suppose, he suggested, the universe began like that. Suppose spacetime, curled up tight in a tiny, dimensionless point, exploded in what we call the Big Bang and expanded until it looks the way it does today. Might it have happened like that? *Must* it have happened like that?

With these questions, Hawking began the intellectual adventure that has continued for more than forty-five years. As he says, 'I started working hard for the first time in my life. To my surprise, I found I liked it. Maybe it is not really fair to call it work.'

1965

In the winter of 1965 Hawking applied for a research fellowship at Gonville and Caius College in Cambridge. Jane came up for the weekend from London, where she was still living while completing her degree at Westfield College, and Hawking reported that 'I was hoping that

Jane would type my application, but she had her arm in plaster, having broken it. I must admit that I was less sympathetic than I should have been. However, it was her left arm, so she was able to write out my application to my dictation, and I got someone else to type it.'

Jane's arm was not the worst setback he encountered applying for the Caius fellowship. He was asked to name two persons as references. Denis Sciama suggested Hermann Bondi. Stephen had attended lectures in Bondi's general relativity course at King's College London, but did not know him well. 'I had met him a couple of times and he had communicated a paper I had written to the Royal Society. I asked him [about giving a reference] after a lecture he gave in Cambridge. He looked at me in a vague way and said, yes, he would. Obviously, he didn't remember me, for, when the College wrote to him for a reference, he replied that he had not heard of me.' That should have doomed Hawking's chances. It might today, with so many applying for research fellowships, but he was fortunate. 'Those were quieter times. The College wrote to tell me of the embarrassing reply of my referee. My supervisor got on to Bondi and refreshed his memory. Bondi then wrote me a reference that was probably far better than I deserved. Anyway I got the fellowship.'

Another boost to Stephen's professional standing in the spring of 1965 was a 'commendation prize' in the privately funded Gravity Prize Competition. He might have done better than 'commendation' had he not missed the deadline for submissions, but, with his wedding coming up, £100 was welcome.[28] During that same spring, at an international conference on general relativity and gravitation in London – the first such event he ever attended – Stephen met Kip Thorne from the California Institute of Technology. Thorne already had his Ph.D. from Princeton. Thorne was deeply impressed with the way this young man who walked with a cane and seemed somewhat wobbly, and spoke with a slight hesitation, was taking techniques introduced into general relativity by Roger Penrose and adapting them to investigate the structure and history of the universe. Their conversation in a tearoom at the conference was the beginning of a lifelong friendship. Thorne is one of a handful of friends, perhaps the only one, with whom Hawking has had frank, matter-of-fact discussions about his bleak expectations for the future.

On 14 July 1965, Stephen Hawking and Jane Wilde were married in a civil ceremony, followed the next day by a religious ceremony in the chapel of Trinity Hall.

Theoretical physics is full of paradoxes. It seems appropriate that one of our great theoretical physicists is a man whose enthusiasm for life was awakened by a tragedy that ought to have embittered and destroyed him, and that his meteoric rise as a scientist started with the practical need for a thesis topic so that he could get a job and marry. With what simplicity Hawking described it: in spite of the Wagner, the tragic hero self-image and the dreams; a year, maybe more, of depression . . . then 'I was happier than I'd been before.'

5

'The big question was, was there a beginning or not?'

AFTER THEIR WEDDING AND A SHORT HONEYMOON IN Suffolk, which was all they could afford, Stephen and Jane Hawking set out across the Atlantic to America for a general relativity summer school at Cornell University in upper New York State. The summer school was another opportunity for Hawking to meet top people in his field. However, he remembers the experience as 'a mistake'. 'It put quite a strain on our marriage, specially as we stayed in a dormitory that was full of couples with noisy small children.'[1]

One summer evening at Cornell, chatting with friends in the night air – chilly in those climes despite the time of year – Hawking suddenly had a choking fit. He knew he should expect such episodes, but since he had resolutely refused to discuss his problems with Jane, she did not, and she had no idea what to do to help him. Finally he signalled her to give him a hard thump on the back. This solved the immediate problem, but the experience left Jane shaken and vividly aware of what faced them. 'The demonic nature of the illness had announced its presence.'[2]

In October, Hawking, aged twenty-three, began his fellowship at Caius. Jane Hawking had another year to go to complete her undergraduate degree at London University. As planned, Hawking would fend for himself during the week. She would join him for the weekends. Since he couldn't walk far or cycle, they needed Cambridge lodgings near his department. Before going to America they had applied for a flat being built in the market square. No one told them that those flats were actually owned by Hawking's college, which might have given their application an advantage. As it turned out, that didn't matter because the flats weren't ready for occupancy that autumn.

The bursar at Caius had earlier informed Stephen that Caius policy was not to help Fellows find housing. Relenting only slightly, he offered Stephen and Jane one room in a graduate student hostel and charged them double because there would be two of them living there at weekends. Then, three days after moving into the hostel, they discovered a small house available for three months in Little St Mary's Lane – one of a group of picturesque cottages that line one side of the lane across from Little St Mary's Church and churchyard garden. The house was only a hundred yards from the new Department of Applied Mathematics and Theoretical Physics (DAMTP) premises in Silver Street, where

Stephen was sharing an office with another young physicist, Brandon Carter. He was able to walk that distance, and he acquired a small three-wheeled car to drive when he needed to get to the Institute of Astronomy in the countryside near town. Later that autumn, when their first three-month lease was approaching its end, the Hawkings learned that another house in the lane was unoccupied. A helpful neighbour located the owner in Dorset and upbraided her for having her house vacant while a young couple had no place to live. The owner agreed to rent.

The choking fits became more frequent. Stephen's sister Mary, still working towards her medical degree at London University, suggested that warmer, drier weather might help. Partly on that advice, in December at the end of the Michaelmas term, the Hawkings took advantage of an opportunity to cross the ocean for a second time. Stephen attended an astrophysics conference in Miami, and from there they went on to Austin, Texas, to spend a week with one of his graduate school friends George Ellis and his wife. They returned to England in time for Christmas and the move into their second, more permanent home in Little St Mary's Lane.

Both Stephen and Jane followed punishing schedules during that first year of their marriage. Hawking was still keenly aware of his lack of mathematical background. Being, as his mother has said, a 'self-educator', he decided to use a time-honoured graduate student method of improving one's own knowledge while also earning some money: if you want or need to learn a subject, teach it. So, in addition to working on his Ph.D. thesis, he supervised undergraduate mathematics for the college.[3] Jane managed her weekly commute, finished her undergraduate degree, engineered the move from one house to another, and typed her husband's Ph.D. dissertation.

They celebrated the completion of Stephen's Ph.D. in March 1966, and there was more to celebrate. Hawking had submitted an essay, 'Singularities and the Geometry of Space-Time', in competition for the prestigious Adams Prize, awarded by St John's College, Cambridge, and named for John Couch Adams, co-discoverer of the planet Neptune. The winner had to be a young researcher who was based in Britain, and the work had to be of international calibre. Hawking's essay was co-winner with Roger Penrose's entry. A proud Denis Sciama told Jane that, in his opinion, Stephen could look forward to a career worthy of Isaac Newton.[4] In spite of his physical difficulties and bleak prospects, these were halcyon days: Cambridge in the 1960s was an extremely stimulating place for someone with Hawking's interests. Everything seemed possible. Surprisingly much was![5]

That spring, Jane Hawking, eager to maintain some intellectual identity and purpose of her own, decided to continue her education and work towards a Ph.D. from London University. For her thesis topic she chose a critical treatment of previously published medieval Spanish texts. This topic allowed her to do her research in libraries rather than from primary sources. Even so, deciding to pursue a Ph.D. was a bold step, for Stephen was requiring more and more care and it was also about this time that the Hawkings decided to start a family. In the autumn of 1966, when Jane's first pregancy began, Stephen's fingers were beginning to curl and writing by hand became almost impossible for him. In an extraordinary move, for which Sciama was responsible, the Institute of Physics funded physical therapy at home for him twice a week.[6]

The Hawkings' first child, Robert, was born on 28 May 1967. It was four years since doctors had told Stephen Hawking he had two years to live. He was still on his feet, and he was a father. Jane recalls: 'It obviously gave Stephen a great new impetus, being responsible for this tiny creature.'[7]

Robert was still an infant when his parents whisked him off to America, on their first visit to the west coast. Hawking attended a seven-week summer school in Seattle, Washington. After this there was a fortnight at the University of California, Berkeley. He was living up to the reputation for 'international calibre' that had helped him win the Adams Prize. They ended the trip with a hop across the continent to spend time with Hawking's childhood friend John McClenahan (the friend who had bet that Hawking would never amount to anything) and Hawking's sister Mary, who now was practising medicine in the eastern United States. After nearly four months in America, Stephen, Jane and their baby returned to Cambridge in October in time for the Michaelmas term. Caius College had renewed Stephen's fellowship for two more years.

People who remember Stephen Hawking in the DAMTP in the mid- to late 1960s recall his making his way around the corridors with a cane, supporting himself against the wall and speaking with what sounded like a speech impediment. More than that, they remember his brashness in sessions involving some of the world's most distinguished scientists. The reputation that had begun when he challenged Fred Hoyle in 1964 was being reinforced regularly. While other young researchers kept a reverent silence, Hawking daringly asked unexpected and penetrating questions and clearly knew what he was talking about. The comments about 'a genius', 'another Einstein', began then. In spite of Hawking's ready wit and popularity, that reputation and his physical problems distanced him from some in the department. One acquaintance told me: 'He was very friendly always, but at the same time, some felt a little shy about asking him out with the gang for a beer at the pub.' It's no wonder Hawking feels it's been a problem preventing people from thinking of him as 'anything less or more than simply human'.[8]

In the late 1960s, Hawking's physical condition began deteriorating again. He had to use crutches. Then it became difficult for him to get about even with crutches. He waged a pitched battle against the loss of his independence. A visitor remembers watching him spend fifteen minutes getting up the stairs to bed on his crutches, determined to do it without help. His determination sometimes seemed to be pigheadedness. Hawking refused to make concessions to his illness, even when those 'concessions' were practical steps to make things easier for him and make him less of a burden to others. It was his battle. He would fight it his way. His way was to regard any concession as caving in, an admission of defeat, and to resist as long as possible. 'Some people would call it determination, some obstinacy,' says Jane Hawking. 'I've called it both at one time or another. I suppose that's what's kept him going.'[9] John Boslough, who wrote a book about Hawking in the early 1980s, called him 'the toughest man

I have ever met'.[10] Even with a bad cold or flu, Hawking rarely missed a day of work. Meanwhile, while Hawking refused to make concessions to his illness, Jane Hawking learned to make no concessions to him. This was *her* way of fighting and part of her campaign to keep his life as normal as possible.

Boslough also described Hawking as a 'gentle, witty man', who quickly made you forget about his physical problems. That 'gentle' wit cut through all nonsense and pretension. Hawking's ability to make light of himself, his problems, and even the science he was so keen on was awe-inspiring. It helped others to like him and most of the time eclipsed the feelings of 'differentness'. For some he became, in the department, the most fun to be around. Hawking seems to have been following, without probably ever having read it, the advice Louisa May Alcott's mother gave her family in times of overwhelming distress: 'Hope and keep busy.'

Hawking's future was more threatened by his communication problems than by his immobility. His speech was becoming more and more slurred and indistinct, so much so that Caius College and the University had to face the fact that he could not give regular lectures. His research fellowship was about to expire again in 1969. Once more Denis Sciama saved the day, this time with help from Hermann Bondi. A rumour got about – no one knows who started it, and maybe it was true – that King's College, just down the street, was going to offer Hawking a Senior Research Fellowship. Caius came up with a way to keep him – a six-year contract for a specially created 'Fellowship for Distinction in Science'. Hawking was becoming an important physicist. He was far too valuable to lose.

Hawking's science continued to occupy his mind far more than concern about canes and crutches and stairs. His almost obsessive enjoyment of his work set the tone of his life. In the late 1960s he was finding out what the universe is like and how it might have begun – what he describes as playing 'the game of universe'. In order to understand the work he was immersed in, we have to go back thirty-five years.

The Game of Universe

Today we take it for granted that we live in a lacy spiral disc galaxy – the Milky Way – one of many galaxies more or less like it in the universe, with vast stretches of space between them. Early in the twentieth century not everyone accepted this picture. It was the American astronomer Edwin Hubble who, in the 1920s, showed that there are indeed many galaxies besides our own. Is there any pattern to the movement of these galaxies? Hubble showed that there is, with one of the most revolutionary discoveries of the century: the distant galaxies are all moving away from us. The universe is expanding.

Hubble found that the more distant a galaxy is, the more rapidly it's moving away from us: twice as far, twice as fast. We observe some extremely distant galaxies receding as fast as two-thirds the speed of light. Does that mean every star in the universe is moving away from us? No. Our near neighbours are milling around, some approaching, some receding. It's between clusters of galaxies that space is expanding. The most helpful way to think of the expansion of the universe is not as things rushing away from one another but as space between them swelling. It's an oversimplification, but imagine a loaf of raisin bread rising in the oven. As the dough swells, the raisins move apart. 'Twice as far, twice as fast' works with raisins as well as with galaxies.

If galaxies are receding from us and from each other, then unless something has changed drastically somewhere along the line, they used to be much closer together. At some moment in the past, wouldn't they all have been in exactly the same place? All the enormous amount of matter in the universe packed in a single point, infinitely dense?

That isn't the only possible history of an expanding universe. Perhaps there was once a universe something like ours, and that universe contracted, with all its galaxies getting closer together, looking as though they were on collision course. But galaxies and stars, and atoms and particles, for that matter, have other motion in addition to the motion that draws them straight towards one another. Planets orbit stars, for instance. The result in that universe might have been that, instead of meeting in a point of infinite density, the galaxies, or the particles that made them up, missed one another, flew past, and the universe expanded again until it looks the way it does today. Could it have happened like that? Which way did it happen? These were questions Hawking had begun to consider in his Ph.D. thesis. 'The big question was,' says Hawking, 'was there a beginning or not?'[11]

His search for an answer began, as we mentioned in Chapter 4, with an idea introduced by Roger Penrose in 1965. Penrose's idea concerned the way some stars may end up – something that three years later was going to be given the spectacular name 'black hole' by John Archibald Wheeler. The concept combined what we know about gravity with what general relativity tells us about the behaviour of light. Hawking's friend Kip Thorne would later remember 1965–80 as a golden age of black hole research. In the enormous achievements made, Hawking was in the vanguard.[12]

What Do We Know about Gravity and Light?

Gravity is the most familiar of the four forces. We all learned early in life that it's gravity you blame when your ice cream cone splats on the rug or when you fall off a swing. Asked to guess whether gravity is a very weak or a very strong force, you might answer 'incredibly strong'. You would be wrong. It's by far the weakest of the four forces. The gravity that's so conspicuous in our everyday lives is the gravity of this great hulk of a planet we live on, the combined gravity of every particle in it. The contribution of each individual particle is infinitesimal. It takes sensitive instruments to detect the faint gravitational attraction between small, everyday objects. However, because gravity always attracts, never repels, it has a talent for adding up.

Physicist John Wheeler liked to think of gravity as a sort of universal democratic system. Every particle has a vote that can affect every other particle in the universe. When particles band together and vote as a bloc (in a star, for instance, or in our Earth), they wield more influence. The very weak gravitational attraction of the individual particles in large bodies like the Earth adds up to a significant force: an influential voting bloc.

The more matter particles there are making up a body, the more mass that body has. Mass is not the same as size. Mass is a measure of how much matter is in an object, how many votes are in the voting bloc (regardless of how densely or loosely the matter is packed), and how much the object resists any attempt to change its speed or direction.

Sir Isaac Newton, Lucasian Professor of Mathematics at Cambridge in the 1600s, the same position Hawking would hold, discovered laws explaining how gravity works in more or less normal circumstances. According to Newton, bodies are not 'at rest' in the universe. They don't just sit still until some force comes along to push or pull them and then later 'run down' and sit still again. Instead, a body left completely undisturbed continues to move in a straight line without changing speed. It's best to think of everything in the universe as being in motion. We can measure our speed or direction in relation to other objects in the universe, but we can't measure them in relation to absolute stillness or anything that resembles absolute north, south, east, west, or up or down.

For example, if our moon were alone in space, it would not sit still but rather move in a straight line without changing its speed. (Of course, if it were truly all alone, there would be no way to tell it was doing this, nothing to which we could relate its motion.) But the moon isn't all alone. A force known as gravity acts on the moon to change its speed and direction. Where does that force come from? It comes from a nearby voting bloc of particles (a massive object) known as the Earth. The moon resists the change. It tries to keep moving in a straight line. How well it's able to resist depends on how many votes are in *it*, how massive it is. Meanwhile, the moon's gravity also affects the Earth. The most obvious result is the ocean tides.

Newton's theory tells us that the amount of mass a body

has affects how strong the pull of gravity is between it and another body. Other factors remaining equal, the greater the mass, the greater the attraction. If the Earth were double its present mass, the attraction of gravity between the Earth and the moon would be double what it currently is. Any change in the mass of either the Earth or the moon would change the strength of the gravitational pull between them. Newton also discovered that the farther apart bodies are, the weaker the pull between them is. If the moon were twice its present distance from the Earth, the pull of gravity between the Earth and the moon would be only one-fourth as strong. Newton's theory is usually stated: Bodies attract each other with a force that is proportional to their mass and inversely proportional to the square of the distance between them.

Newton's theory of gravity is an extremely successful theory. It wasn't improved on for over two hundred years. We still use it, though we now know that it fails in some circumstances, such as when gravitational forces become enormously strong (near a black hole, for instance) or when bodies are moving at near light speed.

Albert Einstein, early in the twentieth century, saw a problem with Newton's theory. If the strength of gravity between two objects is related to the distance between them, then if someone takes the sun and moves it farther from the Earth, the force of gravity between the Earth and the sun should change instantly. Is this possible?

Einstein's theory of special relativity recognized that the speed of light measures the same no matter where you are in the universe or how you're moving and that nothing can move faster than the speed of light. Light from the sun takes about eight minutes to reach Earth. We always see the sun as it was eight minutes ago. So, move the sun farther from the Earth; the Earth won't find out this has happened and feel any effect of the change for eight minutes. For eight minutes we'll continue to orbit just as though the sun hadn't moved. In other words, the effect of the gravity of one body on the other cannot change instantaneously because gravity can't move faster than the speed of light. Information about how far away the sun is cannot move instantaneously across space. It can move no faster than about 186,000 miles (300,000 kilometres) per second.

It's obvious then that when we talk about things moving in the universe, it's not realistic to talk in terms of only the three dimensions of space. If no information can travel faster than the speed of light, things out there at astronomical distances simply don't exist for us or for each other without a time factor. Describing the universe in three dimensions is as inadequate as describing a cube in two. Instead we must recognize the time dimension, admit there are really four dimensions, and talk of spacetime.

Einstein spent several years developing a theory of gravity that would work with what he'd discovered about light and motion at near light speed. In 1915 he introduced his theory of general relativity, requiring us to think of gravity not as a force acting between bodies but in terms of the shape, the curvature, of four-dimensional spacetime itself. In general relativity gravity is the geometry of the universe.

Bryce DeWitt, at the University of Texas, suggested we begin thinking about this curvature by imagining someone who believes the Earth is flat trying to draw a grid on the Earth:

The result can be seen from an airplane on any clear day over the cultivated regions of the Great Plains. The land is subdivided by east–west and north–south roads into square-mile sections. The east–west roads often extend in unbroken lines for many miles, but not the north–south roads. Following a road northward, there are abrupt jogs to the east or west every few miles. The jogs are forced by the curvature of the Earth. If the jogs were eliminated, the roads would crowd together, creating sections of less than a square mile. In the three-dimensional case one can imagine building a giant scaffold in space out of straight rods of equal length joined at angles of precisely 90 degrees and 180 degrees. If space is flat, the construction of the scaffold would proceed without difficulty. If space is curved, one would eventually have to begin shortening the rods or stretching them to make them fit.[13]

According to Einstein the curvature is caused by the presence of mass or energy. Every massive body contributes to the curvature of spacetime. Things going 'straight ahead' in the universe are forced to follow curved paths. Imagine a trampoline (Figure 5.1). In its centre lies a bowling ball, which causes a depression in the rubber sheet. Try to roll a golf ball in a straight line past the bowling ball. The golf ball will certainly change direction slightly when it meets the depression caused by the bowling ball. It will probably do more than that: it may even describe an ellipse and roll back in your direction. Something like that happens as the moon tries to continue in a straight line past the Earth. The Earth warps spacetime as the bowling ball warps the rubber sheet. The moon's orbit is the nearest thing to a straight line in warped spacetime.

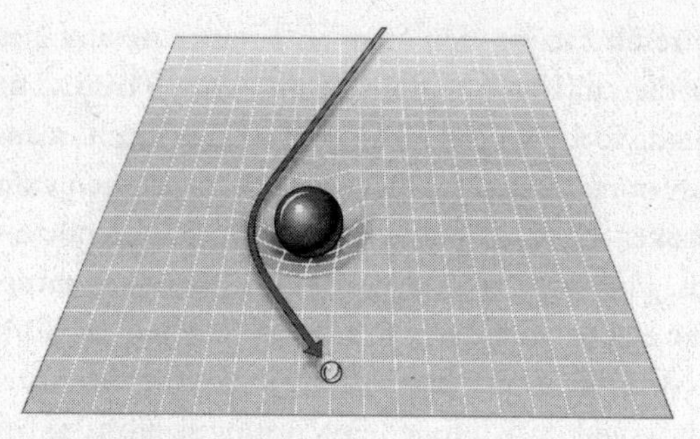

Figure 5.1. A bowling ball bends a rubber sheet where it lies. If you try to roll a smaller ball past the bowling ball, the path of the smaller ball will be bent when it encounters the depression caused by the bowling ball. In a similar manner, mass bends spacetime. Paths of objects in spacetime are bent when they encounter the curvature caused by a more massive object.

Einstein was describing the same phenomenon that Newton described. To Einstein a massive object warps spacetime. To Newton a massive object sends out a force. The result, in each case, is a change in the direction of a second object. According to the theory of general relativity, 'gravitational field' and 'curvature' are the same thing.

If you calculate planetary orbits in our solar system using Newton's theories and then calculate them again using Einstein's, you get almost precisely the same orbits, except in the case of Mercury. Because Mercury is the nearest planet to the sun, it's affected more than the others by the sun's gravity. Einstein's theory predicts a result of this nearness which is slightly different from the result predicted by Newton's theory. Observation shows that Mercury's orbit fits Einstein's prediction better than Newton's.

Einstein's theory predicts that other things besides moons and planets are affected by the warp of spacetime. Photons (particles of light) have to travel a warped path. If

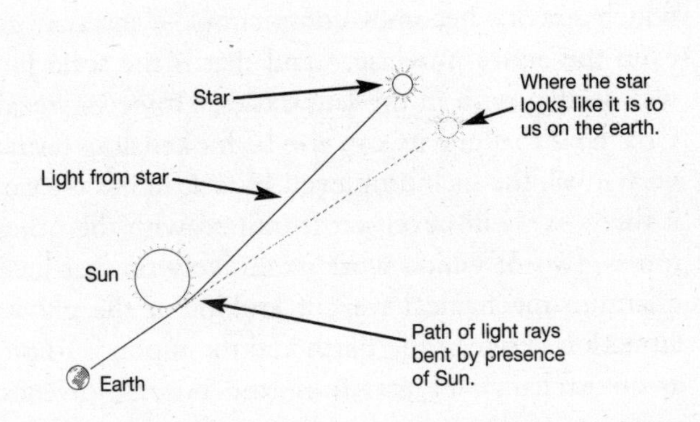

Figure 5.2. Because mass causes curvature of spacetime, the path of light travelling from a distant star bends as it passes a massive body like the sun. Notice the difference between the position of the star as we see it from the Earth and its true position.

a beam of light is travelling from a distant star and its path takes it close to our sun, the warping of spacetime near the sun causes the path to bend inwards towards the sun a bit, just as the path of the golf ball bends inwards towards the bowling ball in our model. Perhaps the path of light bends in such a way that the light finally hits the Earth. Our sun is too bright for us to see such starlight except during an eclipse of the sun. If we see it then and don't realize the sun is bending the path of the star's light, we're going to get the wrong idea about which direction the beam of light is coming from and where that star actually is in the sky (Figure 5.2). Astronomers make use of this effect. They measure the mass of objects in space by measuring how much they bend the paths of light from distant stars. The greater the mass of the 'bender', the greater the bending.

We've been talking about gravity in terms of what we observe on the large scale. That, of course, is the scale on which gravity becomes conspicuous – in stars, galaxies, even the entire universe – and that is the scale Hawking was dealing with in the late sixties. However, recall from Chapter 2 that gravity can also be looked at in terms of the very small, the quantum level. In fact, unless we can study it there, we will never get it unified with the other three forces, two of which work exclusively on that level. The quantum-mechanical way of looking at the gravitational attraction between the Earth and the moon is to picture it as an exchange of gravitons (the bosons, or messenger particles, of the gravitational force) between the particles that make up those two bodies.

With that background, we'll treat ourselves to a little science fiction.

A Disastrous Day for Earth

Remind yourself what the force of gravity feels like on Earth (Figure 5.3a), then pretend you go off on a vacation in space. While you're away something drastic happens to the Earth: it gets squeezed to only half its original size. It still has the same mass, but that mass is pressed together much more tightly. Returning from your vacation, your spacecraft hovers for a while at the place in space where the Earth's surface used to be before the squeezing. You feel as heavy there as you did before you went away. The pull of the Earth's gravity there hasn't changed, because neither your mass nor that of the Earth has changed, and you are still the same distance as before from the Earth's centre of gravity. (Remember Newton!) The moon, out beyond you, still orbits as before. However, when you land on the new surface (a much smaller radius, quite a bit nearer the Earth's centre of gravity), the gravity on that new surface is four times what you remember on the Earth's surface before the squeezing. You feel much heavier (5.3b).

What if something far more dramatic happened? What if the Earth were squeezed to the size of a pea – all the mass of the Earth, billions of tons, squeezed into that tiny space? Gravity on its surface would be so strong that escape velocity would be greater than the speed of light. Even light couldn't escape. The Earth would be a black hole. However, at the radius out in space where the surface of the Earth was before any squeezing, the pull of the Earth's gravity would still feel just the same as it does to us today (Figure 5.3c). The moon would still be orbiting just as before.

As far as we know, that story can't happen. Planets don't become black holes. However, there's a good chance some stars do. Let's retell the story, this time with a star.

Begin with a star that has a mass about ten times that of the sun. The star's radius is about 3 million kilometres, about five times that of the sun. Escape velocity is about 1,000 kilometres per second. Such a star has a life span of about a hundred million years, during which a life-and-death struggle goes on within it.

On one side of the contest is gravity: the attraction of every particle in the star for every other. It was gravity that pulled particles in a gas together to form the star in the first place. The pull is even more powerful now that the particles are closer together. Gravity tries to make the star collapse.

The pressure of the gas in the star opposes gravity. This pressure comes from heat released when hydrogen nuclei in

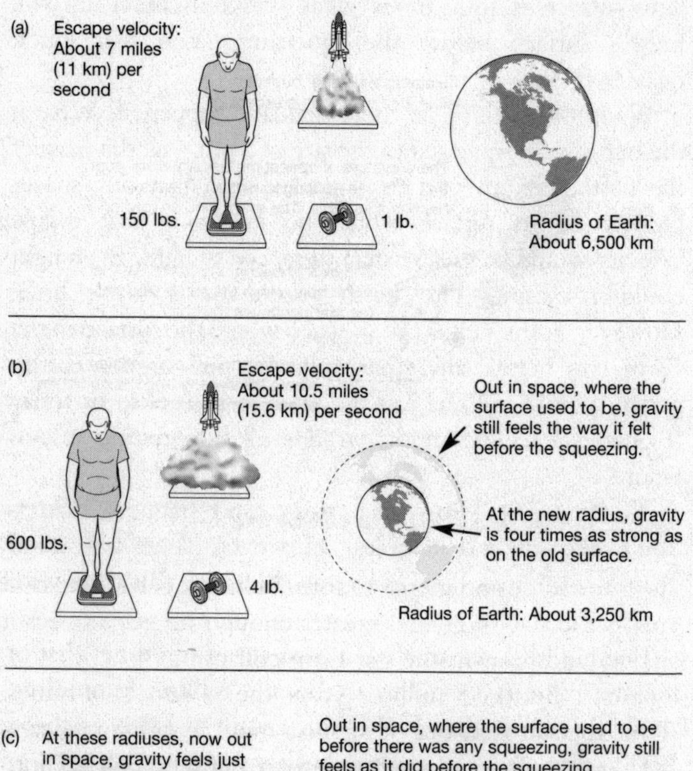

Figure 5.3. The day the Earth gets squeezed.

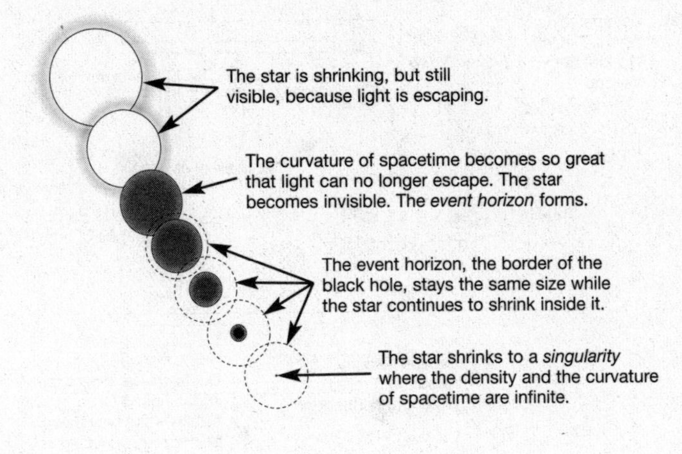

Figure 5.4. A star collapses and becomes a black hole.

the star collide and merge to form helium nuclei. The heat makes the star shine and creates enough pressure to resist gravity and prevent the star from collapsing.

For a hundred million years the contest continues. Then the star runs out of fuel: no more hydrogen to convert into helium. Some stars then convert helium into heavier elements, but that gives them only a short reprieve. When there's no more pressure to counteract gravity, the star shrinks. As it does, the gravity on its surface becomes stronger and stronger, in the same way that gravity on the Earth's surface did in the shrinking Earth story. It won't have to shrink to the size of a pea to become a black hole. When the 10-solar-mass star's radius is about 20 miles (30 kilometres), escape velocity on its surface will have increased to 186,000 miles (300,000 kilometres) per second, the speed of light. When light can

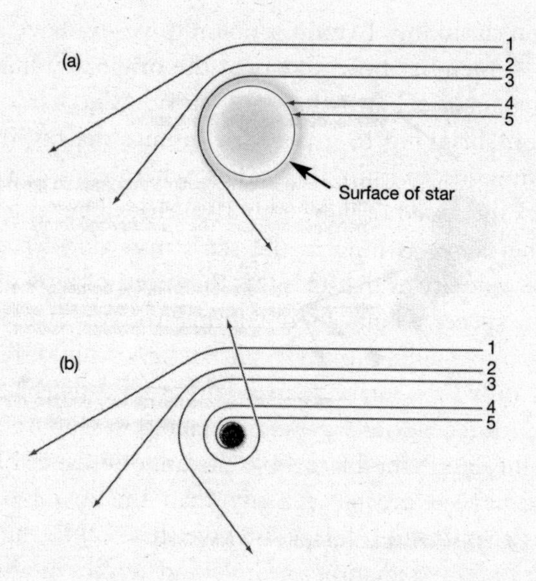

Figure 5.5. In (a), particles from space move towards a star. The paths of particles 1, 2 and 3 are bent as they pass the star. The closer to the star, the greater the bending. Particles 4 and 5 hit the surface of the star. In (b), we see the same particles moving towards the star after it has become a black hole. The paths of particles 1, 2 and 3 are bent exactly as before, because the spacetime outside a star is the same as the spacetime outside a black hole of the same mass. (Recall the shrinking Earth.) Particle 4 circles the black hole and then escapes. It might circle it many times. Particle 5 is captured by the black hole.

no longer escape, the star becomes a black hole (Figure 5.4).*

After the escape velocity on its surface is greater than the speed of light, we don't have to ask whether the star

* Stars less massive than about 8 solar masses probably don't shrink all the way to become black holes. Only more massive stars become black holes.

goes on shrinking. Even if it doesn't, we still have a black hole. Remember how gravity at the original radius never changed in the Earth-shrinking story. Whether our star goes on shrinking to a point of infinite density or stops shrinking just within the radius where escape velocity reaches the speed of light, gravity at that radius is going to feel the same, as long as the star's mass doesn't change. Escape velocity at that radius is the speed of light and will stay the speed of light. Light coming from the star will find escape impossible. Nearby beams of light from distant stars won't only be bent; they may curl around the black hole several times before escaping or falling in (Figure 5.5). If the light enters the black hole, it cannot escape. Nothing can achieve a greater velocity than the speed of light. What a profound 'blackout' we have! No light, no reflection, no radiation of any kind (radio, microwave, X-ray and so on), no sound, no sight, no space probe, absolutely no information can escape. A black hole indeed!

The radius where escape velocity is the speed of light becomes the border of the black hole, the radius-of-no-return: the 'event horizon'. Hawking and Penrose, in the late 1960s, suggested defining a black hole as an area of the universe, or a 'set of events', from which it's impossible for anything to escape to a distance. That has become the accepted definition. A black hole, with its event horizon for an outer boundary, is shaped like a sphere, or, if it's rotating, a bulged-out sphere that looks elliptical when seen from the side (or would, if you could see it). The event horizon is marked by the paths in spacetime of rays of light that hover just on the edge of that spherical area, not being pulled in but unable to escape. Gravity at that radius is strong enough to stop their escape but not strong enough to pull them back. Will you see them as a great orb shimmering in space? No. If the photons can't escape from that radius, they can't reach your eyes. In order for you to see something, photons from it have to reach your eyes.

Classical black hole theory tells us that there are only three secrets a black hole divulges: its mass, its electric charge (if it has any), and its angular momentum or speed of rotation (if it is rotating). John Wheeler, who liked to draw helpful pictures on the chalk-board for his students, drew a television set, a flower, a chair, 'known particles', gravitational and electromagnetic waves, angular momentum, mass and even 'particles as yet undetected' falling into a black hole, shown as a funnel, and nothing coming out at the bottom of the funnel except mass, electric charge and angular momentum. Part of Hawking's work in the early 1970s[14] would help to show that, as Wheeler summed it up: 'Black holes have no hair.'

It's the mass of the black hole that determines its size. If you want to calculate the radius of a black hole (the radius at which the event horizon forms), take the solar mass of the black hole (the same as for the star that collapsed to form it unless that star lost mass earlier in the collapse) and multiply by 2 for miles or 3 for kilometres. You'll find that a 10-solar-mass black hole (that is, a black hole whose mass is ten times the mass of our sun) has its event horizon at a radius of 20 miles (30 kilometres). It's clear that if the mass changes, the radius where the event horizon is also changes. The black hole changes in size. We'll talk more about this possibility later.

Having drawn the curtain at the event horizon, the star

has complete privacy, while any light it emits (any picture of itself that otherwise would be viewed from elsewhere in the universe) is pulled back in. Penrose had wanted to know whether the star would go on collapsing – or just what would happen to it. He had discovered that a star collapsing as we've described has all its matter trapped inside its own surface by the force of its own gravity. Even if the collapse isn't perfectly spherical and smooth, the star does go on collapsing. The surface eventually shrinks to zero size, with all the matter still trapped inside. Our enormous 10-solar-mass star is then confined not just in a region with a 20-mile (30-kilometre) radius (where its event horizon is), but rather in a region of *zero* radius – zero volume. Mathematicians and physicists call that a singularity. At such a singularity the density of matter is infinite. Spacetime curvature is infinite, and beams of light aren't just curled around: they're wound up infinitely tightly.

General relativity predicts the existence of singularities, but in the early 1960s few took this prediction seriously. Physicists thought that a star of great enough mass undergoing gravitational collapse *might* form a singularity. Penrose had shown that if the universe obeys general relativity, it *must*.

6

'There is a singularity in our past'

PENROSE'S DISCOVERY THAT A STAR OF GREAT ENOUGH mass undergoing gravitational collapse must form a singularity set fire to Hawking. With Robert Geroch and Penrose, he began to extend ideas about singularities to other physical and mathematical cases.[1] He was certain the discovery had significant implications for the beginning of the universe. This was exhilarating work, with the 'glorious feeling of having a whole field virtually to ourselves'.[2] Hawking realized that if he reversed the direction of time so that the collapse became an expansion, everything in Penrose's theory would still hold. If general relativity tells us that any star that collapses beyond a certain point must end in a singularity, then it also tells us that any expanding universe must have *begun* as a singularity. For this to be true the universe must be like what scientists call a Friedmann model. What is a Friedmann model of the universe?

A Choice of Universes

Before Hubble demonstrated that the universe is expanding, belief in a static universe (one that isn't changing in size) was very strong, so much so that when Einstein produced his theory of general relativity in 1915, and that theory predicted the universe was not static, Einstein was so sure it was static that he revised his theory. He put in a 'cosmological constant' to balance gravity. Without this cosmological constant the theory of general relativity predicted what we now know to be true: the universe is changing in size.

A Russian physicist, Alexander Friedmann, decided to take Einstein's theory at face value without the cosmological constant. Doing so, he predicted what Hubble would discover in 1929: The universe is expanding.

Friedmann began with two assumptions: (1) the universe looks much the same in whatever direction you look (except for nearby things like the shape of our Milky Way galaxy and our solar system); (2) the universe looks like this from wherever you are in the universe. In other words, no matter where you travel in space, the universe *still* looks much the same in whatever direction you look.

Friedmann's first assumption is fairly easy to accept. The second isn't. We don't have any scientific evidence for or against it. Hawking says, 'We believe it only on grounds of modesty: it would be most remarkable if the universe looked the same in every direction around us, but not around other points in the universe!'[3] Perhaps remarkable, but not impossible, you may argue. Modesty seems no more logical a reason for believing something than pride. However, physicists tend to agree with Friedmann.

In Friedmann's model of the universe all the galaxies

move away from one another. The farther apart two galaxies are, the more rapidly they move away from one another. This agrees with what Hubble observed. According to Friedmann, wherever you travel in space you'll still find all the galaxies moving away from you. In order to understand this, imagine an ant crawling on a balloon that has evenly spaced dots painted on it. You have to pretend the ant can't see the dimension that would allow it to look 'out' from the surface. Nor is it aware that the balloon has an interior. The ant's universe involves only the surface of the balloon. It looks the same in any direction. No matter where the ant crawls on the balloon, it sees as many dots ahead of it as behind. If the balloon is getting larger, the ant sees all the dots move away, no matter where it stands on the surface. The balloon 'universe' fits Friedmann's two assumptions: it looks the same in all directions. It looks the same no matter where you are in it.

What else can we say about the balloon universe? It isn't infinite in size. The surface has dimensions we can measure, like the surface of the Earth. No one would suggest that the surface of the Earth is infinite in size. However, it also has no boundaries, no ends. Regardless of where the ant crawls on the surface, it never comes up against any barrier, finds any end to the surface, or falls off an edge. It eventually gets back to where it started.

In Friedmann's original model, space is like that, with three dimensions rather than two. Gravity bends space around on to itself. The universe is not infinite in size, but neither does it have any end, any boundary. A spaceship will never get to a place in space where the universe ends. That may be difficult to understand, because we tend to think of *infinite* as meaning 'having no end'. The two do not have the same meaning.

Hawking points out that although the idea of circumnavigating the universe and ending up where you started makes great science fiction, it doesn't work, at least with this Friedmann model. You'd have to break the speed limit of the universe (the speed of light) – which isn't allowed – to get all the way around before the universe ends. It's an extremely large balloon. We are extremely small ants.

Time in this Friedmann model, like space, isn't infinite. It can be measured. Time, *unlike* space, *does* have boundaries: a beginning and an end. Look at Figure 6.1a. The distance between two galaxies at the beginning of time is zero. They move apart. The expansion is slow enough and there is enough mass in the universe so that eventually gravitational attraction stops the expansion and causes the universe to contract. The galaxies move towards each other again. At the end of time the distance between them is once again zero. That may be what our universe is like.

Figures 6.1b and 6.1c show two other possible models that would also obey Friedmann's assumptions (that the universe looks the same in every direction and that it looks the same from wherever you are in the universe). In Figure 6.1b, the expansion is much more rapid. Gravity can't stop it, though it does slow it a little. In Figure 6.1c, the universe is expanding just fast enough not to collapse, but not as fast as in Figure 6.1b. The speed at which galaxies are moving apart grows smaller and smaller, but they always continue to move apart. If the universe is like either of these two models, space is infinite. It doesn't curve back around on to itself.

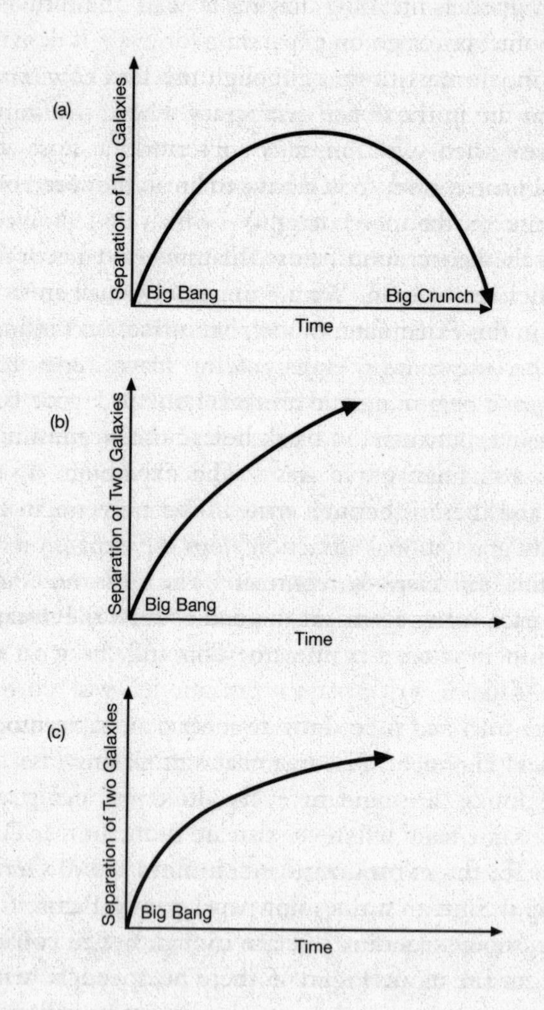

Figure 6.1. Three models that obey Friedmann's assumptions that the universe looks much the same in any direction we look, and the universe looks much the same from wherever you might be in the universe.

Which model fits our universe? Will the universe collapse some day or go on expanding for ever? It depends on how much mass there is in the universe: how many votes there are in the entire democracy. It will take much more mass than we presently observe to close the universe. That is a very simple statement of a more complicated issue, as you'll see later.

Penrose's theory about stars collapsing and becoming singularities only worked with a universe infinite in space that will go on expanding for ever (as in Figures 6.1b and 6.1c), not collapse (as in Figure 6.1a). Hawking first set out to prove that a universe infinite in space not only would have singularities in black holes but also must have begun as a singularity. He was confident enough by the time he finished his thesis to write: 'There is a singularity in our past.'[4]

In 1968, Hawking's and Penrose's essay on the beginning of time won second place in the Gravity Research Foundation Awards, but the question still hung in the balance: what if Friedmann's *first* model was correct, where the universe is not infinite in space and eventually recollapses (Figure 6.1a)? Must that sort of universe also have begun as a singularity? By 1970 Hawking and Penrose were able to show that it must have. Their definitive statement on the subject, in the 1970 *Proceedings of the Royal Society*,[5] was a joint paper proving that if the universe obeys general relativity and fits *any* of the Friedmann models, and if there is as much matter in the universe as we observe, the universe must have begun as a singularity, where all the mass of the universe was compressed to infinite density, where spacetime curvature was infinite.

Physical theories can't really work with infinite numbers. When the theory of general relativity predicts a singularity of infinite density and infinite spacetime curvature, it's also predicting its own breakdown. In fact, all our scientific theories break down at a singularity. We lose our ability to predict. We can't use the laws of physics to say what would emerge from the singularity. It could be any sort of universe. And what about the question of what happened *before* the singularity? It's not even clear that this question has any meaning.

A singularity at the beginning of the universe would mean that the beginning of the universe is beyond our science, beyond anything that claims to be a Theory of Everything. We would simply have to say, time began, because we observe that it did, and that in itself is a very big arbitrary element. A singularity is a door slammed in our faces.

Bedtime Story

Physicists are notorious for being eternally preoccupied with their physics. Even more than most of his colleagues, and partly as a consequence of his disability, Hawking was able to take his work with him anywhere, any time, because it was almost all done in his head. As Kip Thorne described it, he had developed a very unusual ability to manipulate mental images of objects, curves, surfaces, shapes, not merely in three dimensions but in spacetime's four dimensions.[6]

Typical of Hawking's work mode was a bedtime discovery he described in his book *A Brief History of Time*: 'One evening in November of 1970, shortly after the birth of my daughter, Lucy, I started to think about black holes as I was getting into bed. My disability makes this rather a slow process, so I had plenty of time.'[7] Another physicist might have nipped over to his desk and scribbled some notes and equations, but Hawking made one of the most significant discoveries of his career in his head, got into bed, and lay awake for the rest of the night, eager for dawn to break so that he could phone Penrose and tell him about his new insight. Penrose, Hawking insists, *had* thought of it but had not realized the implications.

The idea that had struck Hawking was that a black hole can never get smaller because the area of an event horizon (the radius-of-no-return where escape velocity becomes greater than the speed of light) can never decrease.

To review briefly, a collapsing star reaches a radius where escape velocity is the speed of light. What happens to photons emitted by the star as it collapses past that radius? Gravity there is too strong to allow them to escape, but not strong enough to pull them into the black hole. They stay there, hovering. That radius is the event horizon. After that as the star continues to shrink, any photons it emits are drawn back in.

What Hawking realized was that the paths of light rays hovering at the event horizon cannot be paths of light rays that are approaching one another. Paths of light rays that approach one another would bash into one another and fall into the black hole, not hover. In order for the area of the event horizon to get smaller (and the black hole to get smaller), paths of light rays in the event horizon *would* have to approach one another. But, if they did, they would fall in, and the event horizon would *not* get smaller.

Another way of thinking about this is to realize that a

black hole *can* get *larger*. The size of a black hole is determined by its mass, so a black hole gets larger any time anything new falls in and adds to that mass. If nothing can get *out* of a black hole, its mass can't possibly decrease. A black hole can't get smaller.

Hawking's discovery became known as the second law of black hole dynamics: the area of the event horizon (the border of the black hole) can stay the same or increase but never decrease. If two or more black holes collide and form one black hole, the area of the new event horizon is as big as or bigger than the previous event horizons added together. A black hole can't get smaller or be destroyed or divided into two black holes, no matter how hard it might get zapped. Hawking's discovery had a familiar ring to it. It resembled another 'second law' in physics: the second law of thermodynamics, which is about entropy.

Entropy is the amount of disorder there is in a system. Disorder always increases, never decreases. An assembled jigsaw puzzle put carefully in a box might get jostled, mixing the pieces and spoiling the picture. But it would be very surprising if any jostling of the box caused a mess of unassembled pieces to fall into place and complete the puzzle picture. In our universe, entropy (disorder) always increases. Broken teacups never reassemble themselves. A messy room never straightens itself up.

Suppose you patch the teacup or tidy up the room. Something does become more ordered. Does entropy decrease? No. The mental and physical energy you burn in the process converts energy to a less useful form. That represents a decrease in the amount of order in the universe which outbalances any increase of order you achieved.

There's another way in which entropy resembles the event horizon of a black hole. When two systems join, the entropy of the combined system is as great as or greater than the entropy of the two systems added together. A familiar example describes gas molecules in a box. Think of them as little balls bouncing off one another and off the walls of the box. There's a partition down the centre of the box. Half the box (one side of the partition) is filled with oxygen molecules. The other half is filled with nitrogen molecules. Remove the partition, and oxygen and nitrogen molecules start to mix. Soon there's a fairly uniform mixture throughout both halves of the box, but that's a less ordered state than when the partition was in place: entropy – disorder – has increased. (The second law of thermodynamics doesn't always hold: there is the tiniest of chances, one in many millions of millions, that at some point the nitrogen molecules will be back in their half of the box and the oxygen molecules in the other.)

Suppose you toss the box of mixed-up molecules or anything else that has entropy into a convenient black hole. So much for that bit of entropy, you might think. The total amount of entropy outside the black hole is less than it was before. Have you managed to violate the second law? Someone might argue that the whole universe (inside and outside black holes) hasn't lost any entropy. But the fact is that anything going into a black hole is just plain lost to our universe. Or is it?

One of John Wheeler's graduate students at Princeton, Demetrios Christodoulou, pointed out that according to the second law of thermodynamics, the entropy (disorder) of a closed system always increases, never decreases, and that similarly the 'irreducible mass' (Christodoulou's name for a mathematical combination between a black hole's mass and its speed of rotation) never decreases, no matter what happens to the black hole. Was this resemblance only a coincidence? What connection could Christodoulou's idea or Hawking's more general and powerful statement[8] (the never-decreasing area of the event horizon) have with entropy and the second law of thermodynamics?

Escape from a Black Hole?

In his first announcement to the scientific community of his idea about the event horizon of a black hole never getting smaller, in December 1970 at the Texas Symposium of Relativistic Astrophysics,[9] Hawking insisted that though an increase in the area of the event horizon did indeed resemble an increase of entropy, this was only an analogy.

Another of Wheeler's graduate students at Princeton, Jacob Bekenstein, begged to differ. Bekenstein insisted that the area of the event horizon of a black hole isn't only *like* entropy; it *is* entropy.[10] When you measure the area of the event horizon, you're measuring the entropy of the black hole. You don't destroy entropy if you toss it into a black hole. The black hole already has entropy. You only increase it. When something falls into a black hole, such as a box of molecules, it adds to the mass of the black hole, and the event horizon gets larger. It also adds to the entropy of the black hole.

All of this brings us to a puzzling point. If something has entropy, that means it has a temperature. It is not totally cold. If something has a temperature, it has to be radiating energy. If something is radiating energy, you can't say that nothing is coming out. Nothing was supposed to come out of black holes.

Hawking thought Bekenstein was mistaken. He was irritated by what he thought was Bekenstein's misuse of his discovery that event horizons never decrease. In 1972 and 1973 he joined forces with two other physicists, James Bardeen and Brandon Carter, and seemed to tread close to a concession by coming up with no fewer than four laws of black hole mechanics that appeared to be almost identical to the four well-known laws of thermodynamics 'if one only replaced the phrase "horizon area" with "entropy," and the phrase "horizon surface gravity" with "temperature"'.[11] Nevertheless the three authors continued to stress that these were only analogies, and in the final version of their paper[12] reiterated that their four laws of black hole mechanics were similar to, but distinct from, the four laws of thermodynamics. Although there were many similarities between entropy and the area of the event horizon, a black hole could not have entropy because it could not emit anything. That was an argument that Bekenstein could not gainsay, but even though he was a graduate student up against a trio of established physicists, he was not convinced. It turned out that Hawking, Bardeen and Carter were wrong. It would fall to Hawking himself to show how.

In 1962 when Hawking had begun his graduate studies at Cambridge, he'd chosen cosmology, the study of the very large, rather than quantum mechanics, the study of the very small. Now, in 1973, he decided to shift ground and look at black holes through the eyes of quantum mechanics. It was to be the first serious, successful attempt by anybody to fuse the two great theories of the twentieth century: relativity and quantum mechanics.

Such a fusion, you'll remember from Chapter 2, is a difficult hurdle on the road to a Theory of Everything.

In January 1973, Hawking was thirty-one years old. The new year brought the publication of his first full-length book, co-authored with George Ellis and dedicated to Denis Sciama. Hawking describes *The Large Scale Structure of Space-Time* as 'highly technical, and quite unreadable'.[13] It still appears on the shelves of academic bookshops, and if you pull it down, and are not an accomplished physicist, you will probably agree with him. Although it will never match the sales of *A Brief History of Time*, it has become a classic in the field.

In August and September of that year, during Cambridge's long vacation, Stephen and Jane Hawking travelled to Warsaw for the celebration of the 500th anniversary of the birth of Nicolaus Copernicus, and continued east from there to Moscow. They asked Kip Thorne to go with them, because he had been carrying on joint research with Soviet physicists for five years and knew the ropes in the Soviet Union. Hawking wanted to confer with Yakov Borisovich Zel'dovich and Zel'dovich's graduate student Alexander Starobinsky. These two Russian physicists had been able to show that the uncertainty principle meant that rotating black holes would create and emit particles, produced by the hole's rotational energy. The radiation would come from just outside the event horizon and would slow down the black hole's rotation until the rotation stopped and the radiation ceased. Hawking thought Zel'dovich and Starobinsky were on to something, but he wasn't satisfied with their calculations. After the visit he returned to Cambridge determined to devise a better mathematical treatment.

Hawking expected his calculations to show that rotating black holes produce the radiation the Russians predicted. What he discovered was something far more dramatic: 'I found, to my surprise and annoyance, that even nonrotating black holes should apparently create and emit particles at a steady rate.'[14] At first he thought something had to be wrong with his calculations and he spent many hours trying to find his error. He was particularly eager that Jacob Bekenstein should not find out about his discovery and use it as an argument supporting his idea about event horizons and entropy. But the more Hawking thought about it, the more he had to admit that his own calculations were certainly not far off the mark. The clincher was that the spectrum of the emitted particles was precisely what you'd expect from any hot body.

Bekenstein was right: You cannot make entropy decrease and the universe get more orderly by throwing matter carrying entropy into black holes as though they were great rubbish bins. As matter carrying entropy goes into a black hole, the area of the event horizon gets larger: the entropy of the black hole increases. The total entropy of the universe both inside and outside black holes hasn't become any less.

But Hawking was on to a bigger puzzle now. How can black holes possibly have a temperature and emit particles if nothing can escape past the event horizon? He found the answer in quantum mechanics.

When we think of space as a vacuum, we haven't got it quite right. We've already seen that space is never a complete vacuum. Now we'll find out why.

The uncertainty principle means we can never know both the position and the momentum of a particle at the same time with complete accuracy. It means something more than that: we can never know both the value of a field (a gravitational field or an electromagnetic field, for instance) and the rate at which the field is changing over time with complete accuracy. The more precisely we know the value of a field, the less precisely we know the rate of change, and vice versa: the seesaw again. The upshot is that a field can never measure zero. Zero would be a very precise measurement of both the value of the field and its rate of change, and the uncertainty principle won't allow that. You don't have empty space unless all fields are exactly zero: no zero – no empty space.

Instead of the empty space, the true vacuum, that most of us assume is out there, we have a minimum amount of uncertainty, a bit of fuzziness, as to just what the value of a field is in 'empty' space. One way to think of this fluctuation in the value of the field, this wobbling a bit towards the positive and negative sides of zero so as never to be zero, is as follows:

Pairs of particles – pairs of photons or gravitons, for instance – continually appear. The two particles in a pair start out together then move apart. After an interval of time too short to imagine they come together again and annihilate one another – a brief but eventful life. Quantum mechanics tells us this is happening all the time, everywhere in the 'vacuum' of space. These may not be 'real' particles that we can detect with a particle detector, but they are not imaginary. Even if they are only 'virtual' particles, we know they exist because we can measure their effects on other particles.

Some of the pairs will be pairs of *matter* particles, fermions. In this case, one of the pair is an antiparticle. 'Antimatter', familiar from fantasy games and science fiction (it drives the starship *Enterprise*), isn't purely fictional.

You may have heard that the total amount of energy in the universe always stays the same. There cannot be any suddenly appearing from nowhere. How do we get around that rule with these newly created pairs? They're created by a very temporary 'borrowing' of energy. Nothing permanent at all. One of the pair has positive energy. The other has negative energy. The two balance out. Nothing is added to the total energy of the universe.

Hawking reasoned that there will be many particle pairs popping up at the event horizon of a black hole. The way he pictures it, a pair of virtual particles appears. Before the pair meet again and annihilate, the one with negative energy crosses the event horizon into the black hole. Does that mean the positive energy partner must follow its unfortunate companion in order to meet and annihilate? No. The gravitational field at the event horizon of a black hole is strong enough to do an astounding thing to virtual particles, even those unfortunates with negative energy: it can change them from 'virtual' to 'real'.

The transformation makes a remarkable difference to the pair. They are no longer obliged to find one another and annihilate. They can both live much longer, and separately. The particle with positive energy might fall into the black hole, too, of course, but it doesn't have to. It's free of the partnership. It can escape. To an observer at a distance it appears to come out of the black hole. In fact, it comes from just outside. Meanwhile, its partner has carried negative energy into the black hole (Figure 6.2).

The radiation that's emitted by black holes in this manner is now called Hawking radiation. And with

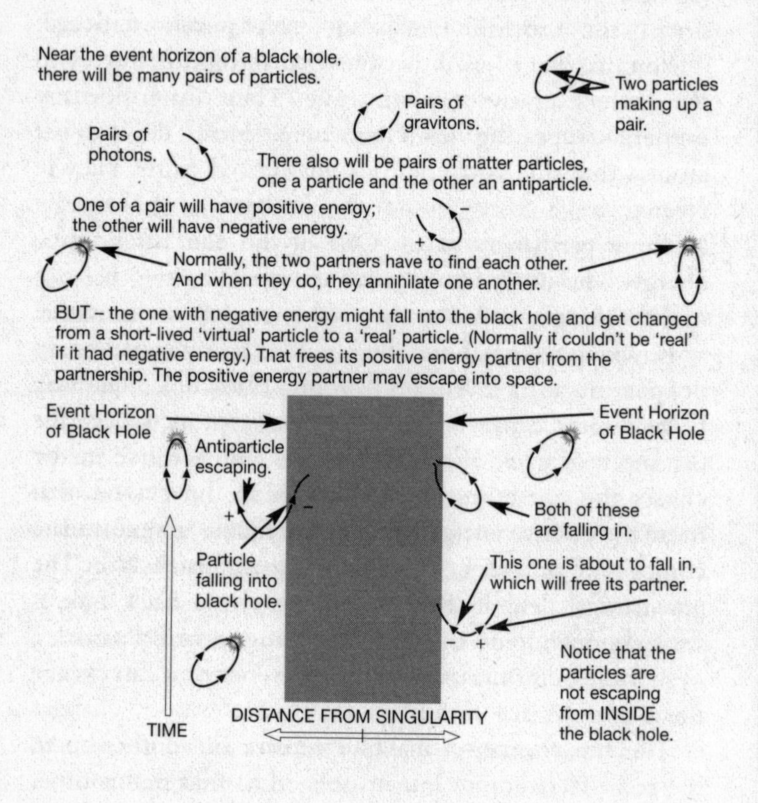

Figure 6.2. Hawking radiation

Hawking radiation, his second famous discovery about black holes, Hawking showed that his first famous discovery, the second law of black hole dynamics (that the area of the event horizon can never decrease), doesn't always hold. Hawking radiation means that a black hole might get smaller and eventually evaporate entirely. It was a truly radical concept.

How does Hawking radiation make a black hole get smaller? As the black hole transforms virtual particles to real particles, it loses energy. How can this happen, if nothing escapes through the event horizon? How can it lose anything? It's rather a trick answer: when the particle with negative energy carries this *negative* energy into the black hole, that makes *less* energy in the black hole. Negative means 'minus', which means less.

That's how Hawking radiation robs the black hole of energy. When something has less energy, it automatically has less mass. Remember Albert Einstein's equation, $E = mc^2$. The E stands for energy, the m for mass, the c for the speed of light. When the energy (on one side of the equal sign) grows less (as it is doing in the black hole), something on the other side of the equal sign grows less too. The speed of light (c) can't change. It must be the mass that grows less. So, when we say a black hole is robbed of energy, we're also saying it's robbed of mass.

Keep this in mind and recall what Newton discovered about gravity: any change in the mass of a body changes the amount of gravitational pull it exerts on another body. If the Earth becomes less massive (not smaller this time, less massive), its gravitational pull feels weaker out where the moon is orbiting. If a black hole loses mass, its gravitational pull becomes weaker out where the event horizon (the radius-of-no-return) has been. Escape velocity at that radius becomes less than the speed of light. There is now a smaller radius where escape velocity is the speed of light. A new event horizon forms closer in. The event horizon has shrunk. This is the only way we know that a black hole can get smaller.

If we measure Hawking radiation from a large black hole, one resulting from the collapse of a star, we'll be disappointed. A black hole this size has a surface temperature of less than a millionth of a degree above absolute zero. The larger the black hole the lower the temperature. Hawking says, 'Our 10-solar-mass black hole might emit a few thousand photons a second, but they would have a wavelength the size of the black hole and so little energy we would not be able to detect them.'[15] The way it works is that the greater the mass, the greater the area of the event horizon. The greater the area of the event horizon, the greater the entropy. The greater the entropy, the lower the surface temperature and the rate of emission.

Exploding Black Holes?

However, as early as 1971 Hawking had suggested that there was a second type of black hole: tiny ones, the most interesting ones about the size of the nucleus of an atom. These would positively crackle with radiation. The smaller a black hole is, the hotter its surface temperature. Referring to these tiny black holes, Hawking declares: 'Such holes hardly deserve [to be called] *black*: they really are *white hot*.'[16]

'Primordial black holes', as Hawking called them, would not have formed from the collapse of stars. They would be relics of the very early universe when there were pressures that could press matter together extremely tightly. A primordial black hole would by now be much smaller even than when it started out. It's been losing mass for a long time.

Hawking radiation would have drastic consequences for a primordial black hole. As the mass grows less and the black hole gets smaller, the temperature and rate of emission of particles at the event horizon increase. The hole loses mass more and more quickly. The lower the mass, the higher the temperature – a vicious circle!

How would the story end? Hawking guessed that the little black hole disappears in a huge final puff of particle emission, like millions of hydrogen bombs exploding. Will a large black hole ever explode? Some models have it that the universe will come to an end long before it reaches that stage.

The idea that a black hole could get smaller and finally explode was so much the reverse of everything anybody thought about black holes in 1973 that Hawking had grave doubts about his discovery. For weeks he kept it under wraps, reviewing the calculations in his head. If he found it so hard to believe, it was fearful to predict what the rest of the scientific world would make of it. No scientist enjoys the prospect of ridicule. On the other hand, Hawking knew that if he was right, his findings would revolutionize astrophysics. At one point he locked himself in the bathroom to think about the problem. 'I worried about this all over Christmas, but I couldn't find any convincing way to get rid of [these findings].'[17]

Hawking tested his idea on his close associates. The reception was mixed. Martin Rees, a friend since their days as graduate students at Cambridge, approached their old thesis supervisor, Denis Sciama, with the exclamation, 'Have you heard? Stephen's changed everything!' Sciama rallied to Hawking's support, urging him to release his findings. Hawking complained that Penrose phoned him, full of enthusiasm, just as he was sitting down to his 1974 birthday dinner, ready to tuck into his goose. He appreciated Penrose's excitement, Hawking said, but, once into the subject, they talked too long. His dinner got cold.[18]

Hawking agreed to present his bizarre discovery in February in a paper at the Rutherford–Appleton Laboratory south of Oxford. Sciama was the organizer of the meeting, the Second Quantum Gravity Conference. Hawking had hedged his bets a little by putting a question mark in the title of his paper, 'Black Hole Explosions?', but, travelling to Oxford, he still agonized over his decision to announce his discovery.

The short presentation, including slides of equations, was greeted with silence that became embarrassing, and few questions. Hawking's arguments had gone over the heads of many in the audience, experts in other fields. But it was more or less obvious to everyone that he was proposing something completely contrary to accepted theory. Those who did understand were shocked and unprepared to argue with him. The lights were snapped back on. The moderator, John G. Taylor, a respected professor from the University of London, rose and declared: 'Sorry, Stephen, but this is absolute rubbish.'[19]

Hawking published this 'rubbish' the next month in the prestigious science magazine *Nature*.[20] Taylor and Paul C. W. Davies disagreed with Hawking in a paper in the same issue.[21] Within a few days physicists all over the world were discussing Hawking's shocking idea. Zel'dovich had reservations at first, but when Kip Thorne was next in Moscow he had an urgent summons to visit the Soviet physicist. When Thorne arrived, Zel'dovich and Starobinsky greeted him with hands in the air, as though they were in the old American West and Thorne had them at gunpoint: 'We give in. Hawking was right. We were wrong.'[22]

Some were calling Hawking's discovery the most significant in theoretical physics in years. Sciama said the paper was 'one of the most beautiful in the history of physics'.[23] John Wheeler, always a master with words, said that talking about Hawking's beautiful discovery was like 'rolling candy on the tongue'.[24] Kip Thorne commented that as Stephen had lost the use of his hands he had developed 'geometrical arguments that he could do pictorially in his head . . . a very powerful set of tools that nobody else really had. And if you are the only master in the world of these new tools, that means there are certain kinds of problems you can solve and nobody else can.'[25] Things were certainly looking up.

Hawking took more time and care putting together a second paper about his discovery. *Communications in Mathematical Physics*, the journal to which he submitted it in March 1974, lost his paper and didn't publish it until April 1975,[26] after he resubmitted it. In the meantime, Hawking and his colleagues went on studying 'Hawking radiation' from many different angles. By the time the next four years had passed – and after a joint paper written by Hawking and Jim Hartle appeared in 1976[27] – Hawking radiation had been generally accepted throughout the theoretical physics world. Most agreed that Hawking had made a significant breakthrough. He had used the activity of virtual particles to explain about something that had arisen from the theory of relativity – black holes. He'd taken a step towards linking relativity and quantum physics.

PART II
1970–1990

7

'These people must think we are used to an astronomical standard of living'

WHEN LUCY WAS BORN ON 2 NOVEMBER 1970, THE Hawkings had recently purchased the house they had been renting in Little St Mary's Lane. Stephen's parents had given them money to fix it up and qualify for a mortgage. The work had finally been completed when Jane was eight months pregnant.

Stephen was still insisting on pulling himself up and down the stairs and dressing himself in the morning and undressing at night. His comment that he had plenty of time to think about photons at the event horizons of black holes while getting ready for bed is one of very few admissions he has made that any of this was extremely slow and arduous. His walking, however, finally became so perilous that he consented to a wheelchair. He'd lost the battle to stay on his feet. Friends watched with sadness, but

Hawking's humour and strength of purpose didn't fail him.

Hawking's loss of the use of his hands, meaning he could no longer write and draw equations and diagrams, did not happen overnight. Over the years of gradual loss, he had time to adapt and train 'his mind to think in a manner different from the minds of other physicists. He thinks in new types of intuitive mental pictures and mental equations that, for him, have replaced paper-and-pen drawings and written equations,' said Kip Thorne.[1] Listening to Hawking himself, you get the impression that he believes he might have chosen this way of working even if he had full use of his hands: 'Equations are just the boring part of mathematics. I prefer to see things in terms of geometry.'[2] The calculations involved in the discovery of Hawking radiation were done almost entirely in his head.

After Lucy's birth, Jane was juggling an all but impossible schedule, trying to finish her Ph.D. thesis while caring for Stephen, her toddler Robert and now her new baby girl. Her mother and a neighbour's nanny helped with the children whenever they could. The cottage in Little St Mary's Lane was a delight. As the children grew and Lucy became a proficient toddler, she joined her brother to play among the flowering plants and ancient stone markers in the Little St Mary's churchyard across the lane. Jane remembers summers with the windows open and the happy voices of her children 'piping in the churchyard'.

When Hawking's January 1971 entry in the yearly Gravity Research Foundation Award, titled 'Black Holes', won top prize, the prize money allowed the Hawkings to buy a new car. Hawking had a salary from Caius and research assistantships from the DAMTP and the Institute of Astronomy. However, the family budget was still tight and not adequate to pay for a private school for Robert when he reached school age. He began instead at the fine local school, Newnham Croft Primary School, which my own daughter would attend fifteen years later. Robert seemed to be following in his father's footsteps, excelling in maths and slow learning to read, but it was a new era, when 'slow learning to read' was not to be accepted without taking proactive steps. Together with Stephen's parents, Jane and Stephen bought half shares in a small terraced house. The rent from it would provide money for Robert's fees at a private school where his mathematical prowess could be particularly respected and encouraged, as well as a retirement income for his grandparents.[3] Robert transferred to the Perse School in Cambridge when he was seven.

The Hawkings continued to try to keep Stephen's illness in the background of their lives and not allow it to become the most important thing about him or about them. They made a habit of not looking to the future. As far as the rest of the world could see, they succeeded so well that it came as a surprise when Jane Hawking mentioned how terrible the difficulties sometimes were. Discussing honours that had come her husband's way, she told an interviewer: 'I wouldn't say [this overwhelming success makes] all the blackness worthwhile. I don't think I am ever going to reconcile in my mind the swings of the pendulum we have experienced from the depths of a black hole to the heights of all the glittering prizes.'[4] To judge from everything Stephen Hawking has written, he barely noticed the depths. Talking about them in any but the most offhand manner, which is the most he allows himself, would be for him a form of giving in, of defeat, and could undermine his resolute disregard of his problems. Most of the time, he continued to refuse to discuss his illness even with Jane, but that didn't, for her, make it any less the gorilla in the room.

Jane remembers that being unable to assist with his children or play with them in an active way was difficult for Stephen. She taught Robert, and later Lucy and Timmy, to play cricket ('I can get them out!' she gloated), and she teased her husband that, unlike other wives, she was not surprised or disillusioned when he proved useless around the house and with the children.

Hawking's practical uselessness became one of the positive side effects of his illness. It might take him a long time to get up and go to bed, but he didn't have to run errands, do home repairs, mow the grass, make travel arrangements, pack his suitcase, draw up lecture schedules or serve in time-consuming administrative positions in the DAMTP or at Caius. Such matters were left to colleagues and assistants and to his wife. He could spend all his time thinking about physics, a luxury which his colleagues envy him.

Jane had anticipated that an overwhelming proportion of these day-to-day responsibilities would fall to her. She had decided even before they were married in the 1960s that only one of them would be able to have a career, and it would have to be her husband. In the 1970s, perhaps partly because attitudes about the role of women were changing, that sacrifice became more difficult for her to accept. She'd thought that providing Stephen with the encouragement and assistance he so badly needed would give her life purpose and meaning. What it was not giving her was an identity. Motherhood wasn't doing that either. As she put it, although she adored her children and 'would not have wanted to farm them out to anybody else, Cambridge is a jolly difficult place to live if your only identity is as the mother of small children'.[5]

To be fair to the university community, whenever you mentioned the name Hawking in Cambridge someone was likely to comment that Jane Hawking was even more remarkable than Stephen. However, Jane Hawking didn't feel that was her reputation. As she saw it, in Cambridge 'the pressure is on you to make your way academically'.[6] That of course was the reason she had decided to go back to university to earn her Ph.D., but her drafts for that thesis were far too often languishing on the shelf.

Jane had much to be proud of in the 1970s. Robert and Lucy were turning out well; Hawking's career as a physicist was skyrocketing; his reputation as a remarkably tough and good-humoured man when the odds were against him was becoming legendary; and she was beginning to make her own mark academically. At the same time she felt increasingly that her enormous and burdensome role in Hawking's success went largely unnoticed. Hers was a problem which is not unusual for persons with a talent for making things look easy: others begin to assume that things *are* easy for them and fail to appreciate the sacrifice and effort involved. Both Jane and her husband knew that none of his success – probably not even his survival – could have happened without her, but she was allowed to share little of the triumph. In photos of him she was sometimes cropped out, thought to be a nurse pushing his wheelchair. Nor could she follow his mathematical reasoning and share his pleasure in that. Nevertheless, 'the joy and excitement of Stephen's success were tremendous',[7] she says. She did not regret the decision she made to marry him, but the rewards 'didn't alleviate the heartrending difficulties of coping day after day with motor neurone disease'.[8]

In spite of the difficulties there were many pleasures the Hawkings shared. Both of them were devoted to their children. They loved classical music and attended concerts and the theatre together. At Christmas they took Robert and Lucy to the pantomime. They also loved entertaining. Don Page, who as a postgraduate researcher would live with the Hawkings for three years as Hawking's assistant, remembers that Jane Hawking was 'very outgoing . . . a great professional asset' to her husband.[9] It wasn't unheard of to find her in the market shopping for a party of sixty people. The Hawkings became renowned for their hospitality.

Both of them were also deeply interested in increasing public awareness of the needs of disabled people and the possibility that they might expect to live normal, even brilliantly successful and active lives. This was not so much a part of our culture in the 1970s as it is becoming today. Britain had passed its Chronically Sick and Disabled Person's Act in 1970, but that was very slow in implementation. Occasionally, Jane Hawking felt indignant enough to protest. She wrote a letter to the board of the National Trust when Anglesey Abbey, a house and gardens open to visitors near Cambridge, insisted the Hawkings park in a car park half a mile away, not near the house.[10] Jane was soon adding campaigning for the rights of disabled people to her already over-long list of activities.

The Hawkings won some battles for wheelchair access. After a lengthy bureaucratic dispute about who should pay for it, a ramp was built at the back entrance into the DAMTP building. The Arts Theatre began reserving spaces for wheelchairs. The Arts Cinema also made room. Elsewhere, the English National Opera at the Coliseum made access possible. Where there was no such access, anyone near at hand was likely to be conscripted to lift Hawking and his chair up and down stairs. At Clare Hall, a graduate college of the University of Cambridge, members of the Astronomy Group were regularly called upon for this duty before and after their meetings. It wasn't always a safe procedure. Attendants at the Royal Opera House in Covent Garden dropped Stephen while carrying him up a flight of stairs to otherwise inaccessible seats.[11]

Faith in God and the Laws of Physics

Looking back in the late 1980s, Jane Hawking attributed her ability to cope for so many years with their unusual and often difficult life – a life with no hope of a long or happy future – to her faith in God. Without that, she said, 'I wouldn't have been able to live in this situation. I wouldn't have been able to marry Stephen in the first place, because I wouldn't have had the optimism to carry me through, and I wouldn't be able to carry on with it.'[12]

The faith which supported her so magnificently was not shared by her husband. It was shared by some of his physics colleagues, but the subject was not one she usually discussed with them. If there has been a religious or philosophical side to Stephen Hawking's confrontation with disability and the threat of early death, he's never spoken about that publicly. However, it seems evident from his books *A Brief History of Time* and *The Grand Design* that God is never far from Hawking's mind. He told an interviewer in the 1980s: 'It is difficult to discuss the beginning of the universe without mentioning the concept of God. My work on the origin of the universe is on the borderline between science and religion, but I try to stay [on the scientific] side of the border. It is quite possible that God acts in ways that cannot be described by scientific laws. But in that case one would just have to go by personal belief.'[13] Asked whether he thought his science is in competition with religion, he answered, 'If one took that attitude, then Newton [who was a very religious man] would not have discovered the law of gravity.'[14]

Hawking said he wasn't an atheist, but he preferred to 'use the term God as the embodiment of the laws of physics'.[15] 'You don't need to appeal to God to set the initial conditions for the universe, but that doesn't prove there is no God – only that he acts through the laws of physics.'[16] However, Hawking definitely did not believe in a personal God who cares for human beings as individuals, relates to them in a powerful and transforming way, and performs miracles. 'We are such insignificant creatures on a minor planet of a very average star in the outer suburbs of one of a hundred thousand million galaxies. So it is difficult to believe in a God that would care about us or even notice our existence.'[17] Einstein shared Hawking's view. Others, including some of Hawking's physics colleagues, would agree with Jane Hawking and call this a sadly limited view of God, pointing out that it's equally difficult to believe that all the intelligent and rational people (a good many scientists among them) who say they have experienced a personal

God are somehow deluded. This enormous difference in outlook could hardly have been illustrated more strikingly than in the views of Stephen and Jane Hawking.

'I used to find Stephen's assertion that he doesn't believe in a personal God quite hurtful,'[18] Jane remembers. She would tell an interviewer in 1988: 'He is delving into realms that really do matter to thinking people and in a way that can have a very disturbing effect on people. There's one aspect of his thought that I find increasingly upsetting and difficult to live with. It's the feeling that, because everything is reduced to a rational, mathematical formula, that must be the truth.'[19] It seemed to her that there was no room in her husband's mind for the possibility that the truth revealed in his mathematics might not be the whole truth. A year later she had changed her outlook somewhat: 'As one grows older it's easier to take a broader view. I think the whole picture for him is so different from the whole picture for anybody else by virtue of his condition and his circumstances . . . being an almost totally paralysed genius . . . that nobody else can understand what his view of God or what his relationship with God might be.'[20]

Truth might have to be mathematical, but physics, for Hawking, was not all there was to life. 'Physics,' he told an interviewer, 'is all very well, but it is completely cold. I couldn't carry on with my life if I only had physics. Like everyone else, I need warmth, love, and affection.'[21]

An Unusual Asset

In the late 1960s it had perhaps been generous of Hawking's college and university departments to keep on a young physicist who hadn't long to live, who might contribute little to his department in terms of lecturing and teaching. The DAMTP had from the beginning exempted Hawking from heavy teaching duties and allowed him to concentrate on his research and a few seminars and graduate students. By the mid-1970s Caius and the University had begun to realize that they'd done themselves a favour. He'd become a considerable asset.

However, at Cambridge extraordinary minds and personalities aren't uncommon. They crop up in one university department or another on a regular basis. It's a healthy environment for genius. No matter how much one is treated with awe in the wider world, within the university community it's generally just business as usual. Even in the late 1970s, when Hawking had become something of a legend, he and all his specialized equipment – gadgets to turn pages for him, computer terminals with special controls so that he could use them like a blackboard – still shared a cramped office with another researcher.

His communication problems were growing more severe. Earlier in the 1970s it was still possible to carry on a normal conversation with Hawking. However, by the late 1970s and early 1980s, when his speech was so slurred that only his family and closest friends could understand him, the job of 'interpreter' frequently fell to a research student. Michael Harwood, who later interviewed Hawking for *The New York Times*, described the process: 'Don Page, sitting beside him, leans close to hear the indistinct words, mouths each phrase to be certain he has caught it, often pauses and asks for a repetition, speaks a phrase back to Hawking sometimes to make certain, corrects himself.'[22] Another interviewer recalls that he often thought Hawking had finished a sentence only to find with the 'interpretation' that he'd spoken just one word. Hawking wrote his scientific papers by dictating them in this tedious fashion to his secretary. But he was learning to state ideas in the fewest words possible and to get to the point in scientific papers and conversations.

What he was saying in those few words was receiving worldwide attention. The procession of awards and recognition had escalated soon after Hawking announced his discovery of exploding black holes. In the spring of 1974 he was inducted into the Royal Society, one of the world's most prestigious bodies of scientists. At thirty-two he was young for the honour. During the rite of investiture, a ceremony dating from the seventeenth century, new Fellows walk to the podium to write their names in a book the earliest pages of which contain the signature of Isaac Newton. Those present when Hawking was inducted remember that the president of the society, Sir Alan Hodgkin, Nobel Prize winner in biology, broke tradition and carried the book down to Hawking in the front row. Hawking could still write his name with great effort, but it took him a long time. The gathering of eminent scientists waited respectfully. When Hawking finished and looked up with a broad grin, they gave him an ovation.

That same spring, the Hawkings enthusiastically accepted an invitation from the California Institute of Technology, where Kip Thorne was a professor, to spend the academic year 1974–5 there.[23] Hawking would be a Sherman Fairchild Distinguished Scholar. The offer included an excellent salary, a house, a car and even a new electric wheelchair. All medical expenses would be covered (the British National Health did not cover expenses out of the country) as well as schooling for Robert and Lucy.

In the spring when that offer came, it had been almost four years since the Hawkings had purchased and renovated their house in Little St Mary's Lane. Stephen's trips up and down the stairs, by this time accomplished by grasping the posts that held up the banister and using only the strength of his arms to pull himself up step by step to the second floor, had for a time been good physical therapy. They were finally proving impossible. Caius College chose to be more helpful with housing than they had been when the Hawkings were newlyweds. There was a new bursar, who offered them a spacious ground-floor flat in a college-owned brick mansion on West Road, not far from the back gate of King's College. The arrangement in this house, with a faculty family having the rather elegant although sometimes a bit run-down-around-the-edges lower floors, with graduate students in rooms above, was not unusual at that time in Cambridge. The flat Caius offered had high ceilings and large windows and required only a little modification and renovation to make it more suitable for the family and convenient for a wheelchair. This work could be completed while the Hawkings were in California. They would be able to move in when they returned. Except for a gravel parking area in front, the house was surrounded by gardens tended by Caius College gardeners, who were also willing to listen to suggestions and gardening plans from Jane Hawking. It would be an ideal childhood home for the Hawking children.

Even though he had given up on stair climbing,

Hawking could still feed himself and get into and out of bed, but these actions too were becoming increasingly difficult. Jane was still managing without outside help, working on overdrive to try to keep his life as normal as it could be with his worsening condition, make it possible for him to continue his work, and at the same time see that Robert and Lucy weren't robbed of a normal childhood. She had also occasionally been finding time to work on her thesis. But the Hawkings both knew that something had to change.

Planning ahead for the trip to California, Jane, over the Easter holiday, suggested a new solution for caring for Stephen – one that he could accept without feeling it represented a concession and courage-sapping defeat. They would begin a custom of asking a graduate student or a postdoctoral research student to live and travel with them. In return for free accommodation and extra attention from Hawking, the student would help him get ready for bed and get up. When the Hawkings departed for California, one of his graduate students, Bernard Carr, would go with them.

Jane Hawking booked flights and, with Bernard Carr's help, packed and moved belongings, two small children, husband and specialized equipment across the world to southern California with an efficiency that awed her friends.

A Place in the Sun

In August 1974, Kip Thorne met the Hawkings at Los Angeles airport in a gleaming new American station wagon that would be theirs for the length of their stay. It had been a long flight, over the North Pole from London, but the family revived in the languid air of southern California as Kip manoeuvred the freeways through the sprawling city with its skyscrapers and amazingly tall palm trees to Pasadena, ten miles north-east of downtown Los Angeles.[24]

They arrived, just at dusk, at the house that had been arranged for them – a lovely home with white weather-boarding, lights gleaming in every window. The house was across the street from the Caltech campus and had a view of the mountains. Jane described it in a letter she wrote on that first day to her parents back in England: 'It is as elegant inside as it is pretty out. These people must think we are used to an astronomical standard of living. If only they knew!'[25]

Stephen, Jane and their children discovered humming-birds on the patio, a huge climbable California oak tree in the garden, a television and several bathrooms. There was a swimming pool nearby on campus. You could easily drive to Disneyland. Stephen's new, state-of-the-art electric wheelchair was waiting for him. Like a racing car driver testing a new, faster, more manoeuvrable model, Hawking zoomed around, finding out what it could do, stopping to allow engineers to make adjustments.

As the Hawkings settled into their California life, a new and rather disturbing experience, which they had known to expect, was the frequent tremors and occasionally a rather startling earthquake. Their neighbours and Hawking's colleagues at Caltech seemed to take these in their stride and assured Jane and Stephen that small, frequent tremors made it less likely there would be a big, dangerous quake. The house and its occupants would make it through the year with no damage.

Robert's and Lucy's school was the Pasadena Town and

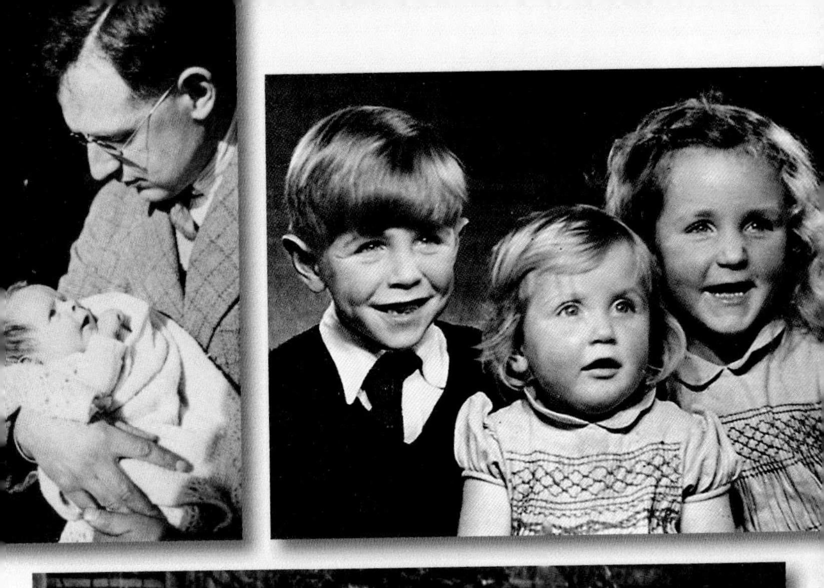

'An ordinary English boyhood' – Highgate and St Albans. (*Clockwise from top left*): Frank Hawking with newborn son Stephen; Stephen and his sisters Philippa and Mary; proud owner of a new bicycle; on holiday with Philippa and Mary on the steps of the family's gypsy caravan.

'Not a grey man' – undergraduate days at Oxford. (*Above*): success on the river as cox of his college boat; (*below*): carousing at a Boat Club party (Stephen is waving the handkerchief).

Hoyle and Narlikar derive their theory from the action:

$$A = \sum_{a \neq b}\sum \iint G(a,b)\, da\, db,$$

where the integration is over the world-lines of partic[les] a, b. In this expression G is a Green functio[n] that satisfies the wave equation:

$$G(x,x')_{;ij}\, g^{ij} + \tfrac{1}{6} R\, G(x,x') = \frac{\delta^4(x,x')}{\sqrt{-g}}$$

where g is the determinant of g_{ij}. Since the do[uble sum] in the action A is symmetrical between all pairs [of] particles a, b, only that part of $G(a,b)$ that [is] symmetrical between a and b will contribute to t[he action,] i.e. the action can be written

$$A = \sum_{a \neq b}\sum \iint G^{*}(a,b)\, da\, db$$

where $G^{*}(a,b) = \tfrac{1}{2} G(a,b) + \tfrac{1}{2} G(b,a)$.

Thus G^{*} must be the time-symmetric Green functi[on ...] be written: $G^{*} = \tfrac{1}{2} G_{ret} + \tfrac{1}{2} G_{adv}$ wher[e]

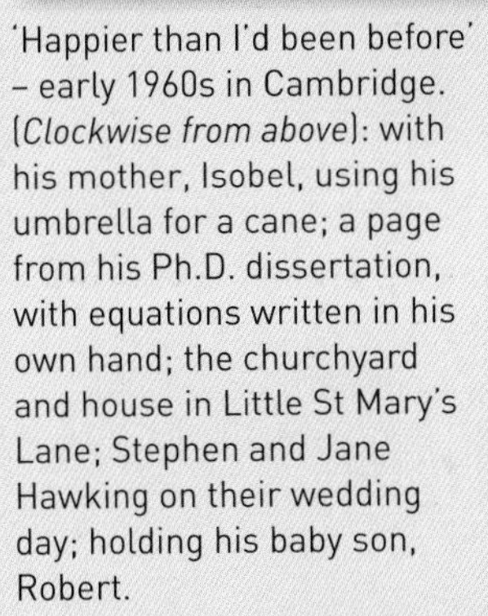

'Happier than I'd been before' – early 1960s in Cambridge. (*Clockwise from above*): with his mother, Isobel, using his umbrella for a cane; a page from his Ph.D. dissertation, with equations written in his own hand; the churchyard and house in Little St Mary's Lane; Stephen and Jane Hawking on their wedding day; holding his baby son, Robert.

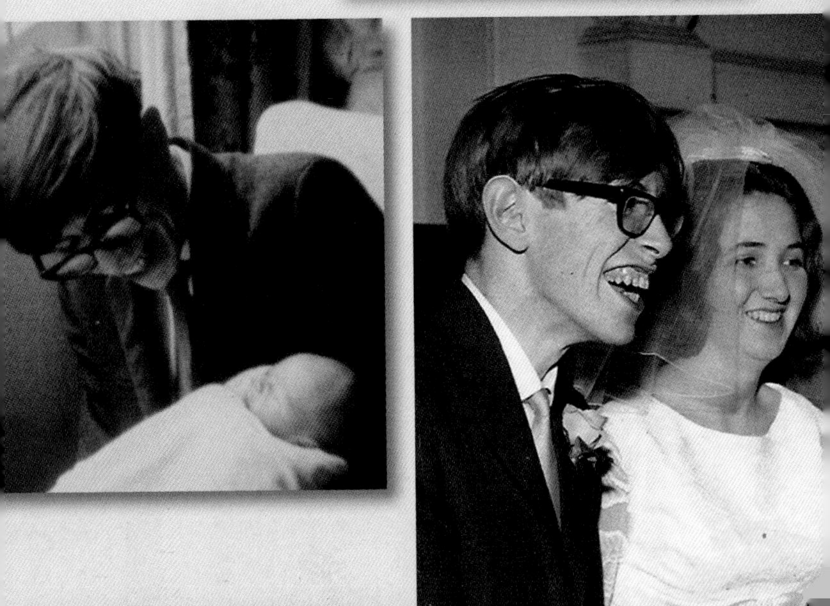

'I have a beautiful family' – 1970s. (*Clockwise from above*): the house in West Road; at the door in a state-of-the-art wheelchair; with Lucy, Jane and Robert; chess with Robert.

(*Above*): Tim, a new addition in the family sandbox, with Robert, Lucy, Stephen and Jane; (*below*): Jane's new passion: music.

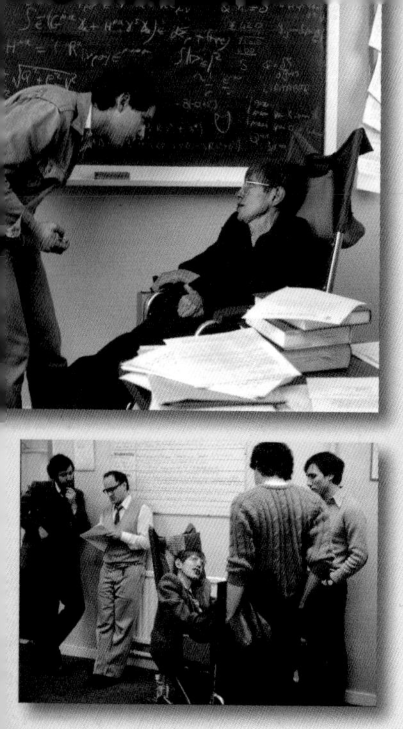

'International calibre.' (*Top left*): in the spring of 1983 shortly before he lost his ability to speak, with research student Chris Hull, who is listening intently to understand his words; (*above*): homeward bound along the path through King's College, accompanied by Don Page; (*left*): tea with colleagues and students in Cambridge.

(*Below*): with his personal assistant Judy Fella in his Cambridge DAMTP office; (*right*): in the rain on King's Parade, with nurse Elaine Mason.

'I have written a bestseller!' – late 1980s. (*Top to bottom*): the Stephen Hawking fan club in Chicago celebrates *A Brief History of Time*; with colleagues in a bar in Berkeley, California, demonstrating the first version (monitor still only taped together) of his computer speech programme; with Roger Penrose and Kip Thorne.

'On the shoulders of giants' – 1980. Lucasian Professor of Mathematics, with the statue of former holder of the title Isaac Newton in Trinity College, Cambridge; (*inset*) receiving the coveted Israeli Wolf prize.

Country School. Three-year-old Lucy liked it so much the first day that she made up her mind she would stay the full day rather than the half-day she was assigned. She was nowhere to be found when her mother came to fetch her. A panicked staff discovered Lucy having her lunch calmly in the lunch room with the older children. Robert found a new role, making himself invaluable to his mother as a master navigator of the Los Angeles freeways with, seemingly, a detailed map in his seven-year-old head. Bernard Carr dived into Caltech student life with a passion and went out to parties nearly every evening after getting Hawking to bed. After the parties he sat up the rest of the night watching horror movies. Fortunately Stephen also was not fond of getting up early in the morning.

Jane was caught up in a social whirl and the Hawkings entertained almost constantly. It is rare in the Cambridge colleges, with the notable exception of Clare Hall, for spouses to be nearly as much a part of the social scene as Jane experienced at Caltech. She found the change exhilarating, if also a little exhausting. In addition to new local friends, there were visitors from abroad, friends and family including Jane's parents and Stephen's mother and aunt. Stephen's sister Philippa visited from New York, where she was living. The Hawking home, so convenient to the campus, was the site of frequent parties for the Caltech Relativity Group.

The spontaneity and directness of Californians were a surprise and, to Jane, a pleasant change from the diffidence and, sometimes, outright avoidance she and her husband were used to encountering in England. It was not easy, because of Stephen's speech problems, for those who did not know them well to carry on a conversation with the pair of them, but Californians seemed more than willing to try. Stephen Hawking was becoming an international celebrity in his field when he arrived at Caltech and he was, accordingly, given star treatment. To be fair, had he been introduced to Cambridge similarly for the first time, his and Jane's reception in University circles might have been characterized by some of the special attention they experienced in Pasadena.

For an academic taking a sabbatical year abroad, the time spent away from home is often not only an enormous boost to his or her creativity and intellectual energy, but also a watershed for other members of the family. So it was with the Hawkings. An eight-year-old computer-wiz friend got Robert excited about the field that would eventually become his career, information technology. Another faculty wife invited Jane to attend an evening choral class that met each week to sing through a major choral work – Jane's first introduction to an avocation that would become her passion and engage her for many years.

Hawking had an air-conditioned office, and ramps sprang up all over the Caltech campus. He basked in the company of other eminent researchers and dined as an honoured, delighted guest in the student 'houses'. The California Institute of Technology was and is one of the greatest centres in the world for the study of physics and physics research. It is a smaller institution than Cambridge or Oxford, but the faculty includes many of the world's most outstanding academics, undisputed leaders in their fields. For Hawking, new colleagues and stimulating ideas abounded, drawing his interest to areas he hadn't explored and to fresh ways of approaching the problems he was already working on. It was here that he first met Don Page, then a Caltech graduate student, who would play a large role in the Hawkings' future. Page and Hawking wrote a paper that year which suggested that the explosion of primordial black holes might be observed as gamma ray bursts.[26] Legendary arch-rivals Richard Feynman and Murray Gell-Mann were also at Caltech, and Hawking attended their lectures. Both were cutting-edge particle physicists, not cosmologists, but Hawking was finding more and more need for expertise in particle physics in his study of black holes, and here was a priceless opportunity. He would soon be employing Feynman's idea of 'sums-over-histories' in a new way as he explored possibilities for describing the origin of the universe. Jim Hartle, whom Hawking had met in Cambridge and who was at the University of California-Santa Barbara, spent some time at Caltech that year, and he and Hawking developed the description of Hawking radiation discussed in Chapter 6.[27]

Hawking didn't spend the entire year, uninterrupted, in Pasadena. Just before Christmas he joined his friend and colleague George Ellis to attend a conference in Dallas, and in April he was invited to Rome to receive, from Pope Paul VI, the Pope Pius XII medal, awarded 'to a young scientist for distinguished work'. Hawking was eager to see the document in the Vatican Library in which Galileo had recanted, under extreme pressure and threat of torture, his discovery that the Earth goes around the sun. Hawking took the opportunity to lobby for a formal apology to Galileo, whom the Catholic Church had treated so badly three and a half centuries before. That apology would not be long in coming.

It was in California that Hawking began to think

seriously about a problem which would set him at odds with some of his colleagues for many years: the loss of information in black holes. We will examine later what 'information' means in this context. For now, think of it as information having to do with everything that went into the making of the black hole when it formed and everything that has fallen in since. How irrevocable is this loss? What might it mean for our ability to understand the universe and make predictions? Could it really represent the breakdown of physics? That was the title he gave a paper he wrote that year – 'Breakdown of Physics in Gravitational Collapse'. When it finally appeared in November 1976, he had changed the title to something less shocking, unless you stopped to think about it: 'Breakdown of Predictability in Gravitational Collapse'.[28]

And, of course, there was Kip Thorne, the dear friend and colleague who had been instrumental in making the whole visit possible. Thorne and Hawking put their signatures (Hawking's thumb print) to a document recording their first famous bet, about whether the binary star system Cygnus X-1 contains a black hole.

Penthouse v. Private Eye

The pre-history of Thorne's and Hawking's bet began in 1964, before John Wheeler had even coined the name black hole. That year Yakov Zel'dovich and his graduate student Oktay Guseinov at the Institute of Applied Mathematics in Moscow began combing the lists of many hundreds of binary star systems that astronomers had previously observed and catalogued. They were looking for stars so extremely massive and compact that they are unlikely to be anything but black holes. The search for black hole candidates had begun, and it was no straightforward undertaking. Such candidates are, by nature, invisible to an optical telescope.

In order to understand what binary star systems are and why they are good places to look for black holes, picture a scene described by John Wheeler. In a dimly lit ballroom, the women are all dressed in white gowns. Some of the men are also dressed in white, but a few of them are in black formal attire. Watching the couples waltz, from a balcony above, we know there are two people in each pair, but in some cases we can see only one of them, the woman in white.

A binary star system consists of two stars circling one another, as the man and woman in one of Wheeler's couples do. In some binary systems, only one star is visible. How do we know that there *are* two stars? In the ballroom, by watching how the visible woman moves, it's fairly obvious that she must have a partner. Likewise, by studying the movement of some stars, it is possible to conclude that they are not alone out there.

Seeing a star apparently alone but moving as though it has a partner does not always signify a black hole. The invisible companion might be a small, dim, low-temperature star – a white dwarf, or a neutron star. Calculating the mass of such stars is complicated, and mass is a vital statistic when you're trying to determine whether something is a black hole or not. Suffice it to say that, again, this was information that astronomers in the 1960s were beginning to find ingenious ways to glean from the movements of the visible star.

In 1966, Zel'dovich and another colleague, Igor Novikov, decided that identifying strong black hole candidates would require using both optical telescopes and X-ray detectors. The observation of X-rays indicates a source of considerable energy powering their emission, and having material fall towards a black hole or neutron star is one of the best ways known to liberate energy. In a binary system, that is what happens when the very compact star or black hole pulls material away from its companion star. So researchers looked for binary systems where one partner shows up brightly in the visible part of the spectrum but is dark in the X-ray part; while the other partner is dark in the visible part of the spectrum but bright in the X-ray part.

Cygnus X-1 was a very promising candidate. Here was a binary system where an optically bright but X-ray dark star orbits with an optically dark but X-ray bright companion. This system is in our galaxy, about 6,000 light years from Earth. The two stars complete one orbit in 5.6 days. An optical telescope reveals what seems to be a blue giant star, too dim to be detected with the naked eye. Studies of the Doppler shift in its light show that it must have a companion. Cygnus X-1 is that companion. It can't be seen at all with an optical telescope, but it is one of the brightest objects in the X-ray sky. The X-ray emission fluctuates violently and chaotically, as expected when matter falls towards a black hole or neutron star. Cygnus X-1's mass is at a minimum 3 solar masses, probably greater than 7 solar masses, and most likely about 16 solar masses. In December 1974, such uncertainty in calculations of its mass made the Hawking–Thorne bet possible. Cygnus X-1 was an excellent black hole candidate, but experts were only about 80 per cent certain that it was a black hole and not a neutron star.

The bet document laid out the terms: If Cygnus X-1 turned out to be a black hole, Hawking would give Thorne a one-year subscription to the magazine *Penthouse*. If it turned out not to be a black hole, Thorne would give Hawking a four-year subscription to *Private Eye*. Hawking called his surprising bet *against* Cygnus X-1 being a black hole an 'insurance policy'. 'I have done a lot of work on black holes, and it would all be wasted if it turned out that black holes do not exist. But in that case, I would have the consolation of winning my bet.' The bet document was framed and hung on the wall of Thorne's office at Caltech, awaiting the advancement of science.

Sadly, as the year in California approached its end, the feelings of depression, inadequacy and low self-esteem that had plagued Jane back in Cambridge caught up with her again. She began to reassess the social whirl of that year in Pasadena and view it as a sort of frenzied escapism from these problems. She felt she had fallen easy prey to the women's liberation movement's insistence that a woman who had no job outside her home should consider herself a failure, lacking in personal fulfilment.[29] Jane concluded this was indeed the prevalent feeling among unemployed faculty wives she had met. Excursions with them to art museums and galleries and plays now seemed only the sad, kindly attempts of these women to make up for the bleakness of their lives and hers. A thoughtful and discerning friend, recognizing Jane for the remarkable woman she was, and perhaps sensing that she was not capable of recognizing this herself, gave her a pearl brooch on the day that Stephen Hawking was awarded his Papal medal, saying that she, Jane, should be given something too.[30]

8

'Scientists usually assume there is a unique link between the past and the future, cause and effect. If information is lost, this link does not exist'

BACK IN CAMBRIDGE AFTER THEIR SOJOURN IN CALIFORNIA, the Hawkings settled into their new home in West Road.[1] For Stephen, after a year in an electric wheelchair that could move fast and be used both indoors and out, there was no going back to an old-fashioned model. His request to the UK Department of Health for a chair like the one he'd had in California was rejected. The Hawkings drained their savings to purchase it themselves.

In this new vehicle, the daily journey to the DAMTP in Silver Street took about ten minutes, not by the shortest route but by a more pleasant and less-trafficked one. He followed a curving tree-lined footpath through King's College that took him across the 'Backs' past meadows with grazing cows and immaculately mown lawns, crossing the River Cam by a hump-backed stone bridge behind King's College Chapel. From there he had a choice. He could use a side entrance from King's and, by way of a not heavily used back street called Queens' Lane, pass in the shadow of the towering medieval gates of Queens' College to arrive at Silver Street. Or he could exit King's by its main front gate into busy King's Parade, and then turn right to reach Silver Street. Either way, the hazardous crossing of this narrow, busy thoroughfare provided a modicum of risk and excitement to climax an otherwise peaceful journey. Around the back of his building was his ramp. He would time his arrival for about 11 a.m. Hawking's new graduate student assistant, Alan Lapades, sometimes made the journey with him, but often Hawking went on his own, the state-of-the-art wheelchair giving him optimum independence.

By the autumn of 1975, Hawking had completed the six years of his Caius Fellowship for Distinction in Science. Though he couldn't lecture, he was a very good mentor, willing to spend so much time in discussion with his students that some colleagues wondered how he ever got his own work done. The University of Cambridge put an end to rumours circulating that he might emigrate permanently to the USA by offering him a 'readership' and also a secretary, Judy Fella. Judy was a vivacious and rather glamorous addition to the DAMTP and a godsend to both Stephen and Jane. His having his own secretary relieved Jane of the burden of doing all Hawking's scheduling and travel bookings. Jane turned once again to her neglected thesis and, following up on her new musical interest, began voice lessons.[2]

In the summer of 1975, BBC television lorries lumbered into the Hawkings' West Road forecourt and snaked cables into the house to film Stephen for a documentary, *The Key to the Universe*, moving to Silver Street to film a seminar in the DAMTP. Media invasions such as this, which would happen many times in the future, ran roughshod over other interests and priorities of colleagues, students, staff and family. Over the years it would become an annoyance, but the first few times it was very exciting.

Jane's plan of having live-in student assistants was working well. Hawking was eager for Don Page, whom he had met at Caltech, to come to fill that position. Page, who was finishing his Ph.D. and applying for postdoc posts, was a close friend and an extremely promising physics colleague. When Gary Gibbons was invited to Munich for a year, that freed funds to hire Page as Research Assistant. Later, a NATO fellowship would pay him. In autumn 1976, this unusually tall young man with a powerfully resonant speaking voice joined the West Road household.

For Hawking, an agnostic bordering on atheism, Don was an unexpected choice – intellectually brilliant, scrupulously moral, devoutly and outspokenly religious. He would not stick to physics topics on journeys to and from Silver Street but often veered off into discussions of what he had read in the Bible that morning. Stephen parried with friendly sceptical banter but respected Don's opinions and faith. Don would remain a personal friend and a highly valued academic collaborator long after his student assistantship was over.

With Judy Fella at the office and Don Page at home, some of the responsibility for Stephen was off Jane

Hawking's shoulders. For the first time in many years, Stephen travelled without her, in the summer of 1977, when Page accompanied him on a revisit to Caltech lasting several weeks.

It was about this time that Hawking and other younger Fellows were invited to come to London for Prince Charles's induction into the Royal Society. The prince was intrigued by Hawking's wheelchair, and Hawking, twirling it around to demonstrate its capabilities, carelessly ran over Prince Charles's toes. Soon rumour had it that running over toes was not always accidental, and eventually it would get around that one of Hawking's regrets in life was not having an opportunity to run over Margaret Thatcher's toes. People who annoyed him, it was said, found themselves a target. 'A malicious rumour,' insists Hawking. 'I'll run over anyone who repeats it.'[3]

In the autumn of 1977, just two years after his 'readership' began, the University of Cambridge promoted Hawking to a chair in gravitational physics, with a welcome rise in salary. He was now a 'professor', a distinction much more rarely achieved at Cambridge and most English universities than in an American one.

It was that December of 1977, when Jane joined the choir of St Mark's Church in Barton Road for the Christmas season of special music, that she first met a young man named Jonathan Hellyer Jones, who was the organist there. Jonathan was a gifted musician, younger than Jane and Stephen by a few years. He had recently lost his wife to leukaemia after only one year of marriage. The Hawking home became a haven for him. Both Jane and Stephen were emotional supports, but Jonathan also more than pulled his own weight in the household – sitting at the piano teaching seven-year-old Lucy, and volunteering to help with Stephen's physical needs. Jane joined the St Mark's congregation. She also was at last able to find the time to begin working seriously on the final chapter of her thesis.[4]

Perhaps it was inevitable that this young man, intimately and generously involved with the family, a god-send to Jane as she struggled to keep up with her tasks as caregiver, homemaker, Ph.D. student and mother, and sharing her faith and her love of music, would become more than a helpful supportive friend. Jane, with characteristic honesty, did not keep it a secret from Stephen when she and Jonathan developed a romantic attachment. Hawking apparently accepted this, saying nothing except that, as Jane has written, 'he would not object so long as I continued to love him'.[5] After that, the subject was rarely mentioned between them, and Jane's relationship with Jonathan remained platonic for a long time. They chose 'to maintain our code of conduct in front of Stephen and the children, suppressing displays of close affection'. They did not move out to live together. 'Jonathan and I had struggled with our own consciences and decided that the greater good – the survival of the family unit, Stephen's right to live at home within that family unit and welfare of the children – outweighed the importance of our relationship.'[6] Their secret was so well guarded that only a small circle of family and friends were aware of that relationship, and whatever grief Jane and Jonathan may, or may not, have caused him, Hawking kept that to himself.[7]

In the autumn of 1978 Jane was pregnant again. With not only Don Page and Judy Fella but also Jonathan on board, she decided to set herself a deadline for finishing her thesis before the baby was born in the spring. It was now or never. In the winter she set that work aside only briefly to organize a charity concert to raise funds for the newly formed Motor Neurone Disease Association. Stephen was a patron. Jane was beginning to be a presence in the music world of Cambridge.[8]

The Hawkings' third child, Timothy, was born in April on Easter Sunday. To commemorate another birthday a hundred years before, Hawking and Werner Israel celebrated the anniversary of the birth of Albert Einstein (14 March 1879) by inviting colleagues to contribute articles to a book reflecting current research related to general relativity. The introduction foreshadowed the topic of Hawking's Inaugural Lecture as Lucasian Professor, commenting on 'Einstein's dream of a complete and consistent theory that would unify all the laws of physics'.[9]

Hawking received several major international awards and honorary doctorates in the mid- to late 1970s, including the Hughes Medal from the Royal Society honouring 'an original discovery in the physical sciences' for 'distinguished contributions to the application of general relativity to astrophysics', and, in 1978, the coveted Albert Einstein Award from the Lewis and Rosa Strauss Memorial Fund in the United States. That award is not given every year and is the most prestigious American award for a physicist. One of the honorary doctorates, in the summer of 1978, was from his alma mater, Oxford. Most significant of all for Hawking's future, in the autumn of 1979 the University of Cambridge gave him the venerable title of Lucasian Professor of Mathematics. He at last got a private office. The impressive tome in which each new university teaching officer inscribes his or her name was brought to him there more than a year later. Somehow, this formality had been neglected earlier. 'I signed with great difficulty. That was the last time I signed my name.'[10]

In 1980, Hawking's battle for independence reached a crisis point. Martin Rees was by then the holder of another prestigious chair at Cambridge, the Plumian Professorship of Astronomy. He had known the Hawkings since before their marriage and had witnessed at close hand both Stephen's soaring success as a scientist and his inexorable physical decline. Some time in the late winter or early spring Rees asked Jane to come in for a chat with him at the Institute of Physics. Earlier that winter, when a bad cold had turned more serious and Jane had also been ill and slow to mend, Stephen, on the family doctor's recommendation, had entered a nursing home for a short while until they both recovered. Rees feared that this would be the first of many times when the Hawkings would not be able to cope without additional help. He offered to find funding to pay for limited home nursing.[11]

To Hawking, this was unthinkable. Agreeing to it would be giving in to his illness and would introduce an unwelcome, impersonal intrusion into his life. He would become a patient. After some consideration, however, he changed his mind. There would be advantages. He would be much freer to travel, not always having to depend on his wife, friends and students. What he had at first seen as a loss of independence might be a gain instead.

N=8 Supergravity

Hawking's Inaugural Lucasian Lecture, 'Is the End in Sight for Theoretical Physics?', with which this book began, took place on 29 April 1980. In that lecture he chose N=8 supergravity as the front-running candidate for the Theory of Everything. It was not one of his own theories, but it seemed to him and many other physicists extremely promising. Supergravity grew out of the idea of supersymmetry, a theory that suggested that all the particles we know have supersymmetric partners, particles with the same mass but different 'spin'.

In *A Brief History of Time*, Hawking suggests it's best to think of spin as meaning what a particle looks like when you rotate it. A particle with spin 0 is like a dot. It looks the same no matter from what direction you look at it and no matter how you turn it. A particle with spin 1 is like an arrow: you have to turn it around a full rotation (360 degrees) to have it look the same as when you started. A particle with spin 2 is like an arrow with two heads (one at each end of the shaft): it will look the same if you turn it only half a rotation (180 degrees). All seems fairly straightforward so far, but now we come to something slightly more bizarre, particles with spin ½. These have to complete two rotations in order to return to their original configuration.

We saw in Chapter 2 that every particle we know in the universe is either a fermion (particles that make up matter) or a boson (a 'messenger' particle). Elementary fermions have spin ½. An electron, for example, has to finish two rotations in order to return to its original configuration. Bosons, on the other hand, all have whole number spins. The photon, W and Z bosons, and the gluon have spin 1, returning to their original configuration after one rotation (the single-headed arrow). The graviton theoretically has spin 2 (the double-headed arrow), returning to its original configuration after half a rotation.

Supersymmetry theory suggests a unification of matter and the forces of nature by proposing 'supersymmetric partners' for each of these particles, with every fermion having a boson for a partner, and every boson a fermion – a little like the procession at the end of an old-fashioned wedding uniting two formerly incompatible families, with every bridesmaid going out on the arm of a groom's attendant from the other clan. There was a Latin flavour to the naming of the bosons' partners. The photon's theoretical supersymmetric partner was christened the photino. The graviton's was named the gravitino. As for the fermions, their supersymmetric partners were all given names by adding an s. And so we have the selectron and the squark.

There were several versions of supergravity around at the time of Hawking's Lucasian Lecture. N=8 supergravity had the advantage of being the only one that worked in four dimensions (three of space and one of time), and though it called for a goodly number of as yet undiscovered particles, it did not call for an infinite number of them, as other attempts at theories of quantum gravity did. The name N=8 came from the fact that in this theory the graviton had not just one but eight supersymmetric partners. A little awkward for processing down the aisle, but it worked well in the theory.

It was not long after his Lucasian Lecture that Hawking and others realized how stupendously difficult it was to do any calculations using what had seemed to be such a promising theory. In addition to the graviton and the eight gravitinos, there were 154 types of particles. The general conclusion was that it would take about four years to finish a calculation, even if you used a computer – making sure to account for all the particles, searching for infinities that might still lurk somewhere, and being sure not to make an error!

Another problem was that none of the supersymmetric partners had ever been observed or looked likely to be. The theory says they have the same mass as their 'normal' partners, but this symmetry (in the world we can observe) is 'broken', leaving the supersymmetric partners hundreds, even thousands of times more massive. The energy required to discover them in a lab is tremendous. 'Symmetry-breaking' is something we'll return to in another context.

The 'Higgs field', proposed in 1964 by Peter Higgs at the University of Edinburgh, theoretically pervades the entire universe and is responsible for the symmetry-breaking that makes the supersymmetric partners so difficult to discover experimentally, and also for the masses of more familiar particles. If the theory is right, the Higgs field should show up as the 'Higgs particle', with spin 0. The Higgs particle itself is very massive and has never been observed, but, if it indeed exists, it may be discoverable with the high energies of the Large Hadron Collider at CERN on the Swiss–French border at Geneva. One of Hawking's famous bets is that the Higgs particle will not turn up at all. He has become more interested in the possibility that the LHC may discover some of the supersymmetric partners, or perhaps produce a mini-black hole.

In the spring of 1980, when Hawking gave his Lucasian Lecture, Jane was working towards her doctoral oral examination, scheduled for June. No one was surprised when she passed. She would not be officially graduated until April 1981, but with her degree now a certainty, she began to tutor local children in French, and then took a part-time position at the Cambridge Centre for Sixth Form Studies, helping students prepare for their sixth-form examinations and gain entrance to university.[12] A little later, degree in hand, she became a teacher at sixth-form level. The beginning of her professional career was something of an anticlimax after her years of labour to finish her thesis. There was still a dispiriting gap between her own considerable intellectual achievements and the much more conspicuous success of her husband. Nevertheless, 'It is fulfilling a part of me,' she said, 'that I feel has been suppressed for a long time and the marvellous thing is that it is totally compatible with what goes on at home.'[13]

Tim was a toddler, Robert was thriving at the Perse School, and Lucy had become a favourite at her primary school. The first I ever heard of any of the Hawkings was a comment from the head teacher at Newnham Croft Primary School that my six-year-old daughter reminded her of Lucy Hawking. This was obviously meant as a compliment. When I met Lucy, I found that it was. She by then had finished at the school but was back for a visit, helping with the small children in the playground. A radiant, clean-cut, blonde sixteen-year-old, she was brimming with intelligence and personality, with a thoughtfulness and composure beyond her years. She said her father was a physicist at Cambridge, but she hoped to go to Oxford instead because she had lived in Cambridge all her life and wanted a change of scene. Lucy told interviewers that she was not like her father. 'I was never any good at science. I even managed to be hopeless at maths as well, which was slightly embarrassing.'[14] But she was a fine student and cellist. It's unlikely anyone has ever wagered a bag of sweets against her success or that of her brothers.

Trouble in the Attic

The year spent at the California Institute of Technology in 1974–5 had been such a success from Stephen Hawking's point of view and that of his hosts that he began a habit of returning for a month almost every year. Caltech renewed and extended his Sherman Fairchild Distinguished Scholar appointment to make this possible.

By 1980, Kip Thorne had noticed a change in Hawking's attitude towards his research, summed up when Hawking told him: 'I would rather be right than rigorous.' This rather enigmatic statement referred to the fact that most mathematicians are satisfied with nothing less than a firm mathematical proof that they are correct. That had been Hawking's attitude in the 1960s and 1970s. Now he was saying that such rigour is not necessarily the best way to arrive at 'right'. It could lead one not to see the forest for the trees. He had become more speculative, happy to be perhaps 95 per cent sure of something and then move on. 'Stubbornly intuitive' was how *New York Times* columnist Dennis Overbye described him.[15] Though Hawking's intuition was often on target, for him the quest for certainty had changed to a quest for 'high probability and rapid movement towards the ultimate goal of understanding the nature of the universe'.[16]

Hawking's experience of California had broadened to include more than Los Angeles and Pasadena. He liked the steep streets of San Francisco and the opportunities they offered for a dare-devil in a motorized wheelchair. He was just as reckless a driver now as he had been when he took Jane Wilde to the Trinity Hall May Ball in 1963. His colleague Leonard Susskind remembers him poised at the top of the steepest section of one of those hills, with an incline so severe that motorists shudder, fearing that their brakes must surely fail or the car somersault forward. Hawking flashed his companions one of his mischievous smiles and then disappeared, barrelling down the hill, almost in free fall. When they caught up with him at the bottom he was sitting there grinning with satisfaction and asked them to take him to a steeper hill.[17]

In San Francisco in 1981, Hawking called attention to an issue that troubled some of his colleagues far more than his reckless disregard for the hazards of San Francisco streets. The venue was more than eccentric, the attic of the mansion of Werner Erhard, founder of a pop psychology called EST. EST targeted people who felt they were sadly lacking in self-confidence and were willing to pay fees of several hundred dollars for help. As many as a thousand at a time gathered for intensive two-week sessions in hotel ballrooms to submit to therapy that could be authoritarian, demeaning, and verbally and physically abusive. The claim was that they would experience at least temporarily a personal transformation and emerge enlightened, self-confident and outgoing. For some, it worked.

This scheme had made a fortune for Erhard. He was also a physics aficionado and he used his wealth to cultivate the friendship of a number of theoretical physicists at the top of their profession and not at all lacking in self-confidence, including Richard Feynman. Erhard hosted elaborate gourmet dinners for them and equipped his mansion's attic to accommodate small, exclusive physics conferences, funded out of his pocket. Though Erhard was the founding father of what his critics regarded as a hokey, violent pop psychology, he was a pleasant, interesting, extremely intelligent man. The elite of physics chose to ignore the first persona and enjoy the second, and did not say no to what he offered them.[18]

When Stephen Hawking made his announcement at a conference in 1981 in Erhard's attic, he was still able to speak with his own voice, but it had been several years since anyone except those who knew him best could understand him. His interpreter on this occasion was Martin Rocek, then a Junior Research Fellow at the DAMTP, later to make his name in physics in the areas of string theory and supersymmetry. Rocek travelled with Hawking on that trip to California, struggling more or less successfully to understand him and repeat his words clearly for others. Film clips made of this process show what a prodigious effort was involved. For most lectures, such as his Lucasian Lecture, Hawking's graduate assistant would deliver the talk and '[Hawking] would just sit around and add brief comments if a student would say something wrong'.[19] Whatever the Byzantine nature of the process involved, Hawking's announcement in Erhard's attic was clear enough.

He had been focusing his attention on black holes for about fifteen years, arriving at equations of such simple clarity and elegance that he felt they had to be correct.

Such results, he insisted, reveal a deep harmony underlying nature.[20] By 1981, hardly anyone was treating Hawking radiation as a dubious concept. But Hawking had begun to realize, at least as early as his sabbatical in Pasadena in 1974–5, that at the heart of the equations surrounding that discovery lurked a paradox that threatened to undermine the whole of physics. It had to do with the loss of information in black holes and the threat this posed to a fundamental tenet of physics, the law of information conservation: information can never be lost from the universe. Don Page, in 1979, had been the first to disagree with Stephen's conclusion, but Stephen had not budged.

It's important to understand what 'information' means in this context. You can think of this lost information as information about everything that went into making the black hole in the first place and everything that fell in later. But what does a theoretical physicist mean by 'information'? The clue is in the words 'information encoded in the particles that make up the universe'.

Here, from the history of the study of black holes, is an example that helps explain what 'information' means to a theoretical physicist: in his book *The Black Hole War*, Leonard Susskind tells of an Einstein-like thought experiment performed by Jacob Bekenstein when he was considering the question of the entropy of a black hole. (Bekenstein's proposals, you will recall, presented a challenge to Stephen Hawking in 1972.) It's hard to imagine that anything could take much less information into a black hole than a single photon falling in, but, in fact, one photon does carry a good deal of information. Most significantly for Bekenstein, it carries into the black hole information about the location where it fell in.

Bekenstein wanted to consider a smaller amount of information than that for his thought experiment. He wanted to reduce the amount of information to one 'bit', a unit of information suggested by John Wheeler that has the smallest possible size in the universe – a quantum distance calculated by Max Planck in the early twentieth century. In order to achieve this, Bekenstein made use of Heisenberg's uncertainty principle and imagined 'smearing out' the location where the photon fell in. He imagined a photon of such long wavelength that the probability of its entry point would be spread out over the entire event horizon, making it as uncertain as possible, so that it would convey only a single 'bit' of information – that the photon was, in fact, inside the black hole. The photon's presence would add to the mass of the black hole and, of course, to the area of the event horizon, by a minuscule amount, which Bekenstein proceeded to calculate.

Obviously, 'information' in this story means something a little more subtle than what you and I normally think of as everyday 'information'. It isn't only information such as what channel John Wheeler's television was tuned to when it fell into the black hole.

The idea of information disappearing into a black hole and becoming inaccessible to those on the outside was not new to anyone at the conference. Locked-away information like that is not a violation of information conservation. The information in a black hole may be inaccessible to anyone on the outside, but it is still in the universe. Hawking was thinking of something more drastic. When a black hole has finally radiated all of its mass away and disappeared, what has actually happened to everything that went into forming the black hole and everything that later fell in?

If you have been following this book carefully, you may be raising your hand to suggest that it would all have been recycled as Hawking radiation. That radiation, of course, would not look anything like, say, a hapless astronaut who fell in, but could it somehow be the solution to the problem? After all, the law of information conservation has it that information encoded in the particles of which everything in the universe is made up can be scrambled, chopped up, destroyed, but if the fundamental laws of physics as we now understand them are correct, it can always be retrieved from the particles that make it up. In principle, with the information in hand, anything can be restored.[21]

For example, let's say you burned this book. You might think you would never be able to finish reading it. However, *in principle*, if you could study the burning process carefully enough to trace all the molecular interactions that turned the book to ashes, then by running the process backwards you could have the book again. It would be a lot easier to go out and buy another copy, but *in principle* you *could* reconstruct it.[22]

Hawking would have none of that. He insisted that Hawking radiation could not serve as the escape vehicle back into the outside universe for information trapped in a black hole. If you tossed this book into a black hole, such reconstruction would be impossible. Hawking radiation is not the 'ashes' or the scrambled or chopped-up remains of what fell into the black hole. Recall that the 'escaping' member of the particle pair in Hawking radiation (the way it was explained in Chapter 6) doesn't come from inside the black hole. It comes from just outside. The escaping particle carries with it no news whatsoever about whether the black hole is full of astronauts, unmatched socks, or Winnie the Pooh's grandmother's honeypot. It doesn't know. Hawking radiation has no direct connection to whatever went into forming the black hole in the first place or what has fallen in. There are some physicists who held out hope that such information is somehow encoded in the Hawking radiation, but Hawking was not one of them. He thought that information does not escape and is completely lost when the black hole evaporates. You can't restore it, not even in principle. Hawking christened the dilemma he had raised 'the information paradox'.

The problem looked likely to extend beyond black holes. In an interview on the BBC programme *Horizon* in 2005, Leonard Susskind, who was there in Erhard's attic, remembers his shock at Hawking's arguments and the realization that if Hawking was right – if information really is irretrievably lost in black holes – then that is not the only place it is going to be lost. Pieces of the universe must be missing. We can forget about predictability. Forget the dependability of cause and effect. Nothing we think we know in science can be trusted.[23]

Hawking perceived that Susskind was probably the only person present who fully realized the implications of his arguments. 'Leonard Susskind got very upset,' he remembers. Scientists, and the rest of us, rely on the link between past and future, between cause and effect. That link is lost if information is lost. 'We wouldn't be able to predict the future. We can't be sure of our past history either. The history books and our memories could be just illusions. It is the past that tells us who we are. Without it we lose our identity.'[24] Kip Thorne pointed out that there was already speculation about the existence of black holes that are smaller than atoms that could be anywhere and everywhere, nibbling away bits of information.[25]

The problem may not seem so disastrous to you or me. Granted, when anything falls into a black hole it carries information with it. The colour and size of a few unmatched socks? Perhaps the dimensions of the unfortunate astronaut? Not information that you or I find interesting or essential. However, this information is necessary, even for a more limited kind of prediction allowed by quantum mechanics.

Hawking traces the discussion about being able to predict the future or the past to Pierre-Simon de Laplace, a mathematician who lived in the late eighteenth and early nineteenth centuries. Laplace's famous proposal was that an omniscient being with unlimited powers of calculation, knowing the laws of the universe and the state of everything in it (that is, the positions and momenta of all the particles in the universe) at any given time, would be able to calculate the state of everything in the universe at any other time in the past or future. Though no one denied the stupefying practical difficulties of acquiring such knowledge and doing all the calculations, Laplace-style scientific determinism remained dogma through the nineteenth and early twentieth centuries. When I heard Hawking lecturing on the subject in Cambridge, he quoted Laplace in French and told us that since we were a Cambridge audience he would not insult our intelligence by offering a translation. One nevertheless soon slyly appeared on his slide screen.

Laplace's omniscient being had to have knowledge of the positions and momenta of *all* particles in the universe. You can't leave out the particles in the toe of the sock. When the sock falls into the black hole, that information is lost from our region of the universe. If black holes exist for ever, well and good, then the lost information isn't entirely lost. It is inaccessible, but still there. If black holes evaporate and disappear from the universe . . . trouble.

Hawking and his colleagues in the attic knew that the 'information paradox' was not the first challenge to Laplace-style scientific determinism. In the mid-1920s when Werner Heisenberg published his 'uncertainty principle', it seemed all bets must be off . . . but only for a while. Arguments having to do with interpretations and implications of the uncertainty principle would go on for years and involve the finest minds in physics, but by the time Hawking studied black holes there was fairly general agreement that even Laplace's omniscient being could not possibly know precisely a particle's position and momentum at the same time.

The uncertainty principle did not, however, ultimately succeed in undermining faith in the dogma of scientific determinism. It soon became clear that laws governing the quantum level of the universe are deterministic in a different way. You can predict what is called the 'quantum state', from which both positions and momenta can be calculated with a certain degree of accuracy. Laplace's omniscient being, knowing the quantum state of the universe at any one moment in time, and the laws of science, could predict the quantum state of the universe at any other time, past or future.[26*]

Now, Hawking had found a new problem, and it seemed to be a serious one. His earlier work had shown that black holes don't last for ever. As Hawking radiation continues, the black hole gets smaller until eventually there is no black hole left. He was insisting that the information about whatever went into forming the black hole in the first place, and whatever fell in, is then irretrievably lost.

Still . . . why all the angst about its loss? Couldn't the universe make do with a little less of this rather arcane information?

No, it seems it cannot. Not and still be the universe we think we know. The law of information conservation is one of the fundamental principles of physics. Information is *never* lost. It can be mixed and scrambled and transformed in ways that make it unrecognizable as the information you started with, but not ever lost. If this law is wrong, then the universe is in effect thumbing its nose at Laplace and all those who have been assuming that he was correct.

Although Hawking's colleagues in Erhard's attic, in Susskind's words, stood there in 'stunned confusion', most of them and other theoretical physicists would go right on

* Hawking's ideas were not the only serious threat to determinism in the 1980s. An equally significant one came from chaos theory. Ilya Prigogine and Isobel Stengers presented this challenge in their 1985 book *Order Out of Chaos*, writing, 'When faced with these unstable systems Laplace's [omniscient being] is just as powerless as we.' (Ilya Prigogine and Isobel Stengers, *Order Out of Chaos*, London: Heinemann, 1985.)

believing that the present has evolved from the past and will go on evolving into the future, that cause and effect continue to operate, that it is meaningful to trace events into the past and into the future, that examining the debris from a collision in a particle accelerator can tell you what happened in the collision – as though Hawking had not hung a Sword of Damocles over all these assumptions. But Hawking stuck to his guns and the information paradox didn't go away. Information, he continued to insist, is truly lost when black holes evaporate, and this means we can predict even less than we thought on the basis of quantum theory.

Was there a fly in the ointment of quantum mechanics? Would the foundations of this well-established, dependable field have to shift? Hawking thought they would. As Kip Thorne has commented, Hawking is stubborn about insisting on his views of how nature works, and he likes to challenge others to show he's wrong.[27] Hawking had laid down the gauntlet. Susskind recalls that in Erhard's attic 'Stephen had a "Stephen" look on his face, a little smile that says, "You may not believe it but I'm right, no mistake about it." We were absolutely sure Stephen was wrong but we couldn't see why.'[28]

9

'The odds against a universe that has produced life like ours are immense'

FOR HAWKING, 1981 WAS A LANDMARK YEAR NOT ONLY for his calling attention to the information paradox. He was turning his attention in a new way to the question of how the universe began and how it would end.

At a conference at the Vatican in September, Pope John Paul II, addressing Hawking and other scientists, said that it was probably futile for humans to try to inquire into the moment of Creation: this knowledge 'comes from the revelation of God.'* Given the state of knowledge and theory at the time, which, thanks largely to Hawking, had the universe beginning in a singularity, no one could

* Hawking's comment in *A Brief History of Time* that the Pope had said scientists 'should not inquire into the moment of Creation' was either a misquotation or a mistranslation of the Pope's words.

gainsay the Pope's words. Most of Hawking's colleagues probably would have reluctantly agreed with the first part of the Pope's statement, while doubting God was ever going to lay down his cards. Hawking himself had recently told author John Boslough: 'The odds against a universe like ours emerging out of something like the Big Bang are enormous. I think there are clearly religious implications whenever you start to discuss the origins of the universe.'[1]

What the Pope and those who briefed him on science had failed to take into consideration was Hawking's propensity for undermining his own previous discoveries. Hawking's title for his presentation at this Vatican conference, 'The Boundary Conditions of the Universe', gave no forewarning that he would propose the possibility that there was no 'beginning' – 'no boundary' to the universe – leaving no necessary role or place for a creator. Had the Pope and his science advisors known, they might have been sufficiently wise and well informed to decide that the Pope should draw a parallel between Hawking's ideas and the Judeo-Christian concept (from the Jewish philosopher Philo of Alexandria and the Christian philosopher St Augustine) of a God existing outside time – the 'I Am' of the Bible – for whom beginnings, endings or anything like our chronological time do not exist. That way of looking at time was to be a major part of Hawking's 'no-boundary proposal'. It was not new to philosophy or religion, but it was to physics.

The work Hawking had done in the late 1960s, in his Ph.D. dissertation and afterwards, seemed to prove that the universe had begun as a singularity, a point of infinite density and infinite spacetime curvature. At that singularity all our laws of physics would break down, and it would be just as useless as the Pope thought to try to investigate the moment of creation. Any sort of universe could come out of a singularity. There would certainly be no way to predict that it would be a universe like ours. It was in this context that Hawking had said that discussing the origins of the universe inevitably had religious implications.[2]

The 'Anthropic Principle'

Most of us have become convinced that the sun, the planets and everything else don't revolve around the Earth. Science also tells us that the universe probably looks the same from any vantage point. Earth, with us as its favoured passengers, isn't the centre of everything.

Nevertheless, the more we discover on both the microscopic and the cosmic levels, the more we're struck with the impression that some careful planning, some incredible fine-tuning, had to occur to make the universe a place where it's possible for us to exist. In the early 1980s Hawking was saying, 'If one considers the possible constants and laws that could have emerged, the odds against a universe that has produced life like ours are immense.'[3]

There are many examples of this mysterious fine-tuning: Hawking points out that if the electric charge of the electron had been slightly different, stars either wouldn't burn to give us light or wouldn't have exploded in supernovae to fling back into space the raw material for new stars like our sun or planets like Earth. If gravity were less powerful than it is, matter couldn't have congealed into stars and galaxies, nor could galaxies and solar systems have formed had gravity not been at the same time the *weakest* of the four forces. No theory we have at present can predict the strength of gravity or the electric charge of the electron. These are arbitrary elements, discoverable only by observation, but they seem minutely adjusted to make possible the development of life as we know it.

Shall we jump to the conclusion that Someone or Something had us in mind when things were set up? Is the universe, as astronomer Fred Hoyle phrased it, 'a put up job', a great conspiracy to make intelligent life possible? Or are we missing other possible explanations?

'We see the universe the way it is because we exist.' 'Things are as they are because we are.' 'If it had been different we wouldn't be here to notice it.' All of these are ways of stating something called the 'anthropic principle'.

Hawking explains the anthropic principle as follows: picture a lot of different, separate universes, or different regions of the same universe. The conditions in most of these universes, or in these regions of the same universe, will not allow the development of intelligent life. However, in a very few of them, the conditions will be just right for stars and galaxies and solar systems to form and for intelligent beings to develop and study the universe and ask the question, why is the universe as we observe it? According to the anthropic principle, the only answer to their question may be that, if it were otherwise, we wouldn't be around to ask the question.

Does the anthropic principle really explain anything? Some scientists say that it doesn't, that all it shows is how what seems like fine-tuning might instead be a random bit of good luck. It's like the old story about giving enough monkeys typewriters so that by the laws of chance one of them would type the first five lines of Shakespeare's *Hamlet*. Even if our sort of universe is highly unlikely, with enough universes around, one of them might very well be like ours.

Does the anthropic principle rule out God? No. However, it does show that the universe could appear tailor-made for our good without there being a God.

John Wheeler thought we might carry the anthropic principle a step further. Perhaps, he suggested, there can be no physical laws at all unless there are observers to work them out. In that case there won't be all those alternate universes, because any universe that didn't allow for the development of observers simply wouldn't exist.

If this is so, does it mean that if we become extinct, so will the universe? Will the stage crew come out and dismantle the set as the last member of the audience leaves the theatre? In fact, if we're not around to remember that it existed, will it ever have existed? Does our having observed a brief slice of its existence give it the power to go on existing after we're gone?

A few physicists like to make a connection between an 'observer-dependent' universe and some of the ideas in Eastern mysticism: Hinduism, Buddhism and Taoism. They get no encouragement from Hawking, who says, 'The universe of Eastern mysticism is an illusion. A physicist who attempts to link it with his own work has abandoned physics.'[4]

Although he didn't invent the idea, the anthropic principle is often associated with Hawking, along with other colleagues and particularly Brandon Carter, Hawking's office-mate in the mid-1960s, who also worked with him trying to refute the ideas of Jacob Bekenstein about black holes and entropy in 1972. Hawking and most other physicists hoped that we wouldn't have to turn to the anthropic principle as the only explanation for why we have the sort of universe we have and not another. 'Was it all just a lucky chance?' Hawking asks. 'That would seem a counsel of despair, a negation of all our hopes of understanding the underlying order of the universe.'[5] Those words would turn out to be prophetic.

Meanwhile, the Pope had said it couldn't be done. The anthropic principle said it was just a roll of the dice (one roll among an almost infinite number) that fell in our favour. Some were arguing that God had the power to change His mind and adjust things, including the laws of the universe, whenever He pleased. But Hawking didn't think an all-powerful God would have any need to change his mind. He believed there are laws that held at the time that we call the beginning, or the Creation – that made our universe the way it is and not some other way – and that we are capable of understanding them. He wanted to know what those laws are. That meant that somehow he had to cut the ultimate Gordian knot: the singularity.

It would be a couple of years before Hawking would fully work out how to perform that heroic feat. Meanwhile, in October 1981, not long after the visit to the Vatican, he was also busy looking at the beginning of the universe through the eyes of a new theory, known as 'inflation theory'.

Big Bang Challenge

Back in the 1960s, everything had seemed to be falling into place for those who favoured the Big Bang theory. In 1964–5 there was a particularly exciting step forward in the quest to understand the history of the universe and decide between the two competing models, the Big Bang and the Steady State. The story has become a classic. It was one of those relatively rare occasions in science when data turns up where no one is looking for it. At Bell Laboratories, in New Jersey, there was a horn antenna designed for use with the Echo I and Telstar communications satellites. The amount of background noise picked up by the antenna hampered the study of signals from space. Scientists working with the antenna had to make adjustments and resign themselves to studying signals that were stronger than the noise. It was an annoyance that most found it possible to ignore, but two young scientists, Arno Penzias and Robert Wilson, took the noise more seriously.

Penzias and Wilson noticed that the level remained the same no matter in which direction they pointed the antenna. That wouldn't be the case if the noise were a result of Earth's atmosphere, since an antenna pointed towards the horizon faces more of the atmosphere than one pointed straight up. The noise had to be coming either from beyond the atmosphere or from the antenna itself. Penzias and Wilson thought pigeons nesting in the antenna might be the source, but evicting the pigeons and clearing away their droppings brought no improvement.

Another radio astronomer, Bernard Burke, heard about Penzias's and Wilson's puzzle with the antenna, and he knew, as they did not, of current work by Robert Dicke at Princeton. Dicke, following up on a proposal made in the 1940s by the Russian-born George Gamow and Americans Ralph Alpher and Robert Herman, was building an antenna to search for radiation surviving from not long after the origin of the universe, from an era when, if the Big Bang theory was correct, the universe would still have been very hot. Gamow, Alpher and Herman had theorized that such radiation should exist, and that by our own era its temperature should have cooled to about five degrees above absolute zero. Burke brought Penzias, Wilson and Dicke together, and they concluded that Penzias and Wilson had discovered by accident the radiation that Dicke had been hoping to find.

The discovery of what soon became known as the 'cosmic microwave background radiation', or CMBR, was dramatic support for the Big Bang theory, for clearly the universe had once been very much hotter and denser than it is now. Hawking and his friend George Ellis wrote a paper in 1968 emphasizing how strongly this evidence supported the Big Bang.[6] However, it also posed a problem for that theory. In repeated measurements, taken out as far as possible in every direction, researchers found that the temperature of the radiation was the same and failed to find small variations in the CMBR that could have resulted in the structure we see today.

Those problems aside, support for the Big Bang kept coming. There was a discovery that quasars, which theorists were realizing might be an early stage of galaxy formation, exist only at enormous distances from Earth. If the Steady State theory is correct, with galaxies continually spreading further and further apart and the emptiness between being filled by the formation of new galaxies, and if quasars are part of the process of galaxy formation, we ought to find quasars fairly evenly distributed near and far throughout the universe. They are not. Quasars' vast distance from Earth in space (and, by virtue of that fact, in time) means they must have existed only when the universe was much younger than it is now.

This particular stage of galaxy formation must have occurred only in the distant past, has not occurred again in later epochs of the universe's history, and is not going on today.

Yet another nail in the Steady State theory's coffin was hammered into place in 1973 when balloon experiments at Berkeley found that the spectrum of the cosmic microwave background radiation was the spectrum that Big Bang theory predicted. And studies of the abundances of various elements in the Milky Way and other galaxies showed that Big Bang predictions of these abundances were on target.

Nevertheless, in the 1970s the Big Bang theory still had stumbling blocks to overcome. While Hawking had diverted his attention to black holes, the question of how the universe began had also never been far from his mind; and how to solve the lingering problems of the Big Bang theory had remained high on the agenda for many of his colleagues around the world. These problems became known as the 'horizon problem', the 'flatness problem' and the 'smoothness problem'.

The horizon problem had to do with the observation that the cosmic microwave background radiation is the same in all directions in areas of the universe too far apart for radiation ever to have passed from one to the other, even in the earliest split seconds after the Big Bang. The intensity of radiation is so remarkably close to identical in those remote areas that it seems they must somehow have exchanged energy and come to equilibrium. How?

The flatness problem had to do with the question of why the universe has not either long ago collapsed again to a Big Crunch or else experienced such runaway expansion that gravity wouldn't have been able to pull any matter together to form stars. A universe somehow poised, as ours seems to be, between those possibilities is so unlikely as to boggle the imagination. The expansive energy (resulting from the Big Bang) and the force of gravity would have had to be so close to equal that they differed from equality by no more than 1 in 10^{60} (1 followed by sixty zeros) at a time less than 10^{-43} seconds after the Big Bang (a fraction with 1 as the numerator and 1 with 43 zeros as the denominator).

The smoothness problem was that, judging by the CMBR, the early universe must have been smooth, without any lumps, clumps, ridges or other irregularities. The question became one of the foremost challenges of astrophysics – a 'missing link' in Big Bang thinking: how did a universe that looked so uniform in the era from which the CMBR comes to us, when the universe was 300,000 years old, become so diverse and clumpy all these years later – with stars, galaxies, galaxy clusters, planets – even such small clumps of matter as you and I? Why was it not possible to see even faint beginnings of that differentiation in the cosmic microwave background radiation?

If that last seems an unlikely problem, think about Wheeler's democracy: the closer to one another the particles are, the stronger they feel each other's gravitational pull. If all particles of matter in the universe are equidistant and there are no areas in which a few particles have drawn together even slightly more densely, then every particle will feel equal pull from every direction and none will budge to move closer to any other particle. It was this sort of gridlock researchers seemed to have discovered in the early universe where matter appeared to have been distributed so evenly that it could never yield and form the structure evident in the universe today.

In the mid-1970s, when Hawking first visited Caltech, no theorist had been successful in overcoming any of these stumbling blocks.

Inflation to the Rescue!

In the late 1970s a young particle physicist at the Stanford Linear Accelerator in California, Alan Guth, came up with a significant revision to the history of the universe as it had been written by cosmologists up to that time. Recognizing right away that he had hit on a good thing, Guth wrote 'SPECTACULAR REALIZATION' in his notebook and drew two concentric boxes around the words. His realization offered a brilliant solution to still stubborn problems in Big Bang theory, and also suggested a way in which the universe could have got to be like the present universe without requiring that its initial state be chosen with such exquisitely precise care.

Guth proposed that the universe might early on have undergone a brief interval of stupendously rapid growth, before settling down to continue expanding at the rate we find in our own era. It was the 'settling down' that particularly set his idea apart. Others had found solutions to Einstein's equations that produced a universe in which the expansion would accelerate during the whole of its existence, or where the expansion began by decelerating and then started to accelerate and never stopped. Guth's universe had only a short surge of accelerated expansion in its extreme infancy.

Guth worked out a process in which the universe, at a time less than 10^{-30} seconds after the Big Bang (a fraction with 1 as the numerator and 1 followed by thirty zeros as the denominator), was subject to a huge repulsive force that for only a short interval behaved like the cosmological constant that Einstein rejected. During a period that lasted only an unimaginably small fraction of a second, this force would have accelerated the expansion, causing violent runaway inflation in the dimensions of the universe from a size smaller than a proton in the nucleus of an atom to about the size of a golf-ball.

In the more than thirty years since Guth first suggested it, physicists have been embroidering on his idea, coming up with new versions, and trying to work out why and how this process might have happened. In order to understand this work, you must get your head around some terminology.

Begin with symmetry-breaking: a simple example is a rod set on end. It could fall in any direction. Gravity, which makes it fall, is 'symmetrical'; it has no preference which direction the rod falls. All directions are equally likely. But when the rod falls it *will* fall one way or another, not every way at once. When the rod falls, the symmetry is broken. Hawking used another example in *A Brief History of Time*: picture a roulette wheel. The croupier spins the wheel and the ball rolls round and round. The situation is 'symmetrical'. Although you, who have placed a bet, may have a preferred outcome for this spin, the physics of the situation has no preferred outcome. The wheel slows and the speed (the high energy) of the ball decreases. Finally it falls into one of the slots in the wheel. The symmetry is broken.[7]

To demonstrate the meaning of 'false vacuum' and 'true vacuum', both of which are relevant in understanding inflation, physicists often like to picture a man's hat, the kind with a brim around the edge and a depression in the crown. Put a marble into the depression in the crown of the hat. It will settle down at the lowest spot available. This is not the lowest it could get on the hat, but it's the lowest on the crown of the hat. Likewise, elementary particles can 'land' in a number of temporary energy levels. These resting places are 'false vacuums'. Jostle the hat or let the marbles bump around against one another and the marbles may roll from the crown of the hat to the brim. In this analogy the brim represents the 'true vacuum'. We can recognize it as the lowest possible energy level in this system.

Something like that could have happened as the universe cooled down. Some of the matter started moving towards a new state with lower energy, in the process liberating the gravitationally repulsive stress that caused the rapid acceleration that was inflation. But, to return to the analogy, the marbles probably didn't all roll down at once. How they rolled, how fast they rolled, how they made it over the edge of the crown, or through the edge of the crown – or whether inflation happened in an entirely different way – these questions have been occupying cosmologists for decades. Nearly all, however, agree that inflation occurred. Inflation has become part of the 'standard model'.

Another useful bit of vocabulary is 'phase transition'. An everyday example of a 'phase transition' is the freezing of water (which also involves a breaking of symmetry). Water as a liquid is symmetrical, the same everywhere and in every direction. Lower the temperature sufficiently and ice crystals form. No longer is everything the same everywhere. The crystals have certain positions and not others, and they line up in some direction and not another. The symmetry is broken. However, if you reduce the temperature of water very carefully, it can go below freezing without ice forming, without the symmetry being broken. The name for this is 'supercooling'. It happens in nature when liquid drops of rain fall during a winter storm and remain liquid even though the air temperature is below freezing, until they encounter something – a tree, a pavement – and immediately freeze.

Alan Guth's suggestion rested on the idea that the universe immediately after the Big Bang would have been extremely hot, with all the particles moving around very rapidly, at high energies. The four forces of nature discussed in Chapter 2 – gravity, electromagnetism, the strong nuclear force and the weak nuclear force – were still united at that time, undifferentiated, as one superforce. The universe expanded slightly, cooled slightly. The particle energies became slightly less, and, as the universe cooled, the forces became separate and distinct from one another. Their initial symmetry was broken. One by one they 'froze out'. It didn't all happen at once, however, because supercooling occurred. There was a phase transition, with bits of the universe going through this transition separately in the form of bubbles where the temperature dropped below some value (analogous to water going below the freezing point) *without* the symmetry between the forces being broken. The result was that the universe entered an unstable, supercooled state, with more energy than it would have had if the symmetry among the forces had been broken.

Bubbles, marbles, hats, frozen water . . . the bottom line

of early inflation theory was, Hawking explains, that during this interval all regions of the universe – including those with more particles than average, and those with less – expanded at an enormous rate, faster than the speed of light. Even where there were more matter particles than average and you might expect gravity to be drawing them together, that could not have happened. As matter particles got farther apart, the result would have been a universe, still expanding, with all its particles few and far between. The expansion would have smoothed out irregularities, meaning that the universe's modern-day smooth and uniform state could have evolved from a great variety of different initial states. The rate of expansion would automatically have become close to the critical rate, solving the 'flatness problem' without the initial rate having been so carefully chosen.[8]

How, then, did the expansion slow down again? Like the rain drops that resist turning to ice but must eventually freeze, the universe had to complete its temporarily arrested phase transition. Guth's original proposal had it that when supercooling occurred, in this overall milieu of *un*broken symmetry, bubbles of *broken* symmetry formed, and these bubbles expanded and joined with one another until everything everywhere was in a new phase of broken symmetry, and the universe was expanding more or less as we discover it doing today.

Initially, there was a problem with this idea. The bubbles would have expanded so quickly that they would have collided with one another, resulting in many irregularities and enormous variations in density and in the expansion rate from one part of the universe to another. This situation could never have developed into our universe.

Nevertheless, Guth went ahead and announced his theory. It promised too much to be allowed to flounder on a glitch he was certain he or others would be able to solve later. It offered solutions to the remaining problems in Big Bang theory: our visible universe could have emerged from a region originally so tiny that it had the opportunity to reach equilibrium before it inflated. The period of runaway inflation could have wiped out the imbalance between the expansive energy and the contracting force of gravity. The prediction that inflation would generate areas of slightly higher and slightly lower density – the seeds of future galaxies, supergalaxies and all the other structure that has evolved in the universe – was particularly promising. Technology for observing the cosmic microwave background radiation had not yet been able to reveal those 'density perturbations', but Stephen Hawking and others had been thinking about them and the smoothness problem since the mid-sixties when Wilson and Penzias had discovered the CMBR. Could inflation provide the answer cosmologists had been looking for?

Hawking, like Guth himself, was not satisfied. Hawking's objection to inflation theory was not that the bubbles would collide and cause havoc instead of a smooth universe, but that in the inflationary phase the universe would have been expanding too rapidly for bubbles of broken symmetry ever to join with one another. He thought they would have scattered apart too quickly, even if they were growing at the speed of light. The result would have been a universe where the symmetry among the four forces was broken in some areas and not in others – again, definitely not our universe. With that in mind in October 1981, Hawking set off for a conference in Moscow.

A Debate in Moscow

The Russian physicist Andrei Linde, aged thirty-three, a graduate of Moscow University and the P. N. Lebedev Physics Institute in Moscow, was about to encounter Stephen Hawking for the first time under rather teeth-grinding circumstances at that Moscow conference.

A few years before Alan Guth developed and published his model of inflation, Linde himself had been thinking along those same lines but recognized that this type of theory had problems. Not so reticent as Linde, Guth of course had recognized the same problems but daringly and, as it turned out, wisely gone ahead and published his paper anyway, beating Linde to the draw. That setback notwithstanding, it would not take long for Linde to catch up and move to the head of the roster of cosmologists working in the field of inflation theory. After 1990 he would also be famous among colleagues at Stanford for his sleight-of-hand magic, acrobatics and hypnosis, but in 1981, when he met Hawking at the conference in Moscow, he was still relatively inexperienced, barely known in the West, and had never been to America or Europe. Stephen Hawking was a distinguished and honoured attendee.

Linde and Hawking both presented papers. In his presentation Hawking spoke of his recent findings that inflation would have generated density perturbations that were too large to result in the universe as we find it today. Linde, in his presentation, explained a way he had worked out the previous summer to solve the problems in his and Guth's original inflation models. Because of the Soviet Union's long delays due to censorship before a paper could be released, Linde's 'new inflation' paper would not be published until early in 1982. Linde had no opportunity at the conference to discuss his ideas with Hawking, but after it ended, circumstances brought them together. At the celebration of Hawking's sixtieth birthday in 2002, Linde vividly described both the trauma and the eventual success of that first meeting.[9]

The Sternberg Astronomy Institute in Moscow had invited Hawking to give a lecture the day after the conference ended, and he had chosen as his subject problems in Alan Guth's inflation theory. At the last minute, Linde, who was fluent in English and Russian, was asked to translate. This was during the time when it was customary for Hawking to have one of his students deliver his talks while he listened and occasionally intervened to make a comment or correction. For some reason this talk had not been prepared in that way. Linde recalled the tedious two-stage translation process. Hawking said something in his garbled voice; Hawking's student struggled to understand and then repeated it clearly in English; Linde translated it into Russian. It all moved at a glacial pace. However, Linde knew the subject well and began adding explanations in Russian. Hawking would make a statement; the student would repeat it; Linde would expound, saving Hawking the trouble of explaining what he had said. Hawking seemed not to object and all went much more smoothly and rapidly, as long as they were talking about the old inflation theory.

There came a moment, however, when Linde was startled to hear the student say, for Hawking, that Andrei Linde had recently 'suggested an interesting way to solve the problems of inflationary theory'.[10] Linde was delighted to translate that announcement into Russian. Top Russian physicists were about to hear Stephen Hawking explain his (Linde's) theory! His future in theoretical physics seemed exceedingly bright, but only for a few seconds. Hawking proceeded to tear apart Linde's new inflation scenario. For a painful and embarrassing half-hour Linde 'was translating for Stephen and explaining to everyone the problems with my scenario and why it did not work'.[11] At the end of the talk, Linde summoned extraordinary courage and told the audience that he had translated but did not agree with Hawking, and he explained why. Then he suggested that he and Hawking continue the discussion in private. Hawking might have interpreted that to mean 'Step outside and say that!' But no, they found an empty office, and while the Institute officials searched in a panic for the 'famous British scientist who had miraculously disappeared',[12] Hawking and Linde talked for two hours and then moved to Hawking's hotel to continue their debate. By then things were looking up for Linde. 'He started showing me photos of his family and invited me to Cambridge. This was the beginning of a beautiful friendship.'[13]

Hawking had not been wrong to be dissatisfied with Linde's 'new inflation'. Linde had suggested a solution to the problem of the bubbles of broken symmetry not being able to join with one another: suppose the bubbles were large enough so that what would later evolve into our region of the universe could be all inside one bubble. In order for this to be possible, the change from symmetry to broken symmetry would have had to happen much more slowly inside the bubble. Hawking objected that Linde's 'new inflation' bubbles would have had to be too large – larger than the entire universe was at the time all this was happening. The theory also still predicted much larger variations in the temperature of the microwave background radiation than had been observed.

Not long after the Moscow conference Hawking made a trip to Philadelphia. In an address there, as recipient of the Benjamin Franklin Medal in Physics from the Franklin Institute, he departed from a strictly scientific topic to talk about something that had been of serious concern to both himself and Jane since the early years of their marriage – the dangerous threat that the USSR's and the United States' escalating nuclear stockpiles were posing to life on Earth. Back in 1962, Diana King had mentioned in Jane Wilde's hearing: 'He goes on Ban the Bomb marches.' Hawking was still marching.

Soon after his return to Cambridge, however, Hawking was back in the inflation discussion. He received in his post a letter from *Physics Letters* asking him to review Linde's inflation paper for publication.[14] Hawking recommended that they publish the paper,[15] in spite of the fact that both he and Linde recognized that there were flaws. The paper was significant and deserved to be widely read, and if Linde were to make the necessary revisions, they would be delayed too long passing through Soviet censorship. At the same time Hawking and a graduate student, Ian Moss, submitted a paper of their own suggesting what they thought was a more satisfactory way to end the inflationary period: if the symmetry had broken (still slowly as Linde had proposed) not just inside bubbles but everywhere at the same time, the result would be the uniform universe that we live in.[16] With all these ideas in the air, Hawking and his DAMTP colleague Gary Gibbons decided to organize a workshop, focusing mainly on inflation, to take place the following summer.

Hawking's super-capable secretary, Judy Fella, got to work.

In January 1982 Hawking turned forty, a birthday he had not expected to reach. There was more to celebrate. Hawking was on the New Year's Honours list, made a Commander of the British Empire. At the investiture ceremony at Buckingham Palace on 23 February, Robert was his father's assistant. Now Hawking could put CBE after his name.

Making Inflation Work

From 21 June to 9 July 1982, the wizards of inflation finally put their heads together in Cambridge at the Nuffield Workshop on the Very Early Universe. Andrei Linde came over from Russia. Alan Guth was there, and so was Paul Steinhardt, a physicist at the University of Pennsylvania, who with his colleague Andreas Albrecht had come up with a 'new inflation' theory independently of Linde and very similar to his, at about the same time.* Hawking's contribution to the meeting was to show how the temperature of the universe during the inflationary period could inevitably lead to small density perturbations.[17]

Later in the summer Hawking flew to California again, this time to Santa Barbara, to spend some weeks at the University of California's new Institute of Theoretical Physics. This was Jim Hartle's home territory, and while Hawking was there the two of them discussed the idea that he had introduced at the Vatican in September 1981.

* Linde, Steinhardt and Albrecht now are given joint credit for the 'new inflation model'.

The 'no-boundary proposal' had taken something of a back seat during all the discussion of inflation theory, but Hawking had not stopped thinking seriously about it. Over the next two years, he and Hartle worked out that proposal.

10

'In all my travels, I have not managed to fall off the edge of the world'

HAWKING'S SUGGESTION THAT BLACK HOLES EMIT radiation had been greeted at first with scepticism in 1974, but we've seen that most physicists soon came to agree it wasn't nonsense after all. Black holes must radiate as any hot body does if our other ideas about general relativity and quantum mechanics aren't badly out. No one has found a primordial black hole, but if one were discovered, physicists would be shocked to find it *not* emitting a shower of gamma rays and X-rays.

Go back to thinking about the particles that are emitted by a black hole in Hawking radiation. A pair of particles appears at the event horizon. The particle with negative energy falls into the black hole. The fact that its energy is negative means we have energy subtracted from the black hole. What happens to that energy? (Recall that we don't think energy can simply disappear from the universe.) It is carried off into space with the positive energy particle (see Chapter 6).

The upshot, you'll remember, is that the black hole loses mass and its event horizon shrinks. For a primordial black hole the whole story may end with the black hole's disappearing completely, probably with an impressive fireworks display. How can something escape from a black hole if nothing can escape from a black hole? It really was one of the great 'locked room mysteries' of all time, solved by 'S. H.'

The idea that matter in a black hole didn't necessarily reach the absolute end of time at a singularity had triggered suspicions about another singularity: the singularity Hawking had decided earlier was at the absolute *beginning* of time. Quantum theory offered a fresh possibility: maybe the Big Bang singularity is, as Hawking terms it, 'smeared away'. Maybe the door isn't slammed in our faces after all.

Hawking points to a similar problem which quantum theory solved early in our century, a problem related to Rutherford's model of the atom: 'There was a problem with the structure of the atom, which was supposed to consist of a number of electrons orbiting around the central nucleus, like the planets around the Sun' (look back at Figure 2.1). 'The previous classical theory predicted that each electron would radiate light waves because of its motion. The waves would carry away energy and so would cause the electrons to spiral inwards until they collided with the nucleus.'[1] Something had to be wrong with this picture, because atoms *don't* collapse in this manner.

Quantum mechanics, with the uncertainty principle, came to the rescue. We can't know simultaneously both a definite position and a definite momentum for an electron. 'If an electron were to sit on the nucleus, it would have both a definite position and a definite velocity,' Hawking points out. 'Instead, quantum mechanics predicts that the electron does not have a definite position but that the probability of finding it is spread out over some region around the nucleus.' The electrons don't spiral inwards and hit the nucleus. Atoms don't collapse.

According to Hawking, 'The prediction of classical theory [that we will find the electrons at the nucleus] is rather similar to the prediction of classical general relativity that there should be a Big Bang singularity of infinite density.'[2] Knowing that everything is at one point of infinite density at the Big Bang or in a black hole is too precise a measurement to be allowed by the uncertainty principle. To Hawking's way of thinking this principle should 'smear out' the singularities predicted by general relativity, just as it smeared out the positions of electrons. There is no collapse of the atom, and he suspected that there was no singularity at the beginning of the universe or inside a black hole. Space would be very compressed there, but probably not to a point of infinite density.

The theory of general relativity had predicted that inside a black hole and at the Big Bang the curvature of spacetime becomes infinite. If that doesn't happen, then Hawking wanted to work out 'what shape space and time may adopt instead of the point of infinite curvature'.[3]

When Time is Time and Space is Space

If you find the following discussion difficult, don't hesitate to skim over it. It isn't necessary to understand every word to appreciate Hawking's theory, but it's more interesting if you can. Of course, the maths Hawking uses to describe it, and that you and I would need in order to understand him completely, is much more complicated than the simple maths you will find here.

Relativity theory links space and time in four-dimensional spacetime: three dimensions of space and one of time. Take a look at what a spacetime diagram is like. Overleaf is one that I once drew showing my daughter Caitlin on her way from her classroom at school to the room where the children ate their lunch. The vertical line on the left represents the passage of time. The horizontal line at the bottom represents all the space dimensions. Any single point on our spacetime diagram represents a position in space and a moment in time. Let's see how this works.

The diagram (Figure 10.1) begins with Caitlin at her desk in her classroom, at 12.00 noon. She sits still, moving forward in time but going nowhere in space. On the diagram a little band of 'Caitlin' moves forward in time. At 12.05 the bell rings. Caitlin moves towards the lunch room. (Her desk still moves forward in time but goes nowhere in space.) Caitlin moves in both time and space. At 12.07 she pauses to retie her trainer. For one minute she moves forward in time but does not move in space. At 12.08 she's off again towards the lunch room, walking a little faster than before so that the food won't all be gone by the time she gets there. At 12.15 she arrives at the lunch room. A physicist would say we have traced Caitlin's 'world-line'.

That spacetime diagram was a very sketchy affair. When physicists draw a spacetime diagram, they often use a common unit for both space and time. They might, for instance, use one yard as the unit of both space and time. (One yard

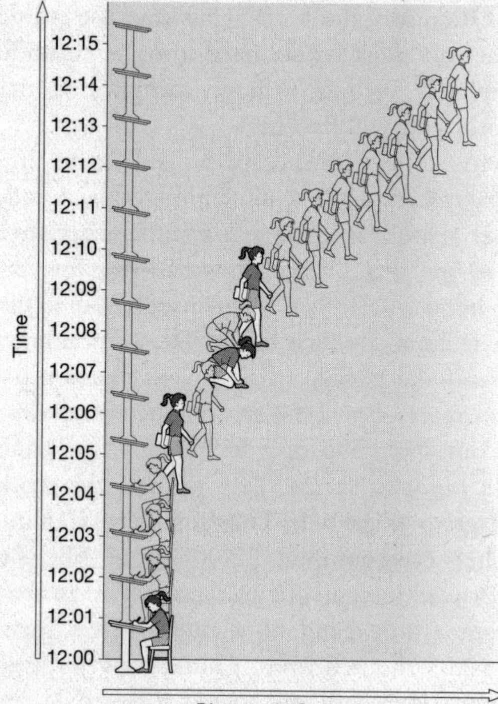

Figure 10.1. Caitlin in Spacetime

of time is very small, only billionths of a second. It's the time it takes a photon, which moves at the speed of light, to travel one yard.) In such a spacetime diagram, if something moves four yards in space and four yards in time, its world-line traces a 45-degree angle. That's the world-line for something moving at the speed of light, a photon, for instance (Figure 10.2).

If something moves three yards in space and four in time, it's moving at three-fourths the speed of light (Figure 10.3a). If something moves four yards in space and

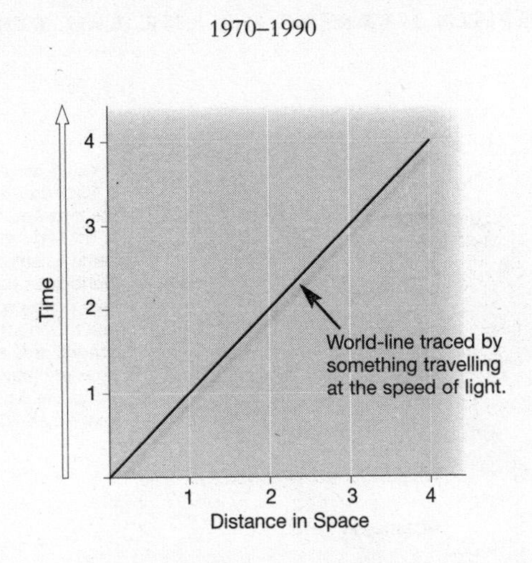

Figure 10.2. A spacetime diagram using one yard as the unit of both space and time. If something travels four yards in space and four yards in time, its 'world-line' traces a 45-degree angle on a spacetime diagram. That's the world-line for a photon, or anything else moving at the speed of light.

three in time, it's exceeding the speed of light, which isn't allowed (Figure 10.3b).

The next diagram (Figure 10.4) shows two events

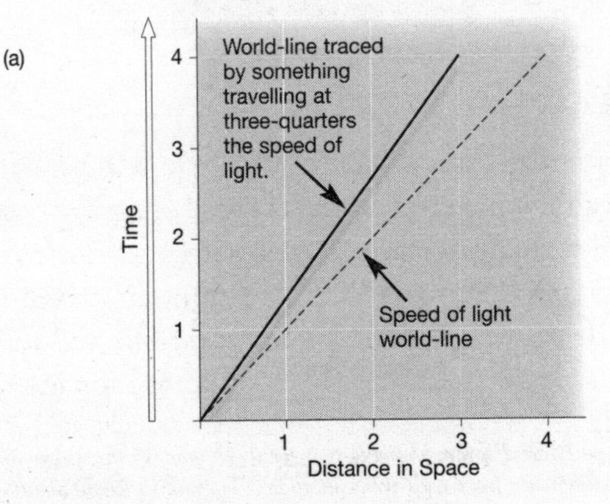

Figure 10.3. (a) A spacetime diagram showing the world-line traced by something moving three yards in space and four yards in time: three-quarters the speed of light.

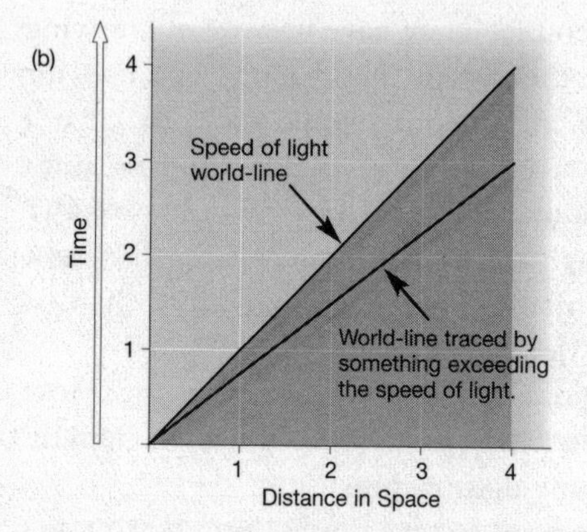

A world-line at more than a 45-degree angle from the time-line, like this one, is not allowed. That's the same as saying that world-lines beginning at zero on the space-line and moving through the shaded area are not allowed. Tracing such a world-line would require faster-than-light travel.

Figure 10.3. (b) World-line traced by something moving four yards in space and three yards in time. When the distance travelled is greater in space than in time, as in this case, the object is exceeding the speed of light (not allowed!).

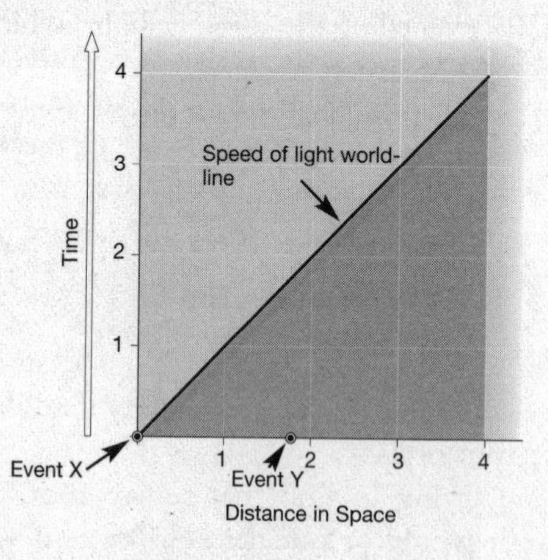

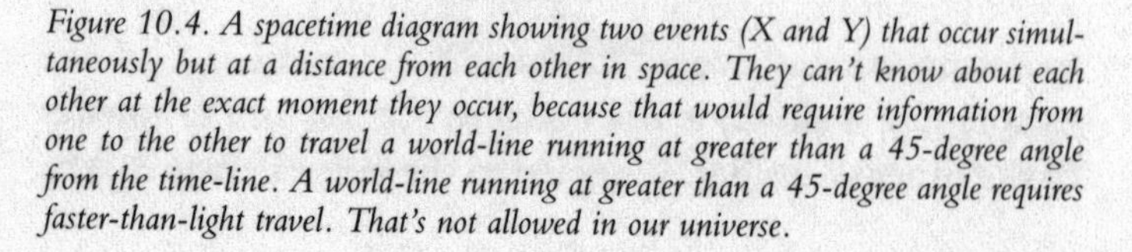
Figure 10.4. A spacetime diagram showing two events (X and Y) that occur simultaneously but at a distance from each other in space. They can't know about each other at the exact moment they occur, because that would require information from one to the other to travel a world-line running at greater than a 45-degree angle from the time-line. A world-line running at greater than a 45-degree angle requires faster-than-light travel. That's not allowed in our universe.

happening simultaneously. They have no way of knowing about each other at the instant they happen because, for them to do so, the information would have to have a world-line running at a 90-degree angle from the time-line. Travelling such a world-line would require faster-than-light travel. Nothing can travel faster than light and light can't manage anything greater than a 45-degree angle on the diagram.

Now we'll talk about the 'length' of a world-line. How shall we say what the length of a world-line is – a 'length' that takes into account all four dimensions?

Let's examine the world-line of something that moves a lot faster than Caitlin. The object in Figure 10.5 moves four yards in space and five in time: four-fifths the speed of light. Think of the distance it moves in the 'space' direction on the diagram as one side of a triangle (side *A*). Think of the distance it moves in the 'time' direction on the diagram as a second side (side *B*). That makes two sides of a right-angled triangle. The world-line of the moving object is the hypotenuse of that triangle (side *C*).

Most of us have learned that the square of the hypotenuse of a right-angled triangle is equal to the sum of the squares of the other two sides. The square of 4 (side *A*) is 16. The square of 5 (side *B*) is 25. The sum of 16 and 25 is 41. The length of side *C*, the hypotenuse, would be the square root of 41.

Never mind trying to find that square root. That is another issue. It would be our next challenge if we were working with our familiar school geometry. However, for spacetime things work differently. The square of the hypotenuse (side *C*) is not equal to the *sum* of the squares of the other two sides. It's equal to the *difference between* the

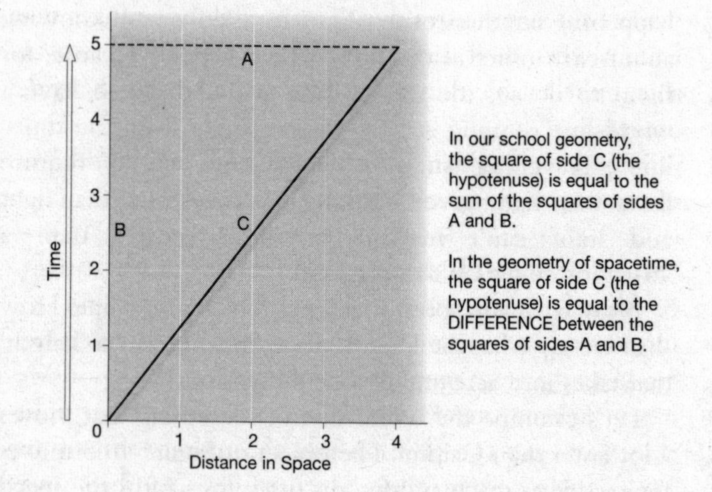

Figure 10.5. A right-angled triangle, using the distance travelled in space as side A, the distance travelled in time as side B, and the world-line travelled in spacetime as side C, the hypotenuse.

squares of the two other sides. Our object travels four yards in space (side *A* of the triangle) and five yards in time (side *B*). The square of 4 is 16; the square of 5 is 25. The difference between 25 and 16 is 9. The square root of 9 is 3. So we know that the third side of the triangle, side C, the world-line of our travelling object, is three yards in length in spacetime.

Let's say, just for the fun of it, that the object is someone wearing a watch. The watch will show that length (three yards) as 'time'. In Figure 10.6 (see p. 206) Lauren remains stationary in space and measures five hours on her watch. Her twin brother Tim, moving at four-fifths the speed of light, meanwhile measures only three hours on his. Tim turns around and returns, again measuring three hours while Lauren measures five. Tim is slightly younger than Lauren when next they meet. This is one of the remarkable, unbelievable things that Einstein taught us about the universe.

Now let's consider spacetime diagrams and world-lines of some smaller objects, elementary particles.

'Sums-Over-Histories', or: The Likelihood of Visiting Venus

Remember the smeared-out positions of electrons in the model of the atom we talked about earlier. Their positions were 'smeared out' because we couldn't measure simultaneously both the position and the momentum of any one of them very precisely. Richard Feynman had a way of dealing with this problem which we now call 'sums-over-histories'.

Imagine that you are considering all the different routes Red Riding Hood might take from home to her grandmother's cottage – not just the quickest way as the crow flies or the safest route avoiding as much as possible the wolf-infested woods, but every possible route she might take. There are billions and billions of possible routes. You ultimately get a gigantic fuzzy picture of her making the journey by all these routes at once. However, some are certainly more likely than others. If you study the probabilities of her taking the various routes, you conclude that she is very unlikely at any time between home and grandmother to be found on the planet Venus, for instance. But according to Feynman you must not completely rule out her passing through there. The probability for that route is extremely low, but not zero.

In a similar manner, with sums-over-histories, physicists

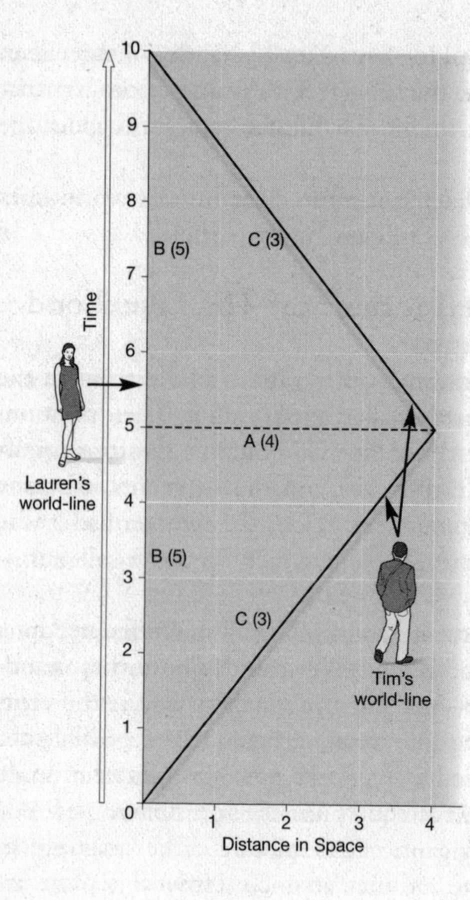

Figure 10.6. The 'twin paradox'

work out every possible path in spacetime that could have been travelled by a certain particle, all the possible 'histories' the particle could have had. It's possible then to calculate the *probability* of a particle's having passed through a particular point, something like calculating how likely Red Riding Hood is to travel by way of the planet Venus. (You don't want to get the idea, however, that particles *choose* a path. That would be carrying the analogy too far.)

Hawking decided to put sums-over-histories to another use, to study all the different histories the universe could have and which are more probable than others.

As we continue, you'll need to know that even though the theory of relativity taught us to think of three dimensions of space and one of time as four dimensions of spacetime, there are still physical differences between space and time. One of these differences has to do with the way we measure the four-dimensional distance between two points in spacetime: the hypotenuse of the aforementioned triangle.

Figure 10.7a (overleaf) shows two separate events (*X* and *Y*) on a spacetime diagram. They are connected by a world-line at more than a 45-degree angle from the time-line. No information can pass between these events without exceeding the speed of light. In a case like this, where the distance between two events is greater in space than it is in time, the square on the hypotenuse (side *C*) of our triangle is a *positive* number. In the language of physics the square of the 'four-dimensional separation' between events *X* and *Y* is positive.

Figure 10.7b also shows two events. The distance between them is greater in time than it is in space. A world-line between these events runs at a less than 45-degree angle from the time-line. Information travelling at less than the speed of light can reach *Y* from *X*. When this is true, the square on the hypotenuse (side *C*) of our triangle is a *negative* number. Physicists say the square of

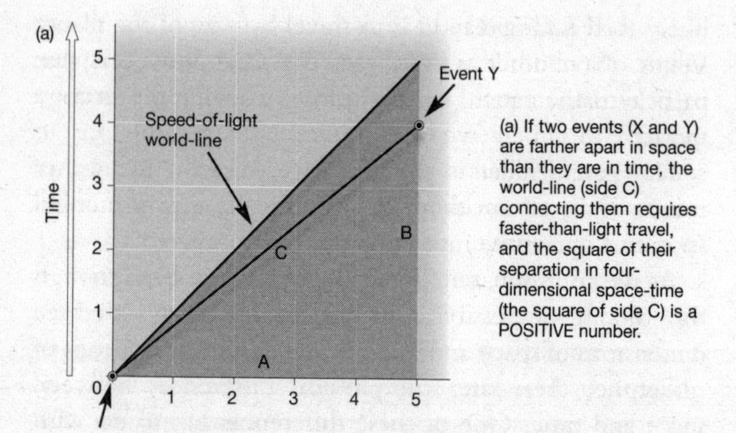

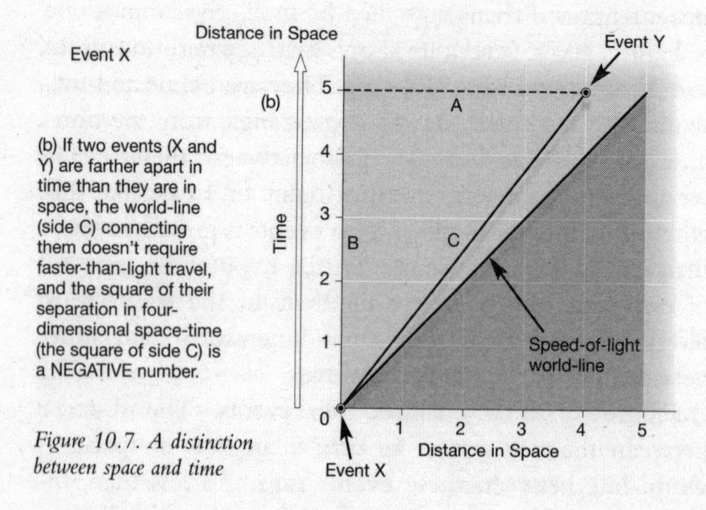

Figure 10.7. A distinction between space and time

the four-dimensional separation between *X* and *Y* is negative.

Perhaps you got lost in those last two paragraphs. If you

didn't, a red light may have flashed on in your brain. The square of a number can't be negative. That doesn't happen in our mathematics. If the square of a number were a negative number, what number could possibly be its square root? What is the square root, for instance, of minus 9? In our mathematics the square of any number (negative or positive) is always positive: 3 squared (3^2) is 9; so is minus 3 squared (-3^2). We can't possibly arrive at minus 9. It's impossible for the square of *anything* to be a negative number.

Stephen Hawking and other mathematicians and physicists have a way around this problem: imagine that there are numbers that do produce negative numbers when multiplied by themselves, and see what happens. Say that imaginary one, when multiplied by itself, gives minus one. Imaginary two, multiplied by itself, gives minus four. Calculate the sums-over-histories of the particles and sums-over-histories of the universe using imaginary numbers. Calculate them in 'imaginary' rather than 'real' time. The time it takes to get from point *X* to point *Y* in Figure 10.7b is imaginary time – the square root of minus nine – imaginary three.

Imaginary numbers are a mathematical device (a trick, if you prefer) to help calculate answers that would otherwise be nonsense. 'Imaginary time' allows physicists to study gravity on the quantum level in a better way, and it gives them a new way of looking at the early universe.

Smearing Out the Speed of Light?

Travelling back into the very early universe, as space becomes more and more compressed, there are fewer possible choices about where a particle is (its position) at a given moment. The position becomes a more and more precise measurement. Because of the uncertainty principle this causes the measurement of the particle's momentum to become less and less precise.

First, let's look at the photon, the particle of light, under more normal circumstances. Photons move at 186,000 miles (300,000 kilometres) per second, making the speed of light 186,000 miles (300,000 kilometres) per second. Now I have to tell you that this might not always be the case. (Having read this far, you are accustomed to such reversals!) We've seen that the probability of finding an electron is spread out over some region around the nucleus of an atom: more likely at some distances than others, but definitely a very smeary affair. Photons, like electrons, can't simultaneously be pinned down precisely as to both position and momentum, because of the uncertainty principle.

Just so, Richard Feynman and others have told us that the probability that a photon is travelling at 186,000 miles (300,000 kilometres) per second may be spread out over some 'region' around that speed. That's the same as saying that, in one way of thinking about it, the speed of a photon fluctuates more or less around what we call light speed. Over long distances probabilities cancel out, so as to make the speed of a photon 186,000 miles (300,000 kilometres) per second. However, over very small distances, on the quantum level, there's a possibility that a photon may move at slightly less or slightly more than this speed. These fluctuations won't be seen directly, but the path of photons on the spacetime diagram, which we've drawn as a 45-degree angle, gets a little fuzzy.

When we're studying the very early universe, when

space is very compressed, that line gets *very* fuzzy. The uncertainty principle means that the more precisely we measure the position of a photon, the less precisely we're able to measure its momentum. When we say that in the very early universe everything was packed to near-infinite density (not a singularity, but nearly there), we're becoming extraordinarily precise about the location of particles such as photons. When we are that precise about position, our imprecision about momentum vastly increases. As we near infinite density we also get near an infinite number of possibilities of what the speed of a photon is. What happens to our spacetime diagram now? Look at Figure 10.8. The world-line of a photon that in more normal circumstances is shown as a 45-degree angle becomes *terribly* smeared out. It fluctuates and ripples wildly.

Here is another way of thinking about what causes this 'rippling', a way which will link it more clearly with other concepts in this book. Travelling back into the very early universe is like shrinking ourselves to a size so unimaginably tiny that we can see what's happening on the level of the extremely small. Imagine it like this: if you look at this page, it seems smooth. You can curl the paper a bit, but it's still smooth. In the same way, although there is some curvature, spacetime around us seems smooth. On the other hand, if you look at this page under a microscope, you see curves and bumps. Similarly, if you look at spacetime on the extremely tiny level, billions on billions of times smaller than an atom, you find violent fluctuations in the geometry of spacetime (Figure 10.9) (see p. 213). We'll discuss this again in Chapter 12 and learn that it might result in something called 'wormholes'. For the time being, the point is that we would find the same violent

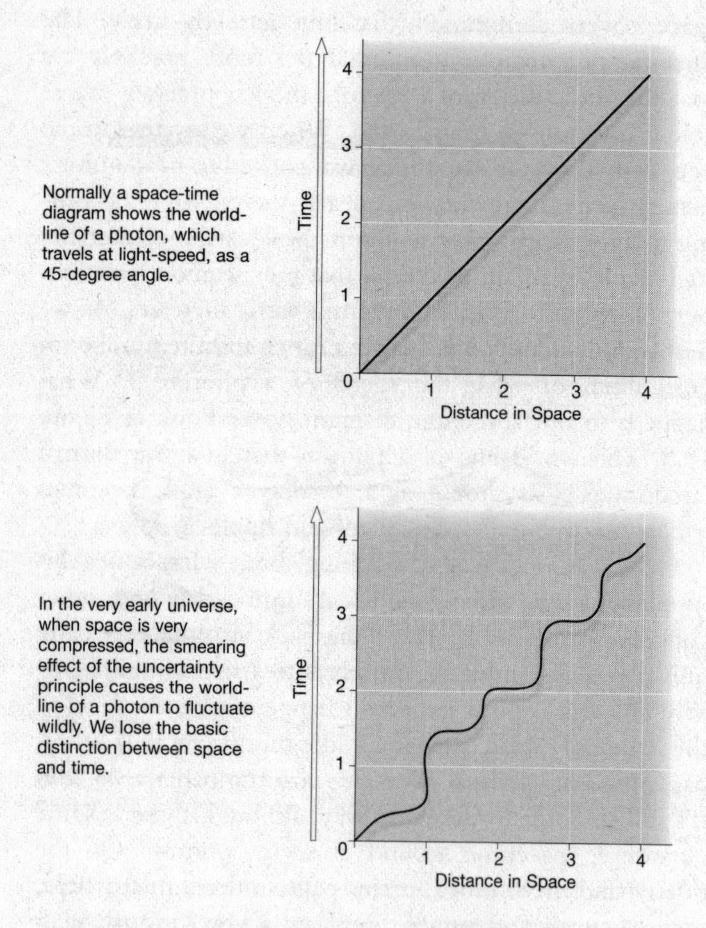

Figure 10.8. The uncertainty principle in the early universe

fluctuations in the very early universe, where everything was compressed to just such extreme smallness.

How can we explain this violent, chaotic scene? Again we turn to the uncertainty principle. We saw in Chapter

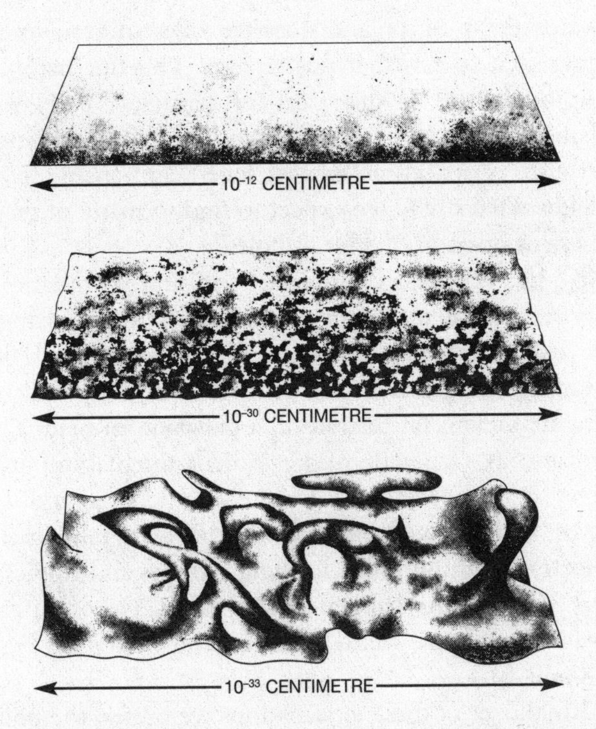

Figure 10.9. The quantum vacuum, as imagined by John Wheeler in 1957, becomes more and more chaotic as you inspect smaller regions of space. At the scale of the atomic nucleus (top), space still looks very smooth. Looking much more closely than that (middle), we see a roughness begin to appear. At a scale 1,000 times smaller still (bottom), the curvature undergoes violent fluctuations.

6 that the uncertainty principle also means that a field, such as an electromagnetic field or a gravitational field, can't have a definite value and a definite rate of change over time. Zero would be a definite measurement, so a field can't measure zero. All fields would *have* to measure exactly zero in empty space. So, no zero, no empty space. What do we have instead of empty space? A continuous fluctuation in the value of all fields, a wobbling a bit

towards the positive and negative sides of zero so as to average out to zero but not *be* zero. This fluctuation can be thought of as the pairs of particles in Hawking radiation. Particle pair production is greater where the curvature of spacetime is most severe and changing most quickly. That's why we expect to find so many of them at the event horizon of a black hole.

In the very early universe we find a situation of extremely great spacetime curvature and rapid change in that curvature. The quantum fluctuations in all fields, including the gravitational field, become very violent. If there are violent fluctuations in a gravitational field, that is the same as saying there are violent fluctuations in the curvature of spacetime. We are not talking about big curves, such as swells on the ocean. We are talking about all sorts of continuously changing crinkles and ripples and swirls. Odd things happen to the world-line of a photon in such a wild and weird environment. Again see Figures 10.8 and 10.9.

Whichever of these explanations we prefer, the point is that the difference between the time direction and directions in space disappears. When time looks like space, we no longer have our familiar situation in which the time direction always lies within the 45-degree angle and space directions always lie outside it.

Hawking sums up what we have just seen: 'In the very early universe, when space was very compressed, the smearing effect of the uncertainty principle can change the basic distinction between space and time.' It's no longer true that if points are farther apart in time than they are in space the square of their separation in four-dimensional spacetime (the square of the hypotenuse of our triangle) is necessarily a negative number. 'It is possible for the square of [that] separation to become positive under some circumstances. When this is the case, space and time lose their remaining distinction – we might say that time becomes fully spatialized – and it is then more accurate to talk, not of spacetime, but of a four-dimensional space.'[4]

When Time Gets Spaced Out

What would this look like? How is this odd situation of four-dimensional space going to join smoothly with spacetime as we know it, in which time flows as time? Using imaginary time it's possible to picture four-dimensional space, where time as we know it is non-existent, curving around and forming a closed surface, a surface without any edge or boundary. If you think you can picture this occurring in four dimensions, either you're mistaken or else you've taken a fresh evolutionary step in brain development. Most of us are doomed to think about it in fewer dimensions. It's easy to picture something with fewer dimensions that hasn't any edge or boundary: the surface of a ball or the surface of the Earth.

In the first Friedmann model of the universe, the universe was finite, not infinite, in size. But in that model it was also unbounded. It had no boundaries, no edges in space. It was like the surface of a ball: no edge, but not infinite in size. Hawking thinks the universe may be finite and unbounded in space *and time*. Time may have no beginning or end. All of it comes around and forms a closed surface, like the surface of the Earth.

This leaves us fairly helpless. We can picture the surface of the Earth, and we can agree that it is finite and unbounded, but what would a *universe* that is finite and unbounded in space and time be like? It's hard to make a mental connection between the shape of a ball and any meaningful concept of a four-dimensional universe. Just trying makes us feel quite blind – groping in the dark. Let's see what else we can say about it that might be helpful.

First, we'll say what it would *not* be like. There would be no 'boundary conditions' – the way things were at the exact point of beginning – because there would be no point of beginning, no boundary, there. The whole thing would just curve around. Hawking suggests we state it exactly like that: the boundary conditions of the universe are that there are no boundaries. There would be no beginning and no end of the universe – *anywhere*. So don't even think of asking, But what about *before* that? That's like asking what's south of the South Pole. A signpost pointing 'south' has no meaning at the South Pole. An arrow of time indicating 'this way to the past' has no meaning when the time dimension has become 'spacey'.

If there were no before and after the universe in the time dimension, would there be any 'elsewhere', any other place, any outside of such a universe – in space dimensions? Hawking's model doesn't say that there isn't. Can you have an outside when you don't have any boundary? In the ball model there is a sense in which we do. It's in the direction the ant on the surface of the balloon in Chapter 6 would see if it could look 'out' from the surface – which, you'll remember, it can't do. That dimension doesn't exist for the ant, but that doesn't necessarily mean it doesn't exist at all. The idea of having 'elsewheres' in space but none in time (no before or after) fits nicely with the idea that the time we live in is only a temporary mutation of what is really a fourth space dimension.

Since all of this may seem too complicated to be meaningful, let's look at it in another, more practical, way. Ask again, What would a universe that is finite and unbounded in space and time be like? The calculations are extremely difficult. However, what they appear to be telling us is that a universe like that could be like our own.

As Hawking described it:

> They predict that the universe must have started out in a fairly smooth and uniform state. It would have undergone a period of what is called exponential or 'inflationary' expansion, during which its size would have increased by a very large factor but the density would have remained the same. The universe would then have become very hot and would have expanded to the state that we see it in today, cooling as it expanded. It would be uniform and the same in every direction on very large scales but would contain local irregularities that would develop into stars and galaxies.[5]

In real time – and that's where we live – it would still appear to us that we have singularities at the beginning of the universe and inside black holes.

Hawking and Jim Hartle presented the physics community with this no-boundary model of the universe in 1983. Hawking liked to emphasize that it was just a proposal. He hadn't deduced these boundary conditions from some other principle. The model appealed to him. He thought 'that it really underlies science because it is really the statement that the laws of science hold everywhere'.[6]

There are no singularities at which they break down. This kind of universe is self-contained. Do we have to explain how it was created? Would it have to be created at all? 'It would just BE,' writes Hawking.[7]

'What Place, Then, for a Creator?'

This raises some sticky philosophical questions. As Hawking puts it, 'If the universe has no boundaries but is self-contained . . . then God would not have had any freedom to choose how the universe began.'[8]

Hawking hasn't said that the no-boundary proposal rules out the existence of God, only that God wouldn't have had any choice in how the universe began. Other scientists disagree. They don't think the no-boundary proposal limits God very much. If God had no choice, we still have to wonder who decided that God would have no choice. Perhaps, suggests the physicist Karel Kuchar, that was the choice God made. Don Page, who reviewed *A Brief History of Time* for the journal *Nature* in England, has a similar viewpoint. Page, of course, was Hawking's graduate assistant in the late 1970s. He had moved on to become a professor at the University of Alberta, Edmondton, Canada. He and Hawking were still good friends and were continuing to collaborate on scientific papers, and Hawking was well aware that Page was likely to come up with some arguments to refute the notion that the no-boundary proposal abolished the need for a Creator. Indeed he did.

To Hawking's question, 'What place, then, for a Creator?', Page answered that in the Judeo-Christian view 'God creates and sustains the entire Universe rather than just the beginning. Whether or not the Universe has a beginning has no relevance to the question of its creation, just as whether an artist's line has a beginning and an end, or instead forms a circle with no end, has no relevance to the question of its being drawn.'[9] A God existing outside our universe and our time wouldn't need a 'beginning' in order to create, but it could still look to us, from our vantage point in 'real' time, as though there had been a 'beginning'.

In *A Brief History of Time* Hawking himself would suggest that there may still be a role for a Creator: 'Is the unified theory so compelling that it brings about its own existence?' If not, 'What is it that breathes fire into the equations and makes a universe for them to describe?'[10] In the book *A Brief History of Time: A Reader's Companion*, the companion book for the film, he would say that if the no-boundary proposal is correct, he will have succeeded in discovering how the universe began. 'But I still don't know why it began.'[11] He intended to find out, if he possibly could.

All of which brings us to a word of caution: although theoretical physicists ask challenging, penetrating questions and present us with mind-blowing proposals and theories, they do not claim to give us 'ultimate answers' – even though the subtitle of a later book co-authored by Hawking might suggest that they do. The best science progresses by suggesting 'answers' and then taking apart and disproving those 'answers'. The most daring and imaginative scientists launch their toy boats and then, it seems, try extremely hard to make them sink.

Hawking's work is a prime example. First he proved that the universe had to start as a singularity. Then with his no-boundary proposal he showed us how there might be no singularity after all. In the meantime he told us that black holes could never get smaller, and then he discovered they could. His work on the Big Bang singularity seemed consistent with a biblical view of Creation, but his no-boundary proposal put the Creator out of a job or at least changed the job description. In *A Brief History of Time* he suggested that we might need the Creator after all, and 'the ultimate triumph of human reason' would be to 'know the mind of God'.[12] Hawking is provocative and open-minded in the way that the greatest thinkers have always been. He reaches clearly defined, well-supported conclusions, and then in the next breath he mercilessly questions and breaks down those same conclusions. He doesn't hesitate to admit that an earlier conclusion was incorrect or incomplete. That's the way his science – and perhaps all good science – advances, and one of the reasons why physics seems so full of paradoxes.

In the process, Hawking has supplied eloquent quotations that can be used to support opposing philosophical points of view. He's been quoted and misquoted by those who believe in God and those who do not. He's been the hero and villain of both camps. However, those who depend on his statements – or statements of other scientists – to support their belief or unbelief risk having the rug swept out from under them at any moment.

Meanwhile, although it may appear to us that Hawking completely reversed himself with the no-boundary proposal, he didn't see it that way. He said that the most important thing about his work on singularities was that it showed that a gravitational field must become so strong that you can't ignore quantum effects. And when you stop ignoring quantum effects, you find out that the universe could be finite in imaginary time but have no boundaries or singularities.

Inflation Gets Chaotic

In 1983, the same year that Hawking and Jim Hartle published their no-boundary proposal, Andrei Linde solved some of the still nagging problems of inflation theory with a new suggestion. The first person in the West to whom he mentioned 'chaotic inflation' was Hawking. Hawking was enthusiastic.

Both 'old' and 'new' inflation had assumed that inflation was just an intermediate short phase in the history of the early universe, that the universe was in a state of thermal equilibrium (meaning it was the same temperature everywhere) before inflation started, relatively homogeneous and large enough to survive until the inflation process got underway. Linde abandoned these assumptions in his chaotic inflation scenario, which didn't require thermal equilibrium, and in which inflation could begin earlier, closer to the Big Bang.

The universe before the inflationary period could have been in a chaotic state. All that was needed was for minuscule parts of that chaos to be able to inflate, becoming smoother and more isotropic* as they did so, like balloons that started out as crinkly pieces of rubber and ended up smooth balls. For all we know, only one tiny part of the chaos did that, though that is probably not likely. In any event, our balloon, as it inflated, pushed other inflating bits far out of range of our potential sight. Maybe other parts of the universe are still chaotic. Or maybe everything

* Isotropy is the quality of being the same in all directions.

is smooth everywhere.

In chaotic inflation there is no phase transition or supercooling at all. Instead, there is a field that has large values in some regions of the universe, while not in others, a kind of 'lucky negligence of the Creator', as Linde puts it.[13] The energy in the regions with large values, Linde was thinking, would be great enough to have a repulsive gravitational effect, causing them to expand in an inflationary way, while in the regions of the field where the values were too small this could not happen. Inflation in the regions where it did happen would create enormous homogeneous islands out of the original chaos, each one much larger than our observable universe. In these regions, the energy of the field would decrease slowly and, in some of them, eventually allow the expansion to reach the rate of expansion that we observe. Have a sufficient number of these regions around and we are highly likely to find one where the conditions were such as to produce a universe like the one we know, with the constants of nature that are arbitrary elements in our theories set just right to allow for the eventual existence of you and me. Maybe only one – in which case that one is ours.

A satisfying resolution to the story, but . . . not the end of it. Chaotic inflation theory also predicted a 'second stage of inflation', occurring much later – a speeding up of the expansion of the universe . . . perhaps in our own modern era. In the early eighties, that idea sounded like science fiction, even to Andrei Linde and Stephen Hawking. We will see that by the end of the century it would no longer be science fiction.

Meanwhile, the most remarkable aspect of inflation theory as it was beginning to be thought of in the early

1980s was that though theorists still had different ideas about the way inflation happened, there *was* agreement that the whole visible universe that we know today could have begun from a much smaller irregularity of mass and energy than anyone had imagined possible. As John Barrow phrased it in *The Book of Universes*: 'Instead of eradicating [irregularities], inflation just swept them beyond the visible horizon in the universe today. They will still be there somewhere far away but the whole of our visible universe reflects the high isotropy and smoothness of a tiny patch of space that underwent inflation.'[14]

Our universe is not, of course, completely smooth. We have solar systems, galaxies and galaxy clusters. Even immediately following inflation, the bit that would eventually expand to be our visible universe must not have been as perfectly smooth as the imagined inflated balloon might suggest. It would be stretched out, but not so much as to avoid still having small variations that would have provided the seeds that could grow into all that astounding structure – large-scale modern variations in density indeed.

As early as the conference Hawking and Gibbons organized in Cambridge in the summer of 1982, the participants realized that inflation would have produced a particular pattern of variations.[15] This would show up as a recognizable pattern of variation in the cosmic microwave background radiation. Observations at that time were not able to show any such pattern. Nevertheless, in the case for and against inflation, though no one could ever be an eye witness to the events with which our universe began, there was hope that there might someday be some evidence to show whether inflation theorists had it right.

11

'It's turtles all the way down'

LUCY WAS ELEVEN YEARS OLD AND COMPLETING HER final year at Newnham Croft Primary School in the spring of 1982. She and her parents had decided that the best choice for the next stage of her education was the Perse School for Girls in Cambridge. Her brother Robert had been attending the Perse School for Boys since he was seven. In the 1960s a practical need for a job, so that he could get married, had sent Hawking off searching for singularities. This time, another practical need – for funds to pay Lucy's school fees – launched him into a new enterprise that was to have a far-reaching impact on the Hawkings and others all over the world. It all began with Hawking's decision to try to earn the money by writing a book about the universe – not an academic book but a book directed to people without a scientific education.

There had, of course, been other popular books about

the universe and black holes. However, Hawking thought none of them spoke enough about the most interesting questions, the questions that had made him want to study cosmology and quantum theory: Where did the universe come from? How and why did it begin? Will it come to an end and, if so, how? Is there a complete theory of the universe and everything in it? Are we close to finding that theory? Is there a need for a Creator?

These were questions that he thought should interest everybody, not only scientists. However, science had become so technical and specialized that the general public was left out of the discussion. The trick in writing the book would be to make it understandable to non-scientists, and that meant using virtually no mathematics. He set to work dictating and completed the first draft in 1984.

Because it was a great deal of trouble dictating a book of this length, Hawking wanted the book to reach as many people as possible. His earlier books had been published by Cambridge University Press, one of the most prestigious academic publishers in the world, but after discussing his proposed new book with them and hearing them predict that it might sell 20,000 copies a year worldwide, Hawking decided he might do better with a publisher more attuned to the mass popular market. He wanted to see his book in airport bookstalls. His American agent discouraged this hope. Academics and students would buy the book, but the popular market, not likely. Hawking thought otherwise.

Several British publishers turned down the proposal, a decision they would regret.[1] But there were some offers. One of the most surprising came from Bantam. Against the advice of his agent, Hawking chose them. Bantam might not have specialized in publishing science books, but they sold many, many books in airports. Bantam in the United States paid $250,000 for the American rights, and Bantam–Transworld in the UK offered £30,000 for the British. Paying that much for a science book was a gamble – one of the best gambles either publisher ever made.

A Year on the Brink

1985 was a difficult year for the Hawking family. That summer, the plan was for Hawking to spend a month in Geneva at CERN. He was looking forward, among other things, to exploring the implications of some recent calculations by Don Page and Raymond LaFlamme having to do with arrows of time. Hawking, his nurses, his secretary Laura Gentry, and some of his students left Cambridge and travelled directly to Switzerland while Jane, Jonathan, Lucy and Tim took a more circuitous and adventurous route, camping across Belgium and Germany. They were to meet Stephen at the Bayreuth Festival and scrub up from campsite to formal standard to attend Wagner's *Ring* cycle. At the time everyone was more concerned about Robert, who was trekking across Iceland and canoeing its northern coastline on a Venture Scout expedition, than they were about Stephen in safe, healthy Switzerland.[2]

On the eve of their arrival in Bayreuth, Jane found a public phone in Mannheim and rang her husband in Switzerland to arrange the schedule for the next day. A very distraught Laura Gentry answered the phone and urged her to come to Geneva at once. Stephen was in hospital with pneumonia. The situation looked grave. Jane arrived there to find Laura's distress well warranted.

He was on a life-support system, in an induced coma, barely hanging on to life.

Knowing Hawking's physical future with ALS, but not aware of his fierce determination to live, doctors gave Jane the choice whether to have him disconnected from life support and allowed to die. This was a heartbreaking situation. The only way to save his life would be to perform a tracheotomy. Afterwards, there would be no more problems with coughing and choking, but he would never again be able to speak or make any vocal sound. That seemed a ghastly price to pay. Hawking's speech was slow and difficult to understand, but it was still speech, and his only possible means of communication. Without it, he couldn't continue his career or even converse. What would survival be worth to him? Nevertheless, Jane remembers that her decision was 'clear and unequivocal, pronounced without a second thought, that Stephen must live, despite the prognosis of likely complications ahead' given her by the doctors, for as his wife she saw herself 'as the agent of life, not death'.[3] 'The future looked very, very bleak,' she recalls. 'We didn't know how we were going to be able to survive – or if he was going to survive. It was my decision . . . but I have sometimes thought – what have I done? What sort of life have I let him in for?'[4]

When Hawking was strong enough, the University of Cambridge paid for an air ambulance to fly him back to Cambridge, where he was admitted to intensive care in Addenbrooke's Hospital. Doctors made a final attempt to avoid the operation, but efforts to wean him off the respirator brought back the choking fits. A tracheotomy was the only way to proceed. Hawking remembers the vivid dreams he had during this time of flying in a hot-air balloon. He decided to take this as a symbol of hope.

Recovering slowly in the hospital, Hawking no longer breathed through his mouth and nose but through a small permanent opening made in his throat at about the height of his shirt collar. The only way he could communicate was to spell out words letter by letter by raising his eyebrows when someone pointed to the right letter on a spelling card.

After several weeks in intensive care, Hawking was allowed to come home on Sunday afternoons. Jane was determined that he would stay with her and their children and Jonathan rather than live in a nursing home. Since 1980, the community and private nurses arranged by Martin Rees had been coming for an hour or two each morning and evening to supplement the care given by Jane, the graduate assistant and Jonathan. However, from now on for as long as he lived, Hawking would need round-the-clock nurses. The cost was astronomical, far beyond the Hawkings' resources. The National Health Service, which in Britain is paid for out of public funds, would have paid for a nursing home but could only offer a few hours' nursing care in the Hawkings' home plus help with bathing. 'There was absolutely no way we could finance nursing at home,' Jane says.[5] Not only Hawking's work as a physicist but any sort of meaningful life at all seemed at an end. It was an end they'd expected to come much sooner, but it was no less bitter for all that.

'At times things have looked absolutely dire for us and then something has come out of those crises,'[6] Jane has commented, recovering some of the optimism with which she had begun their marriage. Kip Thorne in California received word of his friend's plight and immediately got in touch with Jane and suggested that she try to get funding from the John D. and Catherine T. MacArthur Foundation. Another friend, particle physicist Murray Gell-Mann, was on their board. The MacArthur Foundation agreed to help on a trial basis at first, with a grant to cover nursing care. More than three months after Hawking had entered the hospital, he came home to West Road in early November.

An unexpected ray of hope on the bleak horizon came when a computer expert in California, Walt Woltosz, sent a computer program he'd written for his disabled mother-in-law. 'Equalizer' allowed the user to select words from the computer screen and also had a built-in speech synthesizer. One of Hawking's students devised an implement something like a computer mouse, so that Hawking could operate the program by a tiny movement that was still possible for him: squeezing this switch held in his hand. Should that fail him, head or eye movement could activate the switch.

Still too weak and ill to resume his research, Hawking practised with his computer. The first message he produced, after managing to make the computer say 'Hello' in the synthesized voice that was destined to become familiar all over the world, was to ask his graduate assistant, Brian Whitt, to help him finish writing his popular-level book.[7] That would have to wait until he was more proficient with Equalizer, but before long he could produce ten words a minute, not very fast but good enough to convince him that he could continue his career. 'It was a bit slow,' he says, 'but then I think slowly, so it suited me quite well.' Later, his speed improved. He was for a time able to produce more than fifteen words a minute.

Here's how the process worked, and still does, with a few modifications. The vocabulary programmed into the computer contains around 2,500 words, about two hundred of them specialized scientific terms. A screen full of words appears. The top half of the screen and the bottom half are highlighted alternately, back and forth, until Hawking sees the half-screen that includes the word he's looking for highlighted, and squeezes the switch in his hand to choose that half-screen. Then lines of words on that half-screen are highlighted one after the other. When the line with the word he wants is highlighted, Hawking squeezes the switch again. The words on that line are then highlighted one by one. When the word he wants is highlighted, he presses the switch again. Sometimes he misses and the words or lines have to start over. There are a few often-used phrases, such as 'Please turn the page', 'Please switch on the desk computer', an alphabet for spelling out words not included in the program, and, I am told, a special file of insulting remarks, though I haven't seen him use that.

Hawking selects the words one by one to make a sentence, which appears across the lower part of the screen. He can send the result to a speech synthesizer, which pronounces it out loud or over the telephone. (One strange fault with the process is that it cannot pronounce the word photon correctly, but always comes out with foe-t'n.) He can also save something on a disk and later print it out or rework it. He has a formatting program for writing papers, and he writes out his equations in words, which the program translates into symbols.

Hawking writes his lectures this way and saves them on disks. He can listen ahead of time to the speech synthesizer deliver a lecture, then edit and polish it. Before an audience he sends his lecture to the speech synthesizer a sentence at a time. An assistant shows slides, writes Hawking's equations on the board, and answers some of the questions.

Hawking's synthesized computer voice varies the intonation and doesn't sound like a monotonal robot, which to him is an extremely important feature. At first he wished it gave him a British accent, but after a while he became so identified with it that 'I would not want to change even if I were offered a British sounding voice. I would feel I had become a different person.'[8] Just what accent it does give him is uncertain. Some people say it's American or Scandinavian. To me it sounds East Indian, perhaps because of its slightly musical inflection. Hawking can't inject emotion into the voice. The effect is measured, thoughtful, detached. Hawking's son Tim thinks his father's voice suits him. Tim of all the children is least able to remember what Hawking's own voice sounded like. When he was born in 1979, there was little of that left.

Does all of this make conversation with Hawking seem like talking to a machine – like something alien, from science fiction? At first just a little. Soon you forget all about it. Hawking is comfortable with the odd situation and patient when others are not. When he was reading parts of this book while I held the pages, it was his nurse, not he, who suggested that it was unnecessary for me to wait for Hawking to select 'Please turn the page', which involved a number of manoeuvres on the computer screen. As soon as he started clicking, she said, I could turn the page and save him trouble and time. He'd put up with my way of doing things for an hour and a half without indicating that I was in any way inconveniencing him. As it happened, the next time Hawking 'clicked' and I turned the page, he was making a comment, not asking for a page turn.

Hawking's sense of humour is contagious and likely to break out at any moment. However, when one interviewer commented to Hawking that it must be frustrating, telling jokes and having your listeners anticipate the punchline before you have a chance to crack it, Hawking admitted that 'I often find that by the time I have written something, the conversation has moved on to another subject.'[9] Nevertheless, when his face lights up with a smile, it is difficult to believe this man has many problems. The Hawking grin is famous, and it reveals the quality of his love for his subject. It's a grin that says, 'This is all very impressive and serious, but – ain't it fun!'

It is, of course, nothing short of miraculous that Hawking has been able to achieve everything he has, even that he's still alive. However, when you meet him and experience his intelligence and humour, you begin to take his unusual mode of communication and his obviously catastrophic physical problems no more seriously than he seems to himself. That is the way he wants it. He chooses to ignore the difficulty, 'not think about my condition, or regret the things it prevents me from doing, which are not that many.'[10] He expects others to adopt the same attitude.

1985–1986

In the autumn of 1985, with the Equalizer raising Hawking's hopes of getting on with his career and his popular-level book, Jane and Laura Gentry interviewed and hired the twenty-four-hour nursing staff who were going to make life at home possible. There would be three shifts a day, and the nurses had to be trained medical professionals. The tube that had been inserted in his throat had to be suctioned out regularly with a 'mini-vacuum cleaner' so that secretions would not accumulate in his lungs. The 'mini-vacuum cleaner' itself could be a source of infection and cause damage if not used correctly.[11] Not everyone they interviewed wanted this kind of demanding work, and there were also a few false starts.

One applicant who was eager to have the job and willing to dedicate herself to it long term was Elaine Mason, a physically strong, athletic woman with a zany sense of humour and a wonderful taste for colours that showed off her red hair. She impressed Jane as a caring person. Born Elaine Sybil Lawson in Hereford, Elaine was a devout Evangelical Christian whose father, Henry Lawson, had been a clergyman in the Church of England and whose mother had a medical degree. Elaine had worked for four years in an orphanage in Bangladesh, then come back to England and married David Mason, a computer engineer. They had two sons, one about the age of Tim Hawking.

I knew Elaine and David Mason and their sons only as a family with children in the same school as ours, but I remember that on parents' day I competed successfully with Elaine in the egg-and-spoon race. She was reputed to be fiercely competitive, but that wasn't evident in this particular sport. She seemed a refreshingly irrepressible, uninhibited woman.

Stephen's and Jane's hiring of Elaine became a fortuitous choice when her husband adapted a small computer and speech synthesizer and attached it to Hawking's wheelchair. Before that Hawking could run the Equalizer only on his desktop computer. Now his voice could go with him wherever he went. David Mason, like his wife, was devoted to Hawking. 'If he raised an eyebrow, you would run a mile,' he said.[12]

The Hawking household adapted to the new tensions of living a much less private life, in what seemed like a small-scale hospital with strangers abroad in it twenty-four hours a day. Hawking managed to recover enough strength and to handle the Equalizer program well enough to return to his office before Christmas. There were no more solo trips across the Backs; a nurse went with him. In many ways, things were looking up. His son Robert received his A-level exam results and, to everyone's relief, they were excellent. Cambridge would accept Robert for the following autumn, to read natural science as his father had at Oxford.

By the spring of 1986, life had begun to settle into a new, rather optimistic status quo, with one sad break in March when Frank Hawking, Stephen's father, died. Hawking's mother Isobel has said that Hawking was 'very upset by his father's death – it was rather a dreadful thing. He was very fond of his father, but they had grown apart rather and hadn't seen a great deal of each other in the late years.'[13] Hawking, of course, soldiered on. Soon he resumed his travels. His first trip away, to a conference in Sweden, was a success in more ways than one. Murray Gell-Mann was another attendee and witnessed first hand Hawking's ability to take a full part in the conference – evidence of how well the MacArthur funds were being put to work. Jane's application in October for an extension of the funding was approved. Now it would cover medical expenses as well as nurses on a continuing basis.

The Assault on the Airport Bookshops

Having mastered the Equalizer program, Hawking went back to work on his popular-level book in the spring of 1986. It hadn't taken him long, characteristically, to begin to regard the new level of disability as an advantage rather than a calamity. 'In fact,' said Hawking, 'I can communicate better now than before I lost my voice.'[14] That statement is often quoted as an example of raw courage. It was, in fact, the simple truth. He no longer needed to dictate or speak through an 'interpreter'.

Bantam had accepted Hawking's first draft for the book in the summer of 1985, but, with his catastrophic health problems, there had been no opportunity to move forward with the project. In any case, moving forward was not going to be an easy task. Bantam was insisting on some revisions. Hawking ended up almost completely rewriting his first draft.

He knew that even in non-technical language the concepts in his book would not be easy for most people. He claims he is not overly fond of equations himself, in spite of the fact that people compare his ability to handle them in his head to Mozart's mentally composing a whole symphony. It is difficult for him to write equations, even though the Equalizer allows him to express them in words and then rewrites them with symbols. He says he has no intuitive feeling for them. As Kip Thorne pointed out, he likes to think instead in pictures. This, in fact, seemed an excellent method for the book: to describe his mental images in words, helped along with familiar analogies and a few diagrams.

Hawking's and his graduate assistant Whitt's mode of operation fell into a pattern. Hawking would explain something in scientific language and then realize that his readers would not understand. He and Whitt would try to think of an analogy, but neither was willing to use analogies willy-nilly without being certain they were truly valid. Making sure of their validity occasioned lengthy discussions. Hawking wondered just how much to explain. Were some complicated matters better glossed over and left at that? Would explaining too much lead to confusion? Ultimately Hawking explained a great deal.

His editor at Bantam, Peter Guzzardi, wasn't a scientist. He felt that whatever he couldn't understand in the manuscript needed rewriting. He pointed out something that Hawking's students and colleagues had sometimes complained about: Hawking often jumped from thought to thought and came to surprising conclusions, wrongly assuming others could see the connections. Some attributed this to Hawking's need to use few words, but the reason went deeper than that, and his scientific colleagues were experiencing something of the same jumps on a far more advanced level than Peter Guzzardi. Whitt said that sometimes Hawking would tell him that something must be so 'because of what I understand', not because he could prove it or explain how he arrived there. Brian would do the calculations and sometimes have to report to Hawking that he had been wrong, and Hawking would not believe him. Then after some consideration and talking about it, Brian would realize that Hawking was right after all. 'His hunch was better than my calculation. I think that's a very important aspect of his mind: the ability to think ahead rather than go step by step; to jump the simple calculations and just come up with a conclusion.'[15] Nevertheless, for Hawking's editor Guzzardi, jumping the connections between his conclusions wasn't acceptable for a popular-level book. Even when Hawking felt he'd explained simply, Guzzardi often found the explanation unfathomable. At one point, Bantam tactfully suggested having an experienced science writer write the book for Hawking. Hawking vehemently rejected that idea. The revision process became tedious. Each time Hawking submitted a rewritten chapter, Guzzardi sent back a lengthy list of objections and questions. Hawking was irritated, but in the end he admitted that his editor was right. 'It is a much better book as a result,' he said.[16]

Editors at Cambridge University Press who had seen Hawking's book proposal had warned him that every equation he used would cut book sales in half. Guzzardi agreed. Hawking eventually decided he would include only one equation: Einstein's $E=mc^2$. Guzzardi prevailed in a disagreement about the title. When Hawking got nervous about the use of the word 'brief', Guzzardi replied that he liked it very much, it made him smile. That argument won the day. The title would be *A Brief History of Time*. The second draft was finally completed a year after they had begun working on it, in the spring of 1987.

By then, Hawking was also fully involved in the physics world again, continuing his career and gathering more honours and awards. In October 1986 he had received an appointment to the Pontifical Academy of Sciences, and the whole Hawking family had an audience with the Pope. He was awarded the first Paul Dirac Medal from the Institute of Physics. In June and July 1987, after the final draft of *A Brief History of Time* was completed, Cambridge hosted an international conference to celebrate the 300th anniversary of the publication of Isaac Newton's

Principia Mathematica, one of the most significant books in the history of science. Hawking was instrumental in bringing this event about. In connection with it, he and Werner Israel solicited articles from leaders in the fields connected with gravitation and put together the splendid book *300 Years of Gravitation*.[17]

When *A Brief History of Time* was nearing publication, in the early spring of 1988, Don Page was sent an advance copy to review for the journal *Nature*. Page was appalled to find it full of errors: photographs and diagrams in the wrong places and wrongly labelled. He placed an urgent call to Bantam. Bantam editors decided to recall and scrap the entire printing. Then began the intense process of correcting and republishing the book in time to have it in bookshops by the April publication date in the United States. Page now believes he owns one of the few extant copies of the original printing of Hawking's book. That copy is probably quite valuable.

Hawking enjoys pointing out that the American edition of *A Brief History of Time: From the Big Bang to Black Holes* was published on April Fool's Day, 1988. The British edition was launched at a luncheon at the Royal Society on 16 June. The Hawkings watched, astounded, as the book climbed effortlessly to the top of the bestseller lists. There it stayed week after unlikely week, then month after month, soon selling a million copies in America. In Britain the publisher could barely keep enough books on the shelves to meet the demand. Translations into other languages quickly followed. The book was indeed prominently displayed in airport bookstalls, and Hawking had to face the difficulty of getting his speech synthesizer to pronounce the word 'Guinness' when he and his book made the *Guinness Book of World Records*. It insisted on saying 'Guy-ness'. 'Maybe it is because it is an American speech synthesizer,' Hawking quipped. 'If only I had an Irish one . . .'[18]

Perhaps thanks to his persistent editor, Hawking had succeeded in making it possible (though not always easy) to follow him logically from thought to thought, sometimes even to anticipate him. This was a book to be studied, if you didn't have a scientific background, not read quickly. It was well worth the effort, and it was also good entertainment. Hawking's humour makes *A Brief History of Time* in its way a romp through the history of time, not safe to read in any situation where it would be awkward to burst out laughing.

Stephen Hawking rapidly became a household word and a popular hero all over the world. Fans organized a club in Chicago and printed Hawking T-shirts. One member admitted that some of his school friends thought this Hawking on his T-shirt must be a rock star; a few even claimed to have his latest album.

Reviews were favourable. One compared the book to *Zen and the Art of Motorcycle Maintenance*. Jane Hawking was horrified, but Stephen Hawking declared he was flattered, that this meant his book 'gives people the feeling that they need not be cut off from the great intellectual and philosophical questions'.[19]

Did people who bought the book read it and understand it? Some critics suggested that most of those who purchased it never read it and, if they tried, couldn't possibly understand it. They just wanted it to be seen on their coffee table. Hawking lashed back rather forcefully in the foreword to *A Brief History of Time: A Reader's*

Companion: 'I think some critics are rather patronizing to the general public. They feel that they, the critics, are very clever people, and if they can't understand my book completely then ordinary mortals have no chance.'[20] He wasn't overly concerned about its being left on coffee tables and bookcases just for show. The Bible and Shakespeare, he pointed out, have shared that fate for centuries. Nevertheless, he thought lots of people read his book, because he got mounds of letters about it. Many asked questions and made detailed comments. He was often stopped by strangers in the street who said how much they enjoyed it; this pleased him immensely but embarrassed his son Timmy.

Hawking's increasing celebrity status and the need to publicize the book gave him even more opportunities for travel than before. A Hawking visit usually left his hosts exhausted. The Rockefeller Institute in New York was the scene of one such occasion. After a long day of lectures and public appearances, there was a banquet in Hawking's honour. Hawking relished such events and made a show of sniffing the wine and commenting on it. Dinner and speeches over, the party moved to the embankment overlooking the East River. Everyone was petrified lest Hawking roll into the river. To their relief he didn't, and they soon had him safely headed back towards his hotel. In a ballroom opening off the lobby, a dance was going on. Hawking insisted that they not retire yet but crash the party. Unable to dissuade their headstrong honoree, the little group of distinguished scholars hesitatingly agreed, 'although we never do anything like that!' On the dance floor Hawking twirled about in his wheelchair with one partner after another. The band went on playing for him far into the night, long after the original party was over.

Would Hawking write a sequel to his book? Asked that frequently, he answered that he thought not. 'What would I call it? *A Longer History of Time*? *Beyond the End of Time*? *Son of Time*?'[21] Perhaps *A Brief History of Time II* – 'just when you thought it was safe to go back into the airport bookstore!' Would he write his autobiography? Not until he runs out of money to pay his nurses, or so he told me. That was not likely to be soon. *Time* magazine announced in August 1990 that *A Brief History of Time* had so far sold over 8 million copies, and they were still selling. If only he had left out that one equation!

Some accused Bantam and Hawking of exploiting Hawking's condition in marketing the book. They sniffed that his fame and popularity were like a carnival sideshow and blamed Hawking for allowing an overdramatic, grotesque picture to appear on the book cover. Hawking rejoined that his contract gave him absolutely no control over the cover. He did persuade the publisher to use a better picture on the British edition.

On the plus side, the media exposure allowed Hawking to give the world something that may be at least as valuable as his scientific theories and the information that the universe is probably not 'turtles all the way down'.* It

* In his book *A Brief History of Time* Hawking retold the story of an elderly lady who rose at the end of a scientific lecture to take exception with the speaker and insist that the world is a plate supported on the back of a giant turtle. When the speaker asked what she thought the turtle was standing on, she replied that he was very clever to ask that question, but in fact it was 'turtles all the way down'.

brought to millions not only his keen excitement about his work but also the important reminder that there is a profound kind of health which transcends the boundaries of any illness.

For the Hawkings the success of the book brought more than a change in financial status, making Hawking into what *CAM* magazine called 'that rarest of phenomena, a Labour-voting multi-millionaire'.[22] For years, he and Jane and their children had lived with disability and the threat of death. As Jane Hawking described it: 'In a sense we've always been living on the edge of the precipice, and eventually you put down roots at the edge of the precipice. I think that's what we've done.'[23] Now they found themselves threatened in a different way, by the allure and demands of celebrity and the frightening prospect of living up to a worldwide fairy-tale image.

In the mid- to late 1980s, it was often Elaine Mason who accompanied Hawking in his travels, and their growing fondness for one another was evident in a series of photographs taken by a friend of Elaine's, New York photographer Miriam Berkeley. Unfortunately, Elaine's fierce loyalty to Hawking, her protectiveness, her jealously guarded relationship with him and the strength of her personality were beginning not to sit so well with his family, others of his nurses and carers, or with colleagues and staff in the DAMTP. But her relationship with Hawking was a special relationship that was not going to evaporate any time soon. Others were competent and sympathetic, but it was Elaine he preferred to have with him as much as possible.

12

'The field of baby universes is in its infancy'

AS EARLY AS THE 1970S, MAGAZINE ARTICLES AND television specials had told Stephen Hawking's story. In the late 1980s, after the publication of *A Brief History of Time*, virtually every periodical in the world profiled him. Reporters and photographers greeted him everywhere. 'COURAGEOUS PHYSICIST KNOWS THE MIND OF GOD', blared the headlines. His picture was on the cover of *Newsweek* with the words 'MASTER OF THE UNIVERSE' emblazoned across a dramatic background of stars and nebulae. In 1989 he and his family were interviewed for ABC's show *20/20*, and in England a new television special appeared: *Master of the Universe: Stephen Hawking*. He was no longer merely well-known and successful. He'd become an idol, a superstar, in a class with sports heroes and rock musicians.

Jane Hawking spoke of her 'sense of fulfilment that we have been able to remain a united family, that the children are absolutely superb and that Stephen is still able to live at home and do his work'.[1] The world at large knew nothing about Jonathan Hellyer Jones or Elaine Mason and it seemed advisable to keep it that way.

The academic awards kept pouring in: five more honorary degrees and seven more international awards. One of them was the 1988 Wolf Prize, awarded by the Wolf Foundation of Israel and recognized as second in prestige only to the Nobel Prize in Physics. Another Cambridge luminary, Christopher Polge, won the Wolf Prize for Agriculture the same year, and he and his wife Olive and the Hawkings often found themselves feted together. Stephen replied to one interviewer that he did not 'believe in God; there is no room for God in my universe', a statement that Jane found particularly hurtful because they were in Jerusalem, a city with deep spiritual significance for her.

By this time Robert Hawking was a university student, studying physics and rowing for his Cambridge college, Corpus Christi. One of the television specials showed him racing on the river while the rest of the family, including Hawking with his synthetic voice, cheered from the bank. Lucy was considering a career in the theatre. She appeared in the Cambridge Youth Theatre's award-winning production of *The Heart of a Dog*, a Soviet political satire from the 1920s. The production went on to run in Edinburgh and London as well. When the London run conflicted with her entrance examinations for Oxford, Lucy made the daring decision to miss them and allow her application to be judged only on her interview and her A-level exam results. She was admitted to Oxford. As for ten-year-old Tim, Hawking said, 'Of all my children, he is probably the one most like me.'[2] He and Tim enjoyed playing games. Hawking usually won at chess, Tim at Monopoly. 'So we're both quite good at something,' proclaimed Tim.[3] In 1988 the American photographer Stephen Shames had taken pictures of them engaged in an impromptu game of hide-and-seek. Tim excelled at that. He could tell when his father was getting near by the hum of the wheelchair motor.

Lucy told ABC's *20/20* that she and her father 'get on quite well', though both are stubborn. 'I've had lots of arguments with him actually, I must admit, with neither of us willing to give ground. I think a lot of people don't realize just how stubborn he is. Once he gets an idea in his head, he will follow it through no matter what the consequences are. He doesn't let a thing drop . . . He will do what he wants to do at any cost to anybody else.'[4] This sounds harsh, but when I spoke with Lucy, it was very clear she's enormously fond of her father and respects his opinions. In the ABC interview, she said that she thinks he *has* to be stubborn in his situation. It's a necessary mode of survival for him. His strength of will keeps him working day in, day out, grinning and delivering funny one-liners and ignoring a grim physical situation. If it also occasionally makes him appear spoiled and self-centred, it seems entirely reasonable to forgive him. About his health and the fear of his dying, Lucy said, 'I always think, "Oh he's going to be all right", because he's always pulled through everything that's happened to him. You can't help worrying about someone who's so frail. I get quite worried when he goes away.'[5] Lucy had learned to cope with such fears early. Her mother had tried to explain to her what ALS was when Lucy was a child. Then, Lucy had cried, certain that 'he was going to die the next day'.[6]

In the academic world physicists continued to express tremendous respect for Hawking but were a little non-plussed by all the media hype. It didn't take higher maths to multiply book sales figures in the millions and find they amounted to more than Lucy's school fees. There was an occasional hint of sour grapes, a half-suppressed mutter of 'His work's no different from a lot of other physicists'; it's just that his condition makes him interesting.' One colleague commented that 'in a list of the twelve best theoretical physicists this century [the twentieth], Steve would be nowhere near'.[7] That was arguably true, considering the astounding list of physicists who lived in the twentieth century, and Hawking would have agreed, though the 'nowhere near' was perhaps a bit harsh. But there was surprisingly little disparagement. He could more than hold his own in any contemporary company, and everybody knew it. Moreover, his colleagues enjoyed him. Sidney Coleman of Harvard, who rivalled Hawking not only as a physicist but also as a classroom comedian, was pleased that Hawking's celebrity brought him more and more often to America, and frequently to New England. Other physicists who were sometimes unfairly eclipsed by Hawking did not blame him personally.

Nevertheless, it wasn't unreasonable to suggest that Hawking's scientific accomplishments alone would never have made him the celebrity he was or sold millions of books. Were those correct who said he'd exploited his pathetic condition and ridden his wheelchair to fame and fortune? The truth is, although Hawking would almost certainly prefer it otherwise, most of the non-physics world probably appreciates him more for his spirit than for his scientific achievements. He's not the only person who's overcome staggering odds and maintained a positive attitude in adverse circumstances, but who else has done it with quite such brilliant success and engaging style?

For over a quarter of a century Stephen Hawking, perhaps with lapses we'll never know about, had kept up this spirit of optimism and determination for his own benefit. His survival and success depended on his doing so. However, it was a responsibility only to himself and to his family. In the late 1980s, it became a responsibility to millions all over the world for whom he was an inspiration. Many, not only the disabled, expected him and his wife to go on proving that in spite of tragedy, life and people could still be absolutely splendid. We shouldn't be surprised if Hawking was leery about having that larger responsibility thrust on him. He was, he said, no more than simply human. Hawking would later comment that he did not see himself as being tragic and romantic, something like 'a perfect soul living in a flawed body. I am proud of my intelligence, but I have had to accept that the disability is also a part of me.' And for disabled people Hawking had become a superb role model. Nevertheless, the disparity between what he'd achieved and what most could expect was sometimes discouraging. In most things except his illness Hawking has been outrageously lucky.

Jane Hawking pointed out that if her husband were an obscure physics teacher, she couldn't have convinced a foundation to donate over fifty thousand pounds a year for nurses. There would be no computer program. He would be sitting day after pointless day in a nursing facility, away from his home and family, mute, isolated and wasted. Her bitterness about the way the National Health Service failed them led her to campaign for those with similar problems, trying to get the NHS to provide money for home nursing rather than tear apart families. The Hawking image encouraged universities to set up dormitories equipped for students needing round-the-clock nurses in order to attend classes. An abstract glass mini-sculpture sits on a filing cabinet in Hawking's office, a gift from 'Hawking House', a dormitory at the University of Bristol. Cambridge built a similar facility.

Whatever the effect on the rest of the world, in 1989 Stephen Hawking had 'made it', against stupendous odds. The Queen named him a Companion of Honour, making him a member of an Order that consists of the Queen herself and not more than sixty-five other members. It is one of the highest honours she can bestow. The University of Cambridge did what it rarely does and gave one of its own faculty an honorary doctorate. Hawking received his degree from Prince Philip, chancellor of the university, and joined in the pageantry, going to and from the Senate House to the accompaniment of the choirs of King's College and St John's College and the Cambridge University Brass Ensemble. 'This year has been the crowning glory of all Stephen's achievements,' Jane Hawking said. 'I think he is very happy about it.'[8] He loved doing the work he did. 'I have a beautiful family, I am successful in my work, and I have written a bestseller. One really can't ask for more,' he said.[9] He'd earned this fame, and he was enjoying it. For someone who'd thought at the age of twenty-one that he had no reason to go on living, it was heady stuff indeed, a delicious joke on fate.

But fate had its jokes too. There was an obvious down

side to the overwhelming success of his book. Less time for his scientific work. Too many 'extracurricular activities', his students lamented. Too many visitors, but he seldom turned them away. Too many invitations, but he seemed incapable of refusing them. Too much travelling, but he scheduled more and more of it. Too much mail. He had answered the first few letters about *A Brief History of Time* personally, but that had almost immediately become impossible. His graduate assistant and his secretary took on major responsibility for answering his mail.

Notoriety was not all fun. 'It obviously helps me to get things done and it enables me to help other disabled people,' Hawking told a journalist. 'It also means I can't go anywhere incognito in the world. Wherever I go, people recognize me and come up to say how they enjoyed the book and can they have a photograph with me. It is gratifying that they are so enthusiastic, but there are times that I would like to be private.'[10] He came up with a solution, programming his voice synthesizer to say, 'People often mistake me for that man.' Or, 'I am often mistaken for Stephen Hawking.' No one was fooled.

As Hawking juggled his increasingly unmanageable schedule, colleagues began to worry that he would neglect his science. However, Hawking's scientific work did continue. He tied up one small loose end when he was once again visiting Caltech in June 1990. The evidence having to do with Cygnus X-1 that had emerged during the sixteen years since he and Thorne had made their bet gave a 95 per cent certainty that Cygnus X-1 is a black hole. It was time, Hawking decided, to settle with Thorne. When Thorne was away in Moscow, Hawking, with help from 'accessories to the crime', broke into his office where the framed bet was lodged and wrote a note on it conceding the bet. Stephen signed the note with his thumbprint.

While Hawking travelled the globe as a celebrity in the late 1980s, in his head he travelled distances that make these journeys seem paltry by comparison. John Wheeler had earlier (in 1956) introduced the idea of 'quantum wormholes'. Hawking was now attempting adventures through these wormholes into even more exotic climes, into 'baby universes'. Let's stand with him outside space and time, to get a better view.

A New Look at the Cosmic Balloon

Hawking asks us to imagine an enormous balloon, inflating rapidly. The balloon is our universe. Dots on the surface are stars and galaxies. The dots cause dimples and puckers in the surface. As Einstein predicted, the presence of matter and/or energy causes a warping of spacetime.

When we look at the cosmic balloon through a not-very-powerful microscope, the surface, regardless of the puckers, looks relatively smooth. Looking through a much more powerful microscope, we find it isn't smooth after all. The surface seems to be vibrating furiously, creating a blur, a fuzziness (see Figure 10.9, p. 213).

We've seen such fuzziness before. The uncertainty principle causes the universe to be a very fuzzy affair at the quantum level. It's never possible to know precisely both the position and the momentum of a particle at the same time. One way to picture this quantum uncertainty is by imagining that each particle jitters in a sort of random microscopic vibration. The closer we try to look, the more violently it jitters. Scrutinize the quantum level with the greatest possible care, and we at best are able to say only that a particle has *this* probability of being *here* – or *that* probability of moving like *that*. The surface of the cosmic balloon is unpredictable in a similar way. Under high enough magnification the quantum fluctuation becomes so incredibly chaotic that we can say there's a probability for it to be doing – *anything*.

What did Stephen Hawking think this 'anything' might be? In the late 1980s he was pondering the probability that the cosmic balloon will develop a little bulge in it. More familiar balloons, the ones at parties, do that if one point on the surface is weak. Usually party balloons burst immediately when that happens, but on rare occasions a miniature balloon bulges out of the surface. If you could see this happening to our cosmic balloon, you would be witnessing the birth of a 'baby universe'.

It sounds spectacular: the birth of a universe. Will we ever witness such an event? No, first because it happens in imaginary time, discussed in Chapter 10, not 'real' time. Another reason we won't see it, said Hawking, is that if anything can truly be said to start small, it's a universe. The most probable size for the connection between our universe and the new baby – the umbilical cord, if you will – is only about 10^{-33} centimetres across. To write that fraction you use 1 as the numerator and 1 followed by thirty-three zeros for the denominator. That's small! The opening – the wormhole, as it's called – is like a tiny black hole, flickering into existence and then vanishing after an interval too short to imagine. We've spoken of something else with an extremely short life span: in Chapter 6, when we discussed Hawking radiation, we saw that you can think of fluctuations in an energy field as pairs of very short-lived particles. Wormholes similarly are a way of

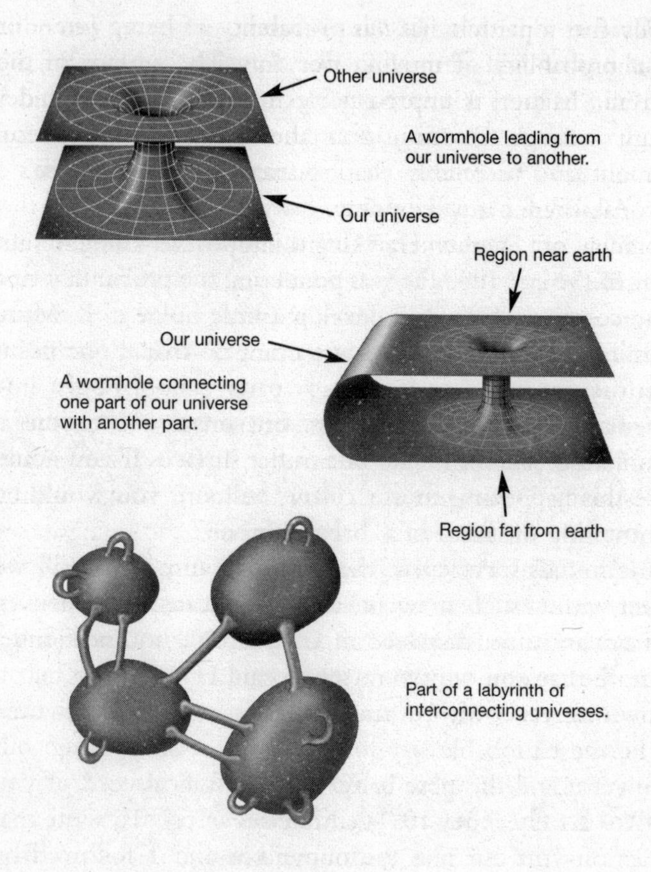

Figure 12.1. Wormholes and baby universes

thinking about fluctuations in the fabric of spacetime: the surface of the cosmic balloon.

Hawking's suggestion was that the baby universe attached to this umbilical cord may *not* be short-lived, and small beginnings don't always continue small. He was thinking that eventually the new universe might expand to become something like our present universe, extending billions of light-years. Like our universe, but empty? Not at all. 'Matter,' Hawking pointed out, 'can be created in any size universe out of gravitational energy.'[11] The result might later be galaxies, stars, planets and, perhaps, life.

Are there many baby and grown-up universes? Do they branch off everywhere? Right inside the kitchen sink? Inside your body? Hawking said yes, it may be that new universes are constantly coming into existence all around us, even from points inside us, completely undetectable to our senses.

Perhaps you're wondering whether our universe began as a bulge from the side of another. It's possible, declared Hawking. Our universe may be part of an infinite labyrinth of universes, branching off and joining one another like a never-ending honeycomb, involving not only a lot of baby universes but adult universes as well. Two universes could develop wormhole connections in more than one spot. Wormholes might link parts of our own universe with other parts of it, or with other times (Figure 12.1).

Life in the Quantum Sieve

Let's stretch our imaginations and look at all of this from the point of view of an electron. If there are quadrillions of wormholes flickering in and out of existence at every point in the universe, an electron is facing something like an enormous, furiously boiling pot of thick porridge. Moving across it is about as tricky as travelling across a giant, continuously changing sieve. An electron trying to move in a straight line in such an environment is almost certain to encounter a wormhole, fall in, and go shooting off into another universe. That sounds suspiciously as though matter will be disappearing from our universe, which isn't allowed. However, according to this theory, there is no danger of such a loss. An identical electron comes back the other way and pops into our universe.

Wouldn't we notice this substitution of electrons? We won't see it that way. To us this event will look like one electron moving in a straight line. Hawking was thinking that the presence of wormholes, however, will make all electrons move as though they have a higher mass than they would if there were no wormholes. Therefore, if we're to try to predict particle masses with any theory, it's important to know whether or not there really are such things as wormholes.

The theory said that if an electron falls into a wormhole accompanied by a photon, it won't appear to be anything out of the ordinary. We will observe only the normal exchange of a messenger particle in an electromagnetic interaction, in which one electron emits a photon and another absorbs it. Hawking was suggesting that perhaps all particle masses and all particle interactions – the ceaseless activity of the four forces, all over the universe – can be explained as this going into and out of wormholes.

You would be right to wonder, at this point, how particles could possibly pass through wormholes. Wormholes would be much smaller than even the smallest particles we know. As with Hawking radiation, what is impossible in any way we try to picture it *is* possible in quantum mechanics.

When Hawking calculated the effect of wormholes on the masses of particles such as electrons, his calculations at first suggested that the masses would be much larger than we actually observe for these particles. He and other researchers later managed to come up with more reasonable numbers. However, at the end of the 1980s Hawking was expressing doubts whether wormhole theory can predict the masses of particles for our universe or any other. As we saw in Chapter 2, when something must be measured directly and cannot be predicted by the theory, that's called an arbitrary element. The masses of particles and the strengths of the forces are, in every theory anyone has come up with so far, just such arbitrary elements. Wormhole theory may not make them any less arbitrary, but it might explain how they happen to be arbitrary. Hawking was thinking that the masses of particles and other fundamental numbers in nature may turn out to be 'quantum variables'. That means they may be uncertain, like the paths of particles or what happens on the surface of the cosmic balloon. These numbers would be fixed at random at the moment of creation in each universe. A throw of the dice, so to speak, and then that's settled for that particular universe – but no way to know from a theory how the dice will fall, or perhaps even to say that one way is definitely more probable than another. Hawking was not sure this was the case in wormhole theory. However, the idea that fundamental numbers in nature – maybe even the 'laws of nature' – might not *be* fundamental to the totality of universes, but different for different universes, was something that he would return to in another context later.

A Severely Warped Universe

'It is a great mystery why quantum fluctuations do not warp spacetime into a tiny ball,' says Hawking.[12] Recall that this is one of the enigmas that theorists must solve in the quest for the Theory of Everything.

Physicists refer to this problem of the energy in the (so-called) vacuum as the cosmological constant problem. You'll remember that Einstein theorized about something called the cosmological constant, which would balance gravity and prevent the universe from changing in size. He later called it 'the greatest blunder of my life'. The term has come to have a related but slightly different meaning. The *cosmological constant*, as scientists now use the term, is a number which tells us how densely this energy in the vacuum is packed: the energy density of the vacuum. Common sense says there shouldn't be any energy there at all, but, as we've seen, the uncertainty principle shows that 'empty' space isn't empty. It seethes with energy. The cosmological constant (the energy density of the vacuum) ought to be enormous, and general relativity theory tells us this mass/energy should be curling up the universe.

However, regardless of what the uncertainty principle and theory of general relativity indicate, we don't have a curled-up universe. Quite the contrary, at the time when Hawking was developing his wormhole theory, the value (the number) of the cosmological constant had been long thought, and observed, to be near zero. We find this out by observing the rate at which galaxies are moving away from one another, and from the fact of our own existence. 'A large cosmological constant either positive or negative would make the universe unsuitable for the development of life,' Hawking points out.[13] The value of the cosmological constant is one of the examples of the 'fine-tuning' we saw in Chapter 9. As we shall see, Einstein was too quick to call it a 'blunder'. But no one knew that in the early 1990s.

How can the cosmological constant possibly be so small, as we observe, when theory tells us it should be enormous? Recall once again the particle pairs in Hawking radiation. In supergravity, the theory Hawking spoke of in his Lucasian lecture, pairs of fermions (matter particles) in the vacuum give negative energy and balance the positive energy of pairs of bosons (messengers). This may indeed be the explanation, or at least part of it, but it's a complicated matter. For one thing, these particles don't only interact with gravity. However, even if we do have a lot of positives and negatives cancelling one another out, for all of these to cancel out to zero is a little too much to swallow. As Sidney Coleman, who shared Hawking's enthusiasm for wormholes, puts it: 'Zero is a suspicious number. Imagine that over a ten-year period you spend millions of dollars without looking at your salary, and when you finally compare what you spent with what you earned, they balance out to the penny.'[14] For the cosmological constant to balance out to zero is even less likely.

Could wormholes solve the mystery? Hawking was thinking that wormholes branching off at every point make the cosmological constant, the energy density of the vacuum, a 'quantum variable' like the masses of particles. It can have *any* value. What's the probability of its being near zero? Imagine the birth of a universe as a 'baby' branching off from an existing universe. Wormhole theory says there are plenty of universes – some more enormous than ours is today, others unimaginably smaller than an atom, and all sizes in between. The infant universe must copy its cosmological constant value from one of these other universes through a wormhole – 'inherit' it, you might say. It isn't important to a human infant whether it inherits a talent for music; it becomes important only when the infant grows larger. It isn't important to a baby universe whether it 'inherits' a cosmological constant value near zero. Its cosmological constant value won't even be measurable until it's quite a bit more grown-up. However, with all those assorted sizes of universes around, the infant is far more likely to inherit its cosmological constant value through wormhole attachments with large, cooler universes of the sort only possible when all those positives and negatives in the vacuum cancel out to zero. Coleman studied the probability of a universe (in wormhole theory) being a universe where the cosmological constant is near zero: our kind of universe. He found that any *other* sort of universe would be highly unlikely.

Wormholes and the Theory of Everything

Wormholes and baby universes fired the imaginations of many physicists. They began responding, disputing this and that, and offering alternative versions. That's always a good sign. 'The field of baby universes is in its infancy,' quipped Hawking, 'but growing fast.'[15] Could wormholes and baby universes contribute to the search for a complete theory of the universe?

First of all, we've seen that the theory seemed to offer a new way of looking at the problem of the cosmological constant, the sticky question of the energy density in the vacuum, which ought to be shrinking the universe but isn't. Did Hawking think wormholes are the theory that will solve this inconsistency between general relativity and quantum mechanics? 'I would not go so far as that,' Hawking said. 'There is no fundamental inconsistency, but there are technical problems which wormholes don't help.'[16]

Second, wormhole theory was a theory that didn't break down if you followed it back to the 'beginning'. With Einstein's theories, if you follow things back to the Big Bang, you reach a singularity where the laws of physics as we know them break down. Hawking's no-boundary proposal showed that in imaginary time there would be no singularity. Wormhole theory suggested that in imaginary time our universe may have begun as a baby universe branching off from another universe.

Third, wormhole theory linked quantum theory and relativity theory in a satisfying, geometric way, allowing us to think of quantum fluctuations, quantum wormholes and baby universes as not too different from the warping of spacetime and black holes on the astronomical level. The fundamental numbers in our universe, such as the masses and charges of particles and the cosmological constant, might be the result of the shape, the geometry of a labyrinth of interconnected universes.

Other theories can't predict the masses and charges of particles. These are arbitrary elements in the theories. An alien who had never seen our universe couldn't take these theories and use them to calculate what these fundamental quantities are, without peeking at the 'real' universe. We've seen that there is argument about whether wormholes might give us a way of understanding and calculating these fundamental numbers, or whether wormholes make their prediction less likely with any theory.

Theorists who work in the field of superstring theory,

which says that the fundamental objects in the universe are not pointlike particles after all but tiny vibrating strings, were hoping their theory might eventually be able to predict particle masses and charges. Hawking was pessimistic. 'If this picture of baby universes is correct, our ability to predict these quantities will be reduced.'[17] If we knew how many universes there are out there and what their sizes are, it would be different, but we don't know that. We can't even see their joining on to or branching off our own. We can't get an accurate picture of the shape of it all. We know only that if universes do join on or branch off, this changes the apparent values of such quantities as particle masses and charges. We end up with a small but definite amount of uncertainty in the predicted values.

Hawking meanwhile wasn't worrying overly much about whether work like this was leading him to the Theory of Everything. His strategy is to concentrate on areas he understands, chipping away at the problem of what happens and how things work when relativity and quantum mechanics are taken together. What he discovers about the universe in this way should hold true, he thinks, regardless of what the theory of everything turns out to be and who finds it. His picture should fit in as part of the larger, or more basic, picture.

Saving History

Science fiction buffs will be disappointed if we don't discuss the possibility that something larger than a particle can travel through a wormhole into another universe or into another part of our universe. There has been a lot of science fiction utilizing the idea. On the face of it this form of travel seems as though it ought to be feasible.

Science fiction and scientific speculation joined hands when Kip Thorne and his graduate students studied the possibility at the request in 1985 of Carl Sagan. Sagan needed a way to get the heroine of his novel *Contact* to a very distant location in space in no time at all. The problem is that a wormhole large enough for you or me or Sagan's heroine to get through would be dangerously unstable. Even so small a disturbance as our presence would destroy the wormhole, and us with it. Thorne eventually thought he had found the answer, a way to hold the throat of a wormhole open using exotic material with a negative energy density – possible perhaps for a civilization much more advanced than our own. Hawking's reaction to Thorne's suggestion was characteristically succinct: 'You're wrong.' 'There is little politeness in our community when one of us believes the other is wrong,' commented Thorne.[18]

Hawking set out to back up his opinion, and the result was something that he called the 'chronology protection conjecture'. His objection was specifically to a wormhole that was a time machine. The 'conjecture' was that nature prevents the trajectory in spacetime that would allow one to travel back in time (a 'closed, time-like curve'). The time-machine wormhole would always explode when you tried to activate it, and that explosion, Hawking declared, would 'keep the universe safe for historians'. No one could travel back in time and change history. Thorne, in a paper written for Hawking's sixtieth birthday in 2002, reminded his readers and listeners that the chronology protection conjecture was just that, a 'conjecture', 'because both he and I were working with the laws of physics in a domain where we are uneasy about whether they really are correct'.[19] Hawking had also argued that 'the best evidence we have that time travel is not possible, and never will be, is that we have not been invaded by hordes of tourists from the future',[20] but he also wryly speculated that it could be that our time in history has become such a notoriously unpleasant tourist destination that visitors from the future always avoid it.

Kip Thorne called Hawking's paper about the 'chronology protection conjecture' a 'tour de force', which didn't of course necessarily mean he agreed with it. For Thorne's sixtieth birthday, Hawking gave him a calculation of the quantum mechanical probability of success for a wormhole time machine. Hawking had not become any more optimistic. He came up with 1 part in 10^{60}.[21]

What about a smaller black hole? When primordial black holes evaporate, what happens to things that fell into them earlier? Wormhole theory suggested that they may not necessarily return to our universe as particles. The particles instead may slip off into a baby universe. The information paradox rears its ugly head! Of course, this baby universe might join on again to our region of space-time. Then it would look like another black hole, which formed and evaporated. Things falling into one black hole would emerge as particles from the other black hole, and vice versa. That's space travel of a sort – if you happen to be a particle – and no information would be lost.

Could wormholes and baby universes offer a solution to the 'information paradox'? If anyone was raising his or her head hopefully, thinking that perhaps the universe also had a way of keeping itself safe from information loss, those hopes were not going to be encouraged by Hawking any time soon.

PART III
1990–2000

13

'Is the end in sight for theoretical physics?'

THE BUILDING THAT HOUSED THE UNIVERSITY OF Cambridge Department of Applied Mathematics and Theoretical Physics from the middle of the twentieth century until 2000 was a grimy behemoth of no architectural distinction whatsoever. One had to conclude that those who worked there happily had to be pretty much oblivious to their surroundings or just loved the old place for other than aesthetic reasons.

The entrance was off Silver Street through a narrow alleyway, an asphalt car park and a red door. The interior of the building was institutional; the floor plan pieced together illogically. A corridor beyond a small reception area made an abrupt right turn past an ancient black metal lift, continued straight for a while, then bent and widened past letterboxes and overstuffed bulletin boards with lecture and seminar notices and some lewd graffiti, narrowed again abruptly, and ended at the door of a large common room.

For decades, it was in this common room that the DAMTP gathered for tea every afternoon at four. For most of the day the room was deserted and dimly lighted. The colour scheme showed a preference for lime green – in vinyl armchairs grouped around low tables, woodwork and the lower halves of the pillars supporting the high ceiling. There was a table with stacks of scientific publications, a rogues' gallery of small photos of present students and faculty on one wall, and formal portraits of former Lucasian professors on another. At the far end of the room enormous windows provided a view of a blank wall across the alley and admitted little light.

Hawking's office and a number of others opened off this common room. On his door there was a small placard: 'QUIET PLEASE, THE BOSS IS ASLEEP'. Probably not true. Hawking spent long hours, over the years, working in that pleasant, high-ceilinged office with his computers, photos of his children, a few plants, a life-size picture of Marilyn Monroe on the door and always, after 1985, one of his nurses in attendance. His one oversized window overlooked the car park.

Hawking's day there usually began at 11 a.m. His secretary reviewed his schedule with him. In the late 1980s, that became something of a joke. They seldom managed to follow it, and anyone who had an appointment with Hawking had to remain flexible.

The day continued with the soft clicking of his hand-held pressure switch. Propped in his chair, Hawking watched the computer screen impassively and selected words to communicate with visitors and interviewers, consult colleagues, advise students, converse over the telephone, write lectures or answer correspondence.

Sometimes you heard the soft hum of his wheelchair motor as he guided it by means of a joystick through the common room and corridors to other rooms in the building for meetings and seminars. A nurse went with him. At intervals the well-modulated computer voice requested his nurse to adjust his position in the chair or suction fluid that accumulates in his breathing passage.

Hawking's nursing staff in the late 1980s was large and competent and varied as to age and sex. They seemed indulgently fond of Hawking and devoted to the task of making him look nice, keeping his hair brushed, his glasses clean, his chin wiped of the saliva that runs from his mouth, and, as they put it, to 'getting him sorted out' many times a day. Hawking had no choice but to be totally dependent on others, but there was never any air of helplessness about him. On the contrary he was vigorous and decisive, unquestionably in charge of his life. His staff said that the strength of his personality made working for him and with him both rewarding and demanding. I was never aware of the unpleasant competition among the nurses that Jane Hawking would later write about.

In the late eighties the mail had become an impossible burden for Hawking's research assistant, his personal assistant Sue Masey and one of his nurses who now helped them. They were struggling valiantly to write thoughtful answers to letters, poems, videotapes from all over the world, many of which told moving stories and deserved a personal response. It was sad to have to resort increasingly to polite preprinted postcards, but it would have taken all Hawking's waking hours to handle even a fraction of his mail.

At 1 p.m., rain or shine, Hawking would propel his

wheelchair with portable computer attached out into the narrow Cambridge streets, sometimes accompanied only by a nurse, sometimes by students, who trotted to keep up with him. It was a short journey through the heart of Cambridge, past the up-market shops in King's Parade, King's College Chapel and the Senate House, to Gonville and Caius, to lunch with other Fellows of his college. There a nurse arranged a bib around his shoulders and spooned food into his mouth. Eating was no deterrent to conversation for Hawking, whose finger moved continually on his hand-held device, choosing words to talk with those near him.

After lunch there was the return journey to the DAMTP. By then Hawking was notorious on at least two continents for his hair-raising wheelchair driving. Students would bound ahead into traffic on King's Parade and Silver Street to stop cars, lorries and bicycles as he recklessly barrelled ahead assuming the right of way. Acquaintances feared he was more likely to be crushed by a lorry than die from ALS.

At 4 p.m. Hawking would emerge again from behind his lime green door. Teatime was a ritual in the department, and the empty, cavernous room would become suddenly deafeningly noisy with voices and the clatter of teacups. Most of the assembled physicists and mathematicians dressed as though they were on a construction site. Someone has commented that Hawking's 'relativity group' looked like a rock group on a bad day. Their talk wasn't small talk. It ranged among wormholes, Euclidean regions, scalar fields and black holes. Equations were scrawled on the low tables. Hawking's wry wit set the tone in his corner of the room, but former students claim that a few remarks from him during tea were often more valuable than an hour's lecture by somebody else. Hawking had mastered the art of packing a lot into a few words. Reading over notes later, you realized how precisely he had chosen his words to say exactly what he meant.

At four thirty the common room would empty as rapidly as it had filled, and all but one of the long, fluorescent lighting fixtures were switched off. Hawking would glide back into his office to work until seven. In the late afternoon his students found him more available to help them.

On some evenings Hawking dined in college, or, in a specially equipped van bought with award money from the 1988 Israeli Wolf Prize in physics, he would be driven to a concert or the theatre. When there was a concert at Tim's school, he would go to hear Tim play the cello with the orchestra. Tim was a good cellist, following in the footsteps of his sister Lucy. On other evenings Hawking would work late in his office.

It was on one of those late evenings, in December 1989, that I went in to talk with him about plans for writing my first book about him. We discussed black holes and I read him a paragraph I'd written to make sure I had it right. When I paused to complain that my writing sounded dull because my editor was opposed to any fun or humour in a book about science, Hawking said, 'It should be fun. Tell him I said so.' I was certain this *would* win the argument with my editor. After all, Hawking's own book was flying off the shelves and had sold millions of copies. At one point, as I watched the words flit across the screen, I was startled that the message was 'Would you please pull me up a little higher in my chair?' Realizing after a moment of confusion that this wasn't meant for me, I glanced over at the young male nurse sitting near us. He came to life, picked up Hawking and set him down in a better position.

By the end of my visit that evening we had sorted out plans. Hawking would tell his personal assistant to supply me with childhood and family photographs and with material he had written about his childhood and his disability that had never been published. In May or June, when I had completed the science sections of the book, he would go over those with me.

The precipice gives way

Life at the focus of as much attention and adulation as Hawking was receiving inevitably takes on an unnatural cast. It isn't easy to keep things in perspective, no matter how levelheaded and grown-up you are or how good a sense of humour you have about yourself. For a quarter of a century Hawking had been convincing people that he was not subhuman. He'd succeeded too well. He'd convinced them he was superhuman. He had never deliberately encouraged this idea. He said he refused to be treated as less *or more* than simply human. But critics pointed out that he'd actually done precious little to discourage the superhero image. To be fair, who would have? It was fun and it sold books. Besides, what good did it do to try to discourage it? When he made statements like 'I get embarrassed when people call it courage; I've just done the only thing open to me in the situation',[1] some took it as false modesty and others as one more example of heroism.

Hawking began openly shouldering, more than previously, the responsibility of being a role model for disabled people. In a speech before an occupational science conference at the University of Southern California in June 1990 he sounded almost militant. 'It is very important that disabled children should be helped to blend with others of the same age. It determines their self-image. How can one feel a member of the human race if one is set apart from an early age? It is a form of apartheid.' He said he counted himself lucky that his disease struck him fairly late, after he'd spent his childhood with able-bodied friends, engaged in normal physical games. He praised the mechanical advances that have helped him. But he went on to say that although 'aids like wheelchairs and computers can play an important role in overcoming physical deficiencies, the right mental attitude is even more important. It is no use complaining about the public's attitude about the disabled. It is up to disabled people to change people's awareness in the same way that blacks and women have changed public perceptions.'[2] Even Hawking's critics couldn't deny that he had gone further than almost anyone else in history towards changing that awareness.

While Hawking ranged all over the world giving talks, receiving honours, holding press conferences, and enjoying the general adulation, more and more frequently accompanied by Elaine Mason, Cambridge friends watched their 'resident supercelebrity' with indulgence and delight, but also with mounting concern. They begrudged him none of the fun, but they worried about him. Was he beginning to believe the 'master of the universe' image? Would celebrity crowd out his scientific work? Mixed with his natural stubbornness, was it making him a wilful prima donna? Would an exalted self-image affect his family? Would the marriage that had endured so much adversity be able to survive? The public likes to own its heroes. Could Stephen ever be just Stephen again? It seemed unlikely.

Jane Hawking's relationship with Jonathan Hellyer Jones was still known to only a handful of very discreet people and news of it had not reached the wider world or the media – an astounding achievement in a town and university community as small as Cambridge. However, Jane had sounded an ominous note in an interview in 1989: 'I started with great optimism. Stephen was then infected with that optimism. His determination has now rather outstripped mine. I cannot keep up with him. I do think he tends to overcompensate for his condition by doing absolutely everything that comes to his notice.'[3] That 'everything' had grown out of all proportion. Jane felt it was a tremendous victory that he was able to live at home and have a fairly normal life. Stephen Hawking wanted much more. There were more doors open to him, more possibilities, than he had ever dreamed of or could ever hope to explore; more demands on his time than he could ever hope to meet.

All these activities and the adulation and awards were distancing him from his family. Increasingly they were carving out lives of their own, separate from his. Robert and Lucy were actively trying to be independent and move out of his shadow. Jane rarely accompanied him in his travels and public appearances. She sought escape in her teaching and her garden and in books and music. Her voice lessons had borne fruit and she had become a valued member of a top-flight Cambridge choir, often singing soprano solos. There were other friends besides Jonathan who shared her religious faith. Her role in Stephen's life had changed. It was, she said, no longer to encourage a sick husband. It was 'simply to tell him that he's not God'.[4]

For twenty-five years Stephen and Jane Hawking together seemed to nearly everyone to have handled adversity magnificently. Again and again Stephen had spoken of their relationship as the mainstay of his life and his success. The *Master of the Universe* television special in 1989 ended with a picture of the two of them watching their sleeping child, Tim, and Hawking saying, 'One really can't ask for more.' The Hawkings' public image had continued to reaffirm that life on the edge of the precipice was, for all its problems, a beautiful life.

In the spring of 1990 the precipice that had been weakening from within for several years crumbled in a way few people had ever expected it would. What had seemed to me a fairly smoothly humming operation on Silver Street when I talked with Hawking the previous December had become, when I visited again with my completed science chapters in the early summer of 1990, frenetic and unhappy. I spent a week conversing with Hawking and having him vet my chapters, aware that for some reason the mood in the department, especially among the staff and those faculty members closest to him, and with Hawking himself, was tense, on a knife edge.

Finally, a mutual friend who knew Jane Hawking well cleared up the mystery. Stephen had told Jane that he was leaving her for Elaine Mason. Even with media attention bordering on paparazzi, the Hawkings had kept their secrets so well that I, who had been interviewing him and was writing a book about him, could hardly believe the news. It seemed a tragic end to what most had thought was a beautiful, heroic marriage. To stalwarts who had been closest to Hawking the break-up had been no surprise. A couple of his most trusted staff had resigned, unwilling to deal with the turmoil of his disintegrating marriage and his new relationship with Elaine. Sue Masey was struggling to keep things moving ahead on an even keel.

The Hawkings separated just short of their twenty-fifth wedding anniversary. Except for a brief mention to the press that autumn that he had left his wife but did not rule out the possibility of a reconciliation, neither Stephen nor Jane Hawking made any public statement. It was symptomatic of the love and respect Hawking's friends and colleagues have for him that in a town where gossip moves like wildfire the news was very slow in spreading. As it did, a widening circle of acquaintances in Cambridge and all over the world reacted to it as a tragedy. The disintegration of marriages was commonplace, but Hawking and the Hawking marriage had seemed so very uncommonplace. At first, opinion turned strongly against him for leaving the wife who had supported him so courageously. Not until Jane Hawking published her memoirs in the late 1990s did a more balanced picture emerge, and it finally became known that the marriage had, in truth, been troubled for years.

Hawking moved out of the house on West Road, and he and Elaine took up residence in Pinehurst, an attractive, upmarket, rather secluded cluster of homes and flats not far away in Grange Road. In the early nineties

Elaine sometimes skateboarded back to that home after delivering her boys to school.

Hawking had relinquished one of the pillars he had always said supported his life: his family. Was another such pillar, his scientific work, also in danger of collapse?

The Lucasian Lecture – Revisited

Though some feared the personal turmoil in Hawking's life would undermine his scientific work, he continued to express his devotion to his science. He said he was 'itching to get on with it'. Was it still possible that he might be the physicist to fit it all together in the Theory of Everything, as the media had been prophesying?

Hawking's work wasn't in the newer mainstream of that effort: superstring theory. However, mainstreams in physics shift overnight, and a mind somewhat set apart may spot the connection that makes several streams converge into one complete theory. There were murmurs that by theoretical physics standards Hawking was already well over the hill. It's young people who usually make the great discoveries. A freshness of mind is required, a passionate, brash approach mixed with a certain amount of naivety. But Hawking certainly still had all of that. It would have been a profound mistake to rule him out on those grounds.

Would he live long enough? His illness was still progressing but very slowly. Did he worry about dying before he finishes his work? In 1990 he replied to that question by saying that he never looks that far ahead. He's lived with the possibility of imminent death for so long that he isn't afraid of it. The kind of work he does is a joint effort, and there are plenty of other physicists to carry on with it.

He's never claimed his presence is necessary for the Theory of Everything to be found. 'But I'm in no hurry to die,' he added. 'There's a lot I want to do first.'[5]

In June 1990, ten years after his inaugural lecture as Lucasian Professor, I asked him how he would change his Lucasian lecture, were he to write it over again. *Is* the end in sight for theoretical physics? Yes, he said. But not by the end of the century. The most promising candidate to unify the forces and particles was no longer the N=8 supergravity he'd spoken of then. It was superstrings, the theory that was explaining the fundamental objects of the universe as tiny, vibrating strings, and proposing that what we had been thinking of as particles are, instead, different ways a fundamental loop of string can vibrate. Superstrings would take a little longer to work out. Give it twenty or twenty-five years, he said.

I asked him whether he believed his no-boundary proposal might turn out to answer the question, what are the boundary conditions of the universe? He answered yes.

Hawking said he thought wormhole theory had important implications for a Theory of Everything. Because of wormholes it was probable that neither superstrings nor any other theory would be able to predict such fundamental numbers in the universe as particle charges and masses.

And if somebody does find the Theory of Everything, what then? According to Hawking, doing physics after that would be like mountaineering after Mount Everest has been conquered. However, Hawking had also said in *A Brief History of Time* that for humanity as a whole it would be only the beginning, because, although a Theory of Everything would tell us how the universe works and why it is the way it is, it won't tell us why it exists at all. It would be just a set of rules and equations. He had wondered: 'What is it that breathes fire into the equations and makes a universe for them to describe?' 'Why does the universe go to all the bother of existing?'[6] Those, he said, are questions that the usual scientific approach of coming up with mathematical models cannot answer.

Hawking still longed to know the answers. 'If I knew that, then I would know everything important',[7] 'then we would know the mind of God'.[8] That was where he had left it at the end of his book, but he had told a television interviewer: 'I'm not so optimistic about finding why the universe exists.'[9] He wasn't considering the question of whether we necessarily need to find the Theory of Everything in order to know the mind of God, whether there are, as Jane Hawking had suggested, other ways to know God besides in the laws of science.

Stardom

In 1990 Hawking received an honorary degree from Harvard University. Those who attended the ceremony and reception remember with fondness that Hawking's fellow honoree Ella Fitzgerald sang especially for him at the reception.

New non-academic faces were showing up outside Hawking's door to read the little plaque, and were being made to wait just as though they were graduate students there for a supervision. It wasn't a local Cambridge or New York photographer, but Francis Giacobetti, photographer of the Pope and Federico Fellini, whose equipment and assistants crowded that side of the common room. Giacobetti believed the subjects of his portraits were best revealed by their hands, the irises of their eyes and in half-profile, and that was the way he was photographing them for an outdoor exhibition that would open in Paris and then travel around the world. Other subjects were Francis Crick, the novelist García Márquez, the architect I. M. Pei.

Nor was it just another interview by a television personality that pumped life to a new level of frenetic high energy in the common room and pushed physics discussions into the corners. Steven Spielberg agreed to produce a film version of *A Brief History of Time*, to be directed by the young Errol Morris.

Morris was an inspired choice, with an unusual, intellectual, eclectic background. As a precocious ten-year-old he had given lectures on the solar system; as a teenager, played the cello and studied music with Nadia Boulanger at Fontainebleau; as an undergraduate, set records climbing in Yosemite; as a graduate student, rather unhappily done graduate work in science history at Princeton (studying with John Wheeler) and philosophy at Berkeley. None of this had appealed to Morris as his life's work, but, along the way, he picked up experience that would help him better understand others who did not fit an ordinary mould, such as Stephen Hawking.[10]

Although the Berkeley faculty had rejected Morris's eccentric thesis proposal on the subject of the insanity plea, monster movies and murderers in Wisconsin prisons, his interest in the subject of 'true crime' continued after he began to make documentary films. Morris made contact with a man named Randal Adams, who was on death row awaiting execution for the murder of a Dallas policeman. Morris found the verdict questionable and took it upon himself to look into the case personally. He was not unqualified, having worked for several years as a private detective in New York when his film-making career was at a low ebb.[11] He documented his investigation of the Adams case on film, and solved it, winning Adams's release. *The Thin Blue Line* premiered in 1988 to enormous critical acclaim and elevated Morris to the level of a major documentary film-maker. With good cause, he dubbed himself a 'director/detective'.[12]

It was this interesting, complicated, brilliant young man whom Spielberg brought together with Stephen Hawking for the filming of *A Brief History of Time.* One of the challenges Morris liked to set himself as a director was 'how to extract a situation's truth without violating its mystery'[13] and it was with that question in mind that he approached the Hawking project.

Spielberg's choice of Morris was partly made in the interest of solving a problem that emerged early on in the development of the film and threatened to doom the project. Hawking envisioned a film that would use all the state-of-the-art science fiction film-making technology and special effects available to Steven Spielberg and his colleagues to bring *A Brief History of Time* spectacularly to the screen. It seemed the project could not have fallen into better hands. He had no intention of allowing his personal life to be featured. The film-makers, however, argued that the film Hawking had in mind would never draw in the mass audience that both they and Hawking hoped to attract. The film must be biographical. Hawking took the matter up personally with Spielberg. One indomitable will had met another, but finally it was Spielberg who prevailed, by bringing Morris into the project and convincing Hawking that Morris could, as Hawking later wrote, 'make a film that people would want to watch, but which doesn't lose sight of the purpose of the book'.[14] Morris saw Hawking's courageous life with severe physical limitations and his bold scientific quest as 'inseparable themes'.[15] He chose to let Hawking narrate the film himself, in his own synthesized voice, and repeatedly filmed him reflected in his computer screen.

Among the most successful hallmarks of Morris's previous work had been his genius as an interviewer and his skilful use of talking heads, and Hawking, in an extraordinary capitulation, gave Morris permission to interview his family, friends and scientific colleagues on film. His permission, however, could not ensure Jane Hawking's permission. She and their three children are not in the film except in photographs. Elaine Mason also refused to be interviewed, but Gordon Freedman, the executive producer, found she was 'a wonderful bouncing nurse' who did 'cartwheels on the soundstage'.[16] Stephen himself would not answer questions or make statements about his personal life in the film. However, Isobel Hawking, Stephen's mother, agreed to appear, and Hawking, at the premiere, thanked Morris for making his mother a film star.

In an Afterword to the book published as a 'companion' to the film, *A Brief History of Time: A Reader's Companion*, Gordon Freedman described the 'very strong working relationship' that developed between Hawking and Morris during the three years' filming first in a London studio and then in Cambridge, when Morris, disappointed at the way the film was turning out, started all over again. 'In the thick of the editing, Stephen

Hawking and Errol Morris could be seen in the edit room for hours on end working toward a single vision for the film.'[17]

A Brief History of Time premiered in New York and Los Angeles in August 1992. It won the 1992 Grand Jury Prize for Documentary Filmmaking and the Documentary Filmmaker's Trophy at the Sundance Festival, and the Filmmaker's Award from the National Society of Film Critics. Philip Gourevitch, writing in *The New York Times Magazine*, commented perceptively: 'Morris's record of Hawking and the people who surround him creates the unexpected impression that he is a normal man who just happens to have the mind of a genius trapped in a devastated body.'[18] David Ansen in *Newsweek* praised the film as 'an elegant, inspirational and mysterious movie. Morris turns abstract ideas into haunting images, and keeps them spinning in the air with the finesse, and playfulness, of a master juggler.'[19] Richard Schickel in *Time* magazine spoke of its 'splendid talking heads' and went on to say, 'that the metaphorical richness of this hypnotic movie has been accomplished by such simple means is a mark of its excellence'.[20]

Morris himself was more than pleased with the final result. He called Hawking 'a symbol to millions of triumph over adversity and man's insignificance in the face of an implacable cosmos out there' and was satisfied that the film successfully conveyed that image. Compared with all previous work he had done, it was, Morris said, 'less cerebral and more moving', in spite of what might have seemed a difficult, esoteric subject, and even though in this film 'everyone is smarter than me'.[21]

In spite of the undeniable beauty of the film and its

critical success, it never reached a mass audience. In fact, it was never marketed or released to theatres in a way that would have allowed it to do that. We shall never know what the result would have been if Hawking had won the argument rather than Spielberg.

The world might have turned upside down when it came to Hawking's personal life: Elaine had become hugely important to him; he had become a film star; but, with all that, when I visited him again a couple of years later in the office and common room in Silver Street, it was as though time there had stood still. The description I'd written in 1990 could have been written that day . . . the small clicking sounds, the words flitting back and forth, up and down, on the computer screen. The synthetic voice enunciated them politely. Students, nurses and colleagues passed in and out. At 4 p.m. the cups were lined up as precisely as a toy army on the counter in the common room. Former Lucasian Professors of Mathematics gazed down from their portraits on the intent little 'rock group on a bad day' as they sipped tea and talked their strange mathematical language. The figure in their midst was pitiful by all normal standards, like a guy on its way to the bonfire on Guy Fawkes Night. He wore a bib, and a nurse held his forehead and tipped his head forward so that he could drink his tea out of the cup she held under his chin. His hair was tousled, his mouth was slack, and his eyes were weary over the glasses that had slipped down his nose a little. But at a disrespectful quip from one of the students his face broke into a grin that would light the universe.

My book in 1990 had ended with the words, 'Whatever the future brings, in this unlikely, paradoxical story, we can hope it will be that grin an artist will capture someday in Hawking's portrait, the portrait that will hang in the empty space still remaining on the common room wall beside his office door. Meanwhile the little plaque is a liar. The boss is not asleep.'

That was more than twenty years ago. Hawking was forty-eight years old.

14

'Between film roles I enjoy solving physics problems'

ANDREI LINDE, WORKING WITH ALEX VILENKIN IN THE late 1980S and early 1990s, discovered that an inflationary universe has an amazing potential: in chaotic inflation theory, it can 'self-reproduce'. The result is an enormous fractal arrangement of universes sprouting one out of another. Hawking's 'baby universe' idea also proposed a multitude of other universes. Our own huge universe was suddenly looking small compared with a possible 'multiverse'.

Inflation Goes 'Eternal'

Imagine again one of those regions in the early universe that is inflating rapidly, while some of its neighbouring regions are not. Within that inflating region, Linde and Vilenkin thought, there would be sub-regions that would expand more rapidly than their parent region, while other sub-regions would not. Nor is that the end of the story. Sub-regions would beget sub-sub-regions, and so on and so forth. In other words each microscopic region that inflated would in turn be made up of microscopic sub-regions, some of which would inflate and be in turn made up of microscopic sub-sub-regions, and so on and so forth – an eternal inflationary universe scheme. 'Eternal inflation' would be a never-ending process of self-reproduction of the universe. As Linde described it, 'the universe is a huge, growing fractal. It consists of many inflating balls that produce new balls, which in turn produce more balls, ad infinitum.'[1] The universe we know and can observe is just one of these regions, or sub-regions, or sub-sub-regions. Cartoons portrayed Linde as a carnival balloon seller.

Are there 'parallel universes' like ours? Not necessarily, though other universes something like ours can't be ruled out. Even though the total picture including what is beyond 'our universe' may be irregular and hugely complicated, in our universe the situation is smooth and relatively simple. Inflation in other regions or sub-regions might also have produced large smooth universes. The whole thing may, in fact, become divided into infinitely many exponentially large regions. However, when energy levels in those other large regions decrease to levels similar to those in our universe today, they will each have different laws of physics. Eternal inflation will produce an infinite variety of universes. In John Barrow's words, 'we find ourselves faced with the likelihood that . . . we inhabit a single, simple patch of space and time in an elaborate cosmic quilt . . . of huge diversity and historical complexity, most of which is totally inaccessible to us.'[2]

And we don't find our patch as it is because it is the most likely kind of universe. It is not.

Barrow used the words 'historical complexity' advisedly, for if the whole picture is as eternal inflation describes, there is no reason to think that our own universe exists at or near the beginning of the eternal process. It might have originated as a sub-region of a sub-region of a sub-region – who knows how far we are down the line? In this model, our own universe has a beginning, but the big picture, the enormous fractal arrangement, perhaps need not have either a beginning or an end. Not everyone agrees that this picture stretches infinitely far into the past. Alan Guth, the founding father of inflation theory, working with Alex Vilenkin and Arvind Borde, thinks that the huge 'eternal' inflation scenario must have a past boundary, a beginning.

If all this is taking place, shouldn't we be able to notice it? The answer is no. The inflation happens so rapidly that the regions and sub-regions and sub-sub-regions, etc., become immediately independent of one another. They move apart faster than the speed of light, becoming separate pocket universes.

In Linde's words, 'According to this scenario, we find ourselves inside a four-dimensional domain with our kind of physical laws, not because domains with different dimensionality and with alternate properties are impossible or improbable, but simply because our kind of life cannot exist in other domains.'[3] Among an infinite variety of universes, it is likely – perhaps inevitable – that one of them, at least, will be the kind where our sort of life can exist. The anthropic principle is alive and well, and not at all the cop-out it once seemed.

If we can't see eternal inflation happening on our doorstep or anywhere else, is there *any* observational evidence – or even *potentially possible* evidence – for eternal inflation? Hawking was going to show that this is not a hopelessly far-out question, though it might have seemed so when eternal inflation first appeared on the scene.

Global celebrity

As the last decade of the twentieth century got underway, Hawking kept up the exhausting level of international travel that had begun soon after the surgery that cost him his voice. Plane flights and hectic schedules were possible for him as long as he had indefatigable, intrepid nurses with him. Travel included more than giving lectures, taking part in conferences and being escorted by dignitaries around tourist sites.

Trips to Japan, in particular, where Hawking went no fewer than seven times in the 1990s, seemed to produce unusual adventures. Joan Godwin, who was with him as a nurse on most of these eastern jaunts, recalls that on one visit Hawking expressed a strong desire to see the northern part of the country. His hosts warned him that this was an area where earthquakes were expected. They suggested a safer destination. Sendai. Sendai is, of course, where the devastating earthquake and tsunami and the subsequent deadly problems with nuclear reactors occurred in 2011, but Joan recalls that they had a delightful visit there. Once the subject of earthquakes had come up, however, Joan did confer with Hawking about what she should do if one occurred. 'Save yourself,' he said. 'Don't worry about me.'

During another visit to Japan, an incident occurred that, though it represented no danger to Hawking physically, was perhaps a threat to his ego. On the bullet train platform he was surrounded as he often was by cameras and reporters clamouring for statements and pictures. Then, in the blink of an eye, they inexplicably turned their backs and sprinted off. Hawking was alone. Investigation revealed that a famous Sumo wrestler had alighted from a train on a nearby platform. Hawking was *not* the most important celebrity in the universe.

Japan was also the setting for a story told by string theorist David Gross at Hawking's sixtieth-birthday party conference in 2002:

> When you travel with Stephen you get to meet all sorts of people you would never meet otherwise. We didn't get to meet the Emperor, which I regret, but we did get to meet somebody who I gather in Japan is even more popular and more famous, that is, the Green Tea Master, and we got to meet geisha girls, etc. But the most interesting experience was when Stephen insisted we all go to a karaoke bar. He actually got us to get up there and sing 'Yellow Submarine', which, if I were to try to reproduce, you would run screeching from the room. Every time the chorus came up Stephen would pipe in 'Yellow Submarine'; he probably still has a 'Yellow Submarine' button that he can push![4]

Staying close to home was much less interesting. It was also no guarantee of safety. On 6 March 1991, the news spread quickly in Cambridge that Stephen Hawking had been knocked over by a taxi the evening before, crossing

Grange Road, where he and Elaine were living in Pinehurst. It had been dark and raining, but the wheelchair had front and rear bicycle lights and should not have been difficult to see. His nurse screamed 'Look out!', and the speeding car caught his wheelchair from behind as he was crossing. He was thrown out on to the road, landing with his legs over what remained of the destroyed chair. It would have been a serious accident even for someone in a state of perfect health and fitness. Nevertheless, he was out of hospital back in his office two days later, with a broken arm in a sling and stitched-up cuts on his head. His personal assistant Sue Masey, his graduate assistant, students and friends had spent those two days frantically bringing in a new wheelchair from another part of the country and finding the necessary parts to repair his computer system, making sure his equipment was restored to normal as rapidly as Hawking.

In 1992, Stephen and Elaine built a large modern house not far from the centre of Cambridge. From there, Hawking took a new route to his office, through an attractive old passage known as Maltings Lane, emerging to skirt a pond and cross a low-lying, wild, green area of grass and trees and small bridges known as Sheep's Green. Then he crossed the River Cam where a weir separates the upper river from the part that runs through the Backs, and from there it was a straight shoot into Mill Lane and to his ramp at the DAMTP back entrance.

Anyone familiar with that footpath through Sheep's Green might expect that the narrow bridges would have presented a problem. They are barely wide enough to allow bicycles to cross, if cyclists are sufficiently skilful to avoid scraping their hands on the wooden railings built at handle-bar height. Hawking in his wheelchair 'threaded the needle' at speed, even in the dark. On one evening, making the trip home late with him, Joan Godwin stumbled off the ragged edge of the path and fell. Hawking, unaware she wasn't still behind him, blazed merrily ahead. When a kindly gentleman helped Joan up, announcing that he was an orthopaedic surgeon and hoped she wouldn't require his services, she asked him please to 'stop that wheelchair'.

The Inflation of Wrinkles

In April 1992, George Smoot, an astrophysicist at Lawrence Berkeley Laboratory and the University of California at Berkeley, and his colleagues at several other institutions, announced that data coming in from the Cosmic Background Explorer (COBE) satellite had revealed 'ripples' in the CMBR. This was a hugely significant discovery. Those ripples were the first evidence of the elusive variations that astrophysicists and cosmologists had been looking for in vain since the 1960s. These tiny differences in the topography of the universe when it was only an estimated 300,000 years old were evidence of a situation that would have given gravity a fingerhold and allowed matter to attract matter into larger and larger clumps, eventually forming planets, stars, galaxies and clusters of galaxies. The no-boundary proposal stood to gain credibility from Smoot's discovery. It had predicted both the overall smoothness of the universe and the deviations from that smoothness that COBE had found.

Hawking recognized that the COBE findings might even be indirect observational evidence of Hawking radiation. As we have seen, according to inflation theory, long before the era in which the cosmic microwave background radiation originated – in fact, when the universe was far less than one second old – it went through a period of runaway inflation. Hawking pointed out that during that period the universe would have expanded so enormously and so rapidly that light travelling towards us from some faraway objects would *never* be able to reach us. It would have to travel at greater than the speed of light to do so. Hearing of some light being able to reach us, while other light can never reach us, suggests the event horizon of a black hole. Hawking proposed that there may indeed have been an event horizon in the early universe that was similar to the event horizon of a black hole, separating the region from which light reaches us from the region from which it doesn't. From that ancient horizon, there would be radiation just as there is from a black hole, and thermal radiation like this has a characteristic pattern of density fluctuations. In the case of the early universe event horizon, these density fluctuations would have expanded with the universe, but then become 'frozen in'. We would observe them today as a pattern of minuscule variations in temperature – the 'ripples' that Smoot found in the cosmic microwave background radiation. Those 'ripples' did indeed turn out to have the characteristic pattern of density fluctuations from thermal radiation such as Hawking radiation.

Star of Stage and Screen

In the autumn of 1992, opera-lover Stephen Hawking found himself portrayed on the stage of the Metropolitan Opera in New York City, or rather as a figure suspended somewhat above it. The opera was not Wagner. It was *The Voyage*, a new work by Philip Glass, who had composed the score for the film *A Brief History of Time*. The Met commissioned *The Voyage* to celebrate the 500th anniversary of Columbus's journey to the new world, but Glass chose not to retell the story of Columbus. Instead he let Columbus symbolize the human longing to explore and discover.[5] In the opera's prologue, a wheelchair-bound figure, clearly meant to suggest Stephen Hawking, floated above the stage, intoning, 'The voyage lies where the vision lies.' A sky full of planets appeared as though he had conjured it, and he flew away.

The following year, Hawking was more personally involved in one short escapade that remains a favourite memory not only for him but for many of the rest of us, travelling – in imagination – far more boldly into space than the real-life sub-orbital flight he still hopes to take. He was not, for a change, conveyed on the wings of theoretical physics.

It began in the spring of 1993, at a party celebrating the release of the home video version of the film *A Brief History of Time*.[6] Leonard Nimoy, who played the Vulcan Spock on *Star Trek*, was among the guests and had the honour of introducing Hawking. As *People Magazine* described it, 'Vulcan's most famous son and Earth's most celebrated cosmologist instantly melded minds'[7] – an allusion that any faithful *Star Trek* viewer would appreciate. When Nimoy learned that Hawking was an avid fan of *Star Trek* and, like nearly every other fan, longed to be on the show, he contacted the executive producer Nick Berman. Berman lost no time arranging for a three-minute scene to be added at the beginning of one of the regular weekly episodes, titled 'Descent'.

The scene was the 'holodeck' of the Starship *Enterprise* – a part of the ship that uses holographic technology to turn the fantasies of crew members into 'reality'. This was the android Data's fantasy – a poker game with Einstein, Newton and Hawking. Not surprisingly, Hawking was the only one who played himself. He was sent the script well ahead of time and programmed his lines into his voice synthesizer. In the 1990s and for most of the first decade of the twenty-first century, Hawking was still capable of a number of facial expressions, and these he put to good use playing his role. 'Everyone was astonished at how much mobility his face has. The vitality behind it is very evident,' commented the episode director Alex Singer; and John Neville, who played Isaac Newton, added, 'When you get that smile in response to something you've said, it's worth the whole day's pay, really.'[8] Brent Spiner, who as Data was the host of this remarkable poker game, summed it up: 'When Rick Berman and I are in the old folks' home sitting in rocking chairs, we're going to be talking about the Hawk.'[9] Hawking also seemed to be getting his priorities right: 'Between film roles,' he commented on the set, 'I enjoy solving physics problems.'[10] His one regret was that 'unfortunately there was a red alert, so I never collected my winnings' although he had 'beaten them all'.[11]

The cameo role on *Star Trek* brought Hawking back into the public limelight to a degree that almost exceeded the fame won him by *A Brief History of Time* and was an asset in publicizing his *Black Holes and Baby Universes and Other Essays*, which came out the same year, 1993. There soon were other appearances in pop culture. The song 'Keep Talking' on Pink Floyd's *The Division Bell* album featured his computerized voice.

Advocate and role model

The *Star Trek* episode also brought him to the attention, far more than his books had, of young people with various forms and degrees of disability. A *Time* magazine article in September 1993 described him speaking in Seattle for more than an hour to a rapt, 'completely focused' audience of teenagers in wheelchairs.[12] After the talk, they crowded around him, asking questions most of which had to do with the practicalities of living with disability and political issues pertaining to the disabled, rather than with science or the cosmos. 'As they wait for Hawking to tap out his answers, they can't stop grinning. Here's a famous scientist, a best-selling author, a *Star Trek* star – and he's disabled, just like they are.'[13] The truth of Hawking's statement that his fame, while being a mixed blessing, 'enables me to help other disabled people',[14] was evident.

He would help in other ways. In the summer of 1995, he lectured to a capacity crowd at London's Royal Albert Hall, no mean amount of crowd-drawing power for a venue seating 5,000. The proceeds went to an ALS charity. He helped to publicize an exhibit of technological aids for disabled people, called 'Speak to Me', at London's Science Museum. His presence or sponsorship could assure a sell-out crowd nearly anywhere in the world. A January 1993 article in *Newsweek* described the public and media frenzy at lectures in Berkeley, California, where many of the audience showed up more than three hours early to get seats.[15] As he rolled to centre stage, photographers jostled one another for good positions and there was a blizzard of flash bulbs. Harnessing that sort of excitement for the good of disabled people was well worth doing.

Over a decade Hawking had become a master manipulator of his public . . . or was it just an accident that he always seemed to come up with attention-getting statements whenever public and media attention appeared to require a boost. As one of his personal assistants once commented to me, 'He isn't stupid, you know.' Whichever it was, it proved an advantage not only for himself but for advocates for the disabled and for science in general.

A Challenge for the 'Prime Directive'

An uncharacteristically sour pronouncement at the Macworld Expo in Boston in August 1994 made the news all over the world: 'Maybe it says something about human nature that the only form of life we have created so far is purely destructive. We've created life in our own image.'[16] Hawking was talking about computer viruses.

Are computer viruses a form of life? Hawking thought they should 'count as life' and, with those words, initiated an uproar. In a recent episode of *Star Trek*, Captain Picard, confronting a super-intelligent virus, had negotiated with it rather than destroy it, to avoid violating the Star Fleet 'prime directive' that forbids interfering in the internal development or social order of any alien society. Destruction of the virus, in this case, would have constituted such a violation. Clearly the writers of *Star Trek* agreed with Hawking. There were plenty of fans of both *Star Trek* and Hawking ready to join the debate, on one side or the other.

'A living being usually has two elements,' Hawking argued. 'First, an internal set of instructions that tell it how to sustain and reproduce itself. Second, a mechanism to carry out the instructions.' Life as we know it is biological life, and these two elements are the genes and the metabolism. But 'it is worth emphasizing that there need be nothing biological about them'. A computer virus copies itself as it moves into different computers and infects linked systems. Though it does not have a metabolism in the usual sense, it uses the metabolism of its hosts, like a parasite. 'Most forms of life, ourselves included, are parasites in that they feed off and depend for their survival on other forms of life.'

Because in biology it is by no means a settled matter what is life and what is not, biologists asked to comment were unwilling to say that Hawking was wrong. Computer viruses do certainly fit some definitions of life.

Hawking closed his speech with yet another startling suggestion for what 'life' might include. Human life spans are too short for long-distance interstellar and intergalactic travel, even at the speed of light. However, the necessary longevity would not be difficult for mechanical spaceships that could land on distant planets, mine their resources and then produce new spaceships. The voyage could go on for ever. 'These machines would be a new form of life based on mechanical and electronic components rather than macro molecules [like biological life],' said Hawking. A bleak prophecy!

With everything else he was managing to fit into his schedule, Hawking found time in 1993 to co-edit a volume of technical papers on Euclidean quantum gravity with Gary Gibbons.[17] Hawking wrote or co-wrote sixteen of the thirty-seven papers himself. That same year he also published a collection of his own papers on black holes and the big bang.[18]

Arrows of Time

Another subject Hawking was lecturing on in his public lectures in the early 1990s was far less disturbing than viruses as a form of life. It was something that had intrigued him for many years: 'arrows of time'. The increase of entropy (disorder) and the human perception of past and future seem to be linked with the expansion of the universe. Why should this be so? As a doctoral student he had considered writing his thesis on this mysterious topic but decided he wanted something 'more definite and less airy-fairy'. Singularity theorems were 'a lot easier'.[19] However, when he and Jim Hartle were developing their no-boundary proposal, Hawking had recognized that this work had interesting implications for arrows of time. He returned to the subject in a paper he wrote in 1985 and intended to work on it more at CERN that summer when he instead ended up so disastrously in hospital.

In the early nineties, with an increasing demand for public lectures, Hawking found that this was indeed a topic that interested his non-expert audiences and that he could explain fairly simply and succinctly. It was also a subject that allowed him to show that eminent scientists are capable of changing their minds and admitting mistakes.

With very few exceptions, the laws of science make no distinction between forward and backward directions of time. The laws are *symmetrical* with respect to time. You could make a film of most physical interactions and reverse the direction of the film and no one who saw it could say which way it ought to run. Strange, then, that this is not our experience at the level of the everyday world. We have a well-defined future and past. We nearly always can tell if a film is running backwards. It would be difficult to mistake one direction for another. How this 'symmetry-breaking' occurs is still one of the great mysteries, but we do know that in the universe as we experience it, our perception of the passage of time seems to be linked with the fact that, in any closed system, disorder (or entropy) always increases with time. The road from order to disorder is a one-way street. Broken pottery does not pull its scattered pieces together and hop back on to the shelf. Entropy, disorder, never decreases.

There are three 'arrows of time': the 'thermodynamic arrow' (the direction in which disorder, or entropy, increases); the 'psychological' or 'subjective' arrow (the way human beings experience time passing); and the 'cosmological arrow' (the direction of time in which the universe is expanding, not contracting). The question that interested Hawking was why these three arrows exist at all, why they are so well defined and why they point in the same direction. Disorder increases, and we experience the passage of time from past to future, while the universe expands. He suspected the answer lay in the no-boundary condition for the universe, with some help from the anthropic principle.

The thermodynamic arrow (having to do with the increase of disorder or entropy) and the psychological arrow (our everyday perception of time) *do* point, always, in the same direction. It is common experience that as time moves forward, disorder, or entropy, increases. Hawking admitted that this is a tautology, concluding that 'entropy increases with time because we define the direction of time to be that in which entropy increases'[20] – but he was satisfied that the psychological arrow and the thermodynamic arrow are essentially the same arrow.

Why then does it point in the same direction as the cosmological arrow of time, with the expansion of the universe? Must it? Enter the no-boundary proposal. Recall that in the classical theory of general relativity all physical laws break down at the Big Bang singularity. It is impossible to predict whether or not the beginning of time would have been orderly or a situation of complete disorder in which there was no possibility of disorder increasing. However, if Hawking's and Hartle's no-boundary proposal is correct, the beginning was 'a regular, smooth point of spacetime and the universe would have begun its expansion in a very smooth and ordered state'.[21] As the universe expanded, the gradual development of all the structure we observe today – galaxy clusters, galaxies, star systems, stars, planets, you and me – represented a continuous, enormous increase in disorder, and this trend continues. Hence, in the universe as we know it, the thermodynamic arrow, the psychological arrow and the cosmological arrow all point in the same direction.

But consider what might happen if Friedmann's first model of the universe is correct (see Figure 6.1, p. 110), the model in which the universe eventually stops expanding and begins to contract. When expansion changes to contraction, the cosmological arrow of time reverses direction. The big question was, would the thermodynamic and psychological arrows of time also reverse direction? Would disorder start to *decrease*? Hawking thought there were all sorts of interesting possibilities for science fiction writers here, but he also pointed out that it was 'a bit academic to worry about what would happen when the universe collapses again, as it will not start to contract for at least another ten thousand million years.'[22]

Nevertheless, the no-boundary condition did seem to mean that disorder would decrease in the collapsing universe, and Hawking at first concluded that when the universe stopped expanding and started to collapse, not only the cosmological arrow but *all three arrows* would reverse direction and all three continue to point in the same direction as one another. Time would be reversed and people would live their lives backwards, 'youthening', as the magician Merlin did in T. H. White's Arthurian novel *The Once and Future King*. Broken teacups would reassemble.

Don Page, by then in the physics faculty at Pennsylvania State University, begged to differ. In a paper that eventually appeared in the same issue of *Physical Review* as Hawking's arrow of time paper, Page argued that the no-boundary condition did not mean that all three arrows would have to be reversed when the universe was in its contracting phase.[23] Raymond LaFlamme, one of Hawking's students, found a more complicated model, and the three argued and sent calculations back and forth. Page, with more experience of working with Hawking, suggested to LaFlamme that it would be best not to tell Hawking their conclusion but first to lay out all their assumptions in such a way that Stephen would arrive at the same result without their having told him what it was.[24] They finally convinced their mentor that he had been wrong. Though the cosmological arrow of time would reverse when the universe stopped expanding and began to contract, the thermodynamic and psychological arrows would not. It was too late to change Hawking's paper, but he was able to insert a note admitting, 'I think that Page may well be right in his suggestion.'[25]

What, then, is the answer to the question, why do we observe the thermodynamic, psychological, and cosmological arrows pointing in the same direction? Because, even though we would not find ourselves 'youthening', we could not survive in the universe when it is collapsing, when the cosmological arrow of time has reversed. At that distance in the future, the universe will be in a state of nearly total disorder, all the stars burned out, the protons and neutrons in them decayed into light particles and radiation. There will no longer be a strong thermodynamic arrow of time at all. We couldn't survive the death of our sun, but even if we could, we also require a strong thermodynamic arrow of time in order to exist. For one thing, human beings have to eat. Food is a relatively ordered form of energy. The heat into which our bodies convert food is more disordered. Hawking had concluded that the psychological and thermodynamic arrows of time are for all intents and purposes the same arrow, and if one fizzles out, so does the other. In the contracting phase of the universe there could be no intelligent life. The answer to the question of why we observe the thermodynamic, psychological and cosmological arrows pointing in the same direction is: because, if things were different, there would be no one around to ask those questions. If that sounds familiar, it is none other than the anthropic principle. As time (in all three senses) passed, Hawking was thinking less and less of the anthropic principle as a cop-out, 'a negation of all our hopes of understanding the underlying order of the universe', and increasingly regarding it as a powerful principle indeed.

More Conjuring at the Event Horizon

Hawking had suspected in 1981 that Leonard Susskind was 'the only one in the room who fully appreciated the implications of what I had said', in Werner Erhard's attic. In the years since then, Susskind had never been able to leave the information paradox problem alone. 'Just about everything I have thought about since 1980 has in one way or another been a response to [Hawking's] profoundly insightful question about the fate of information that falls into a black hole. While I firmly believe his answer was wrong, the question and his insistence on a convincing answer have forced us to rethink the foundations of physics.'[26] In 1993, referring back to work Hawking had done in the 1970s, Susskind came up with a new way of dealing with a contradiction that defied common sense at the event horizon of a black hole.

It will come as no news to anyone who has read even the most rudimentary book about black holes that if someone (let us call her Miranda) falls into one, the experience for her will be radically different from what it appears to be from the vantage point of someone (let us call him Owen) who is watching from a spaceship at a distance outside the black hole. Einstein showed that if two people are moving rapidly relative to one other, each one sees the other's clock slow down and sees the other being flattened out in the direction of motion. Also a clock that is in the vicinity of a massive object (and a black hole is a very massive object) will run more slowly compared with one that is not.

The upshot is that from the vantage point of distant observer Owen, Miranda as she falls towards the black hole seems to be falling more and more slowly with her body squashed to a thinner and thinner pancake. Finally, when Miranda reaches the event horizon, Owen sees her come to a stop. He never sees her fall through the horizon, in fact, never sees her quite get there. Meanwhile, Miranda's own experience is that she falls through the event horizon intact. From Owen's point of view, she is stuck and flattened; from Miranda's point of view, she is still falling.

Susskind had set himself to find out how both could be true and pointed out that although he and you and I, who are neither falling in nor watching from a distance, can agree that both scenarios in our example have occurred, and be troubled by the contradiction, none of us is actually on the spot. Suppose instead that you and I are part of the action. This time I will be the observer at a distance. You fall into the black hole. The crux of the matter is that in a real-life playing out of this story, neither I, the observer at a distance, nor you, who fall into the black hole, ever observes or experiences the contradiction. And you who have experienced an uneventful fall through the event horizon are absolutely incapable of going back and comparing notes with me or of sending me a message. If I happen to fall in later (this possibility stumped Susskind for a while), you would still be so far ahead on the way to the singularity that I would never catch up. It would be impossible for either of us, ever, to know about the version of the story that contradicted our own.

Susskind, and colleagues Lárus Thorlacius and John Uglum, called this principle that neither observer ever sees a violation of the laws of nature 'horizon complementarity'.

Take a moment to recall what 'complementarity' means. It is using two different, perhaps mutually exclusive descriptions in order to gain a better understanding than either description alone provides. Early in the twentieth century it was physicist Niels Bohr's way of addressing a problem in physics known as wave-particle duality. People experimenting with the way light propagates (the way it travels) found that it acts as though it were waves. The description of it as particles is ruled out. However, when they studied the way light interacts with matter, they found that it acts as though it must be particles. The model that describes it as waves is ruled out. By 1920 it was clear that light could be conceived of either in terms of waves or in terms of particles, but that neither model by itself could explain the experimental data, and this odd situation could not be resolved by saying that light is sometimes particles and sometimes waves, or that light is both particles *and* waves. The problem applies to matter as well as to radiation. Bohr wrote to Einstein in 1927, concluding that it was possible to live with what looked like a contradiction 'as long as we don't allow our intuitive feeling that matter and radiation must be *either* wave or particle to "lead us into temptation"'.[27] The descriptions were incompatible but both necessary, and both correct.

The same could be said in the case of horizon complementarity. As Susskind summed it up, 'The paradox of information being at two places at the same time is apparent and yet a careful analysis shows that no real contradictions arise. But there is a weirdness to it,' he admits.[28] Gerard 't Hooft of Utrecht had in 1993 introduced what he called 'dimensional reduction'. Susskind rechristened it the 'holographic principle'.

Go back to thinking about Miranda falling towards the event horizon, as viewed by Owen, the distant observer in the spaceship. From the spaceship, because of time dilation, Miranda appeared to freeze and spread out at the event horizon. Susskind points out that, by the same token, Owen will also see everything else that originally went into the formation of the black hole, and everything that has fallen into it, likewise frozen at the horizon. 'The black hole consists of an immense junkyard of flattened matter at its horizon,' says Susskind.

The 'holographic principle', then, is the idea that information is in some sense stored on the boundary of a system instead of inside it. Think of the holographic image on a credit card, where a three-dimensional image is stored on the two-dimensional surface of the card. Susskind elaborated on that idea to compare a black hole with a giant cosmic projector that takes a three-dimensional person and turns her into a two-dimensional surface on the event horizon. The bottom line was that all that information, stretched out on the edge of the black hole, is not destroyed. It is all there. Not lost.

A very interesting suggestion about how this all happens came from string theory, in which, you will recall, particles do not look like points but rather like tiny loops of vibrating string. The way a loop of string vibrates determines which type of particle it is. First, think of a single string falling into a black hole. You are watching it from a spaceship at a distance. As it approaches the event horizon, its vibration appears to slow down. The string spreads, ending up with whatever information it carries smeared over the entire event horizon. As each string spreads out, it overlaps others, resulting in a dense tangle.

Since everything is made of strings, everything that falls towards a black hole is smeared out like that. The resulting giant tangle of strings covering the surface of the black hole is able to hold the entire enormous amount of information that fell when the black hole formed and afterwards. Here, then, at the horizon, is everything that 'fell into the black hole'. As far as the observer at a distance is concerned, it didn't fall into the hole at all. It stopped at the horizon and was later radiated back into space.

Susskind visited Cambridge in 1994 and looked upon it as a golden opportunity to talk with Hawking and convince him that horizon complementarity could solve the information paradox. Unfortunately Hawking was ill at the time and they were unable to connect. Finally, Hawking attended a lecture that Susskind delivered about black hole complementarity. Susskind remembers the occasion: 'This was the last chance for a confrontation with Stephen. The lecture room was full to capacity. Stephen arrived just as I was starting and sat in the back. Normally, he sits up front near the blackboard. He was not alone; his nurse and another assistant were in attendance, just in case he needed medical attention. It was obvious that he was having trouble, and about half way through the seminar, he left. That was it.'[29] Susskind's ideas would wait until early in the twenty-first century for someone to give them a rigorous mathematical treatment.

15

'I think we have a good chance of avoiding both Armageddon and a new Dark Age'

IN THE SPRING OF 1995, SEVEN YEARS AFTER IT FIRST appeared, *A Brief History of Time* finally went into paperback. Normally this would happen for a book about a year after its initial publication, but because the hardcover continued to sell at such a phenomenal rate, Bantam had repeatedly put off that decision. An interviewer told Hawking that with 600,000 copies sold in the UK, more than 8 million worldwide, and more than 235 weeks on the *Times* bestseller list, it was difficult to imagine there were still people who had waited seven years to learn the secrets of the universe, just so they could buy his book in paperback and save £8.[1] Hawking disagreed: 'It has sold one copy for every 750 men, women and children in the world, so there are 749 to go.' 'Hawking logic!' chimed in his nurse.[2]

Stephen and Jane Hawking finalized their divorce that spring. In July, Stephen made the first public announcement of his engagement and upcoming marriage to Elaine at the Aspen Music Festival in Colorado, at a concert to benefit the Festival and Music School and the Aspen Center for Physics.[3]

Hawking's short, joyful speech introduced a performance of Richard Wagner's *Siegfried Idyll.* Unlike most of Wagner's compositions, this is intimate chamber music and requires only a small ensemble of musicians. It has a romantic history. Wagner composed it to be played at his villa, on the staircase outside the bedroom of his wife Cosima as a surprise on Christmas morning 1870, her birthday. The couple had married the summer before. The *Idyll* was an inspired choice for Hawking's engagement announcement. It combines gentle tenderness and passion in a way that is almost unique in the literature. Elaine fondled Stephen's shoulder lovingly as he exited the stage and the music began. The physicist David Schramm, chairman of the board of the Aspen Center for Physics, commented: 'There is a warmth – a caring expression in Stephen's eyes when he looks at Elaine. There is a very special relationship between them.'[4] Two months later, on 16 September 1995, Stephen and Elaine were married in a ceremony in a Cambridge registry office, followed by a church blessing and a celebration. None of his three children or her two boys were present. Hawking had a statement programmed and ready in his computer: 'It's wonderful – I have married the woman I love.'[5]

The press reaction at the time of their marriage was not kind, questioning Elaine's motives for marrying this extremely wealthy man who was probably not going to live very long. Interviewers must have hoped that Elaine's former husband, David Mason, would provide some snide quotation, but he came to Elaine's defence. All Elaine really wanted, he said, was someone who needed her.[6] Elaine apparently needed Hawking too, for he replied to a question about his reasons for marrying Elaine that 'It's time I helped someone else. All my adult life people have been helping me.'[7] After his wedding, Hawking consistently declined to answer press questions – and those from curious audiences – about his marriage. 'I would rather not go into details of my private life' was his standard reply.[8] Amidst the concerned rumours and the less well-meaning gossip that would surface about Stephen's and Elaine's life together, there would be one consistent theme expressed by those who know him best. The bottom line was 'He loves Elaine.'

Jane Hawking was in Seattle visiting her son Robert when the engagement announcement was made. When she got back to Cambridge, she began to reconsider an earlier decision not to write her memoirs as Hawking's wife. She had tried in vain to find a publisher for a book she was writing about 'Le Moulin', a home that she bought and restored in France. That book offered invaluable advice and practical information for others who might contemplate doing the same. Publishers, however, wanted a book about her personal life with Hawking, not about Le Moulin. One unscrupulous agent tricked her into signing a contract that promised a 'tell-all' later, if the publisher agreed to take the book she was writing now. Jane had waited out the term of that contract and then, in 1994, self-published *At Home in France*.

In the summer and autumn of 1995, with everything changed, Stephen married to Elaine, and Jonathan and herself living openly together in Cambridge with Jane's son Timothy, it seemed to her that the time had come to tell the entire 'Hawking story' from her own much less up-beat point of view. A letter came from an editor at Macmillan Publishers, asking whether she would consider writing an autobiography. This time, Jane said yes.[9]

Tea and a Lecture

When I visited Hawking for tea in the DAMTP in the spring of 1996, his book with Roger Penrose, *The Nature of Space and Time*, had recently appeared. One particular statement in the book had annoyed some critics.[10] Hawking had re-emphasized something he'd been saying for at least a decade, that a theory in physics 'is just a mathematical model and it is meaningless to ask whether it corresponds to reality. All that one can ask is whether its predictions agree with observation.'[11] The rest of us may be curious, but Hawking was insisting, and would continue to insist, that it is meaningless to discuss such questions as whether wormholes really exist.

Over tea, I pursued those ideas a little further with him: all right, granted, it is pointless to ask whether this theory corresponds to reality. But, *is* there, actually, an answer to that question? Is there reality, inaccessible to us perhaps, but solid reality, all the same? To say there is none at all, on any level, is to take a very postmodern view of things. Hawking's reply was interesting: 'We never have a model-independent view of reality. But that does not mean there is no model-independent reality. If I didn't think there is, I could not continue to do science.'[12] In *The Nature of*

Space and Time, a collection of six lectures that Hawking and Penrose had delivered at the Newton Institute in Cambridge in 1994, giving insights into the two men's differing philosophical and scientific points of view, Hawking commented: 'I think Penrose is a Platonist at heart but he must answer for himself.'[13] Hawking's comment to me would certainly make him a Platonist too.[14]

We had our tea in the common room surrounded as usual by the bustling herd of students and physicists, their dress still casual to the point of scruffy, their language still a mixture of English and mathematics, conversations ranging across the span and history of the universe, equations scribbled on the surfaces of the low tables. In this company, it occurred to me, Hawking has never been treated as anyone extraordinary, though his colleagues and students are willing to outwait the long pauses while he composes his sentences, and consider what he says worth the wait. Because of his lack of body language and vocal nuance, the synthesized voice conveys only one mood: infinite, thoughtful patience, lending an oracular air to his statements. His humour, whether he intends it that way or not, comes across as dry wit.

Tea was cut short that afternoon because Hawking was scheduled to give a public lecture. The tickets had all been distributed through the university weeks ago. Someone suggested that I come along with his graduate and post-doctoral students, who would be admitted without tickets. It was a particularly kind suggestion, since I was at least twenty-five years older than any of them – the lecturer's age, not theirs.

The lecture was a media event: sound trucks outside the

building, cables snaking in, spotlights focused on the stage and the audience. The lecture hall was modern and large, though not nearly so capacious as others around the world that Hawking appearances routinely packed to the rafters. About 500 people sat on long, curved, desklike benches, others crowded into the balconies above us. There was a hush as Hawking rolled on to the platform. Something about that calm, ordinary, inexorable progress to centre stage gave it the aura of a visitation from another dimension. He had pre-programmed the lecture into his computer. An assistant worked a slide projector. Even when the lecture and slides included formulas and diagrams that few could follow, Hawking commanded rapt attention.

At that time, Hawking was also busy with another television project, to be aired the next year. *Stephen Hawking's Universe* and a companion book to go with it was a six-part collaboration between the BBC and Public Television in America. This time he got his way when he insisted the series stick to science.

Jane and Jonathan Hellyer Jones were married in July 1997. Earlier that year, in March, Lucy had told the family that she and her fiancé, Alex Mackenzie Smith, a member of the United Nations Peace Corps in Bosnia, were expecting a child. They planned to live together in London, and were married later that spring. They named Hawking's first grandchild William, which is of course Stephen's middle name.[15]

Censorship on a Cosmic Level

It was time to concede another bet. The background of this one began in 1970, when Hawking was first thinking about the light rays at the event horizon of a black hole and what would happen if they approached each other, collided, and fell into the black hole. The question was, could a black hole ever end up with *no* horizon, with the singularity left 'naked', exposed to view? Roger Penrose had proposed a 'cosmic censorship conjecture' – that a singularity would always be clothed inside a horizon. The discussion had gone on through the years, and Hawking had bet Kip Thorne and John Preskill (also of Caltech) that Penrose was right. The loser would reward the winner or winners with 'clothing to cover the winner's nakedness, to be embroidered with a suitable concessionary message'. Since 1991, when the bet was made and signed, Demetrios Christodoulou of Princeton, using computer simulations by Matthew Choptuik at the University of Texas, had done some theoretical calculations which suggested that a singularity without an event horizon might be created under very unlikely special circumstances such as a collapsing black hole. The situation was about as likely as balancing a pencil upright on its sharpened tip, said Choptuik, but it was theoretically not impossible.

Hawking conceded the bet at a public lecture in California in 1997. The message in the 'embroidery' on the T-shirts he gave Thorne and Preskill indicated that although a naked singularity *could* happen, it probably wouldn't – or shouldn't! The cartoon image showed a shapely woman barely concealing her nakedness behind a towel with the words 'Nature Abhors a Naked Singularity'. When Christodoulou redid his calculations, he found that Hawking's concession had probably been premature, so a new wager was made. This time it was clearly spelled out that the singularity would have to happen without any unlikely special conditions and the loser's message on the garment would have to be unambiguously concessionary. Preskill had meanwhile made the remark that we do know of one naked singularity, the Big Bang.[16]

Hawking made a particularly memorable journey in 1997. He, with Kip Thorne and several of their colleagues, visited Antarctica. Photographs show him bundled up in his wheelchair against a background of ice and snow. He didn't actually get to the South Pole, however, and he's never been to the North Pole, which means he can't say he has *personally observed* that there are no boundaries there.

Speeding Up!

In January 1998, at a meeting of the American Astronomical Society, a young astronomer named Saul Perlmutter made an announcement that rivalled in significance Hubble's discovery that the universe is expanding: the expansion of the universe is speeding up! Cosmologists' jaws dropped. The media soon got the news out that something stunning had turned up, something that ran completely counter to all expectations. In the afterword to the new 2010 edition of his book with Roger Penrose, *The Nature of Space and Time*, Hawking exclaimed about the excitement and importance of this startling development.

Two teams of astronomers made the discovery independently: Perlmutter and his Supernova Cosmology Project of the Lawrence Berkeley National Laboratory in California had been studying supernovae to find out whether the expansion of the universe was slowing down.

They had discovered quite the opposite. It was difficult to believe they were not mistaken, but in March another research group, this one led by Brian Schmidt of the Mount Stromlo and Siding Spring Observatory in Australia, reported similar findings.

Inflation theory was predicting a flat universe, while this new data seemed to be hinting that we might instead have an open universe (Friedmann's second model, see Figure 6.1, p. 110); but another implication of the discovery wasn't so bad for inflation theory. What Perlmutter had found could be taken as the first strong observational evidence that there is a repulsive force operating in the universe, that the type of antigravity acceleration which inflation theory had suggested really does exist. The universe is getting an antigravity boost from somewhere.

Were we seeing evidence of the cosmological constant Einstein had put into his equations of general relativity when he didn't believe what they were implying? He had, of course, taken it out again. Now Perlmutter suggested that there is, after all, a small positive cosmological constant, and Hawking and many others agreed that was the simplest explanation.[17] But there was some unease with this conclusion. Things might not be quite that simple. Perhaps there is a more exotic antigravity stress in the universe. There was talk of a mysterious 'quintessence' (named after a fifth element suggested by Aristotle).

'Dark energy' entered the physics vocabulary to describe the mysterious energy source. And energy, we know from Einstein's most familiar equation, has an equivalence with matter. One suggestion was that, added to ordinary matter and 'dark matter' (whose makeup is still mysterious but whose presence is well confirmed), dark energy might actually produce precisely the flat universe that inflation theory was predicting. In the afterword to the 2010 edition of *The Nature of Space and Time*, Hawking suggested that the presence of enough of this energy might even produce the positive curvature necessary for the closed universe consistent with the original no-boundary proposal.[18] However, in 1998, Hawking was beginning to think he might take another look at that proposal in the light of this unexpected discovery.

Nearing the millennium

In 1998, President Bill Clinton announced the Millennium Evening series – eight lectures and cultural showcases to be hosted by the White House and carried live over the internet – and invited Hawking to be one of the lecturers. His lecture 'Imagination and Change: Science in the Next Millennium', was the second event in the series, on 6 March. He used the opportunity to warn about what he saw as serious dangers – overpopulation and unchecked energy consumption. Hawking thought there was a possibility that we will destroy all life on Earth or 'descend into a state of brutalism and barbarity'. He also expressed his grave doubts that any laws or bans would stop all attempts to redesign human DNA in the next millennium. No matter that most people would approve a legal ban on human genetic engineering, that would not prevent someone somewhere from doing it. Perhaps because he didn't want to leave his listeners completely devastated, he ended in a more upbeat mood: 'I'm an optimist, I think we have a good chance of avoiding both Armageddon and a new Dark Age.'[19]

America seemed to bring out both a darker and a more

light-hearted side of Hawking. Back in California again in 1999, he flew from Monterey, where he was staying, to Los Angeles to do the voice of himself in an episode of *The Simpsons*. This assignment was of truly vital importance to him, as witness the fact that when his wheelchair broke down two days before the flight, his graduate assistant Chris Burgoyne worked for a straight thirty-six hours to repair it in time. If Hawking couldn't rescue the entire human race from Armageddon, he could at least get to Los Angeles in time to 'Save Lisa's Brain'. The line best remembered from the episode was Hawking telling Homer that Homer's theory of a doughnut-shaped universe was 'interesting' and that 'I may have to steal it.' Hawking asked the producers whether it might be possible to create an 'action figure' of him. That went on the market and became a bestseller in toy stores. Hawking (again doing the voice-over) also came to the rescue of the whole universe in an episode of *Dilbert* in which a machine accidentally created a black hole. Dogbert kidnapped Hawking in order to get him to repair spacetime. Needless to say, Dr Hawking cured the universe. Hawking went in for repairs himself that year. Surgery to reroute his larynx to prevent food falling into his lungs made eating a less hazardous and more enjoyable activity.

In the late 1980s, when I first wrote about Stephen Hawking, it seemed inappropriate – indeed impossible – to describe details of the Hawkings' private lives that only a very few people knew. All of this hidden information became public knowledge in 1999 with the publication of Jane Hawking's *Music to Move the Stars*. She was unsparingly frank about her memories of the physical and emotional turmoil of their lives and about her relationship with Jonathan Hellyer Jones. It surprised hardly anyone that Jane's book caused a sensation in the media, but all the furore did seem to come as something of a shock to Jane. Hawking made no public comment except that he never read biographies about himself. His humour survived, as witness his reply to a reporter asking him about Jane's book, who also asked whether he had willed his DNA to science for cloning: 'I don't think anyone would want another copy of me.'[20]

Hawking had appeared in an opera and on television, and had a film made about him. He had not yet been a character in a play. When he received an advance script of Robin Hawdon's *God and Stephen Hawking*, his reaction was to ignore it in the hope it would go away and never reach the stage. When Hawdon added details from Jane's book, Hawking considered taking legal action but decided that would just attract more attention to a play that was 'stupid and worthless'.[21] God, the Pope, the Queen, Jane Hawking, Einstein and Newton were also characters. Lucy saw a performance and, watching her family portrayed on stage, found herself both 'horrified and mesmerized', fighting an 'insane urge to climb on the stage and join them'.[22]

Meanwhile, Hawking was broadening his horizons in a different direction at the behest of his son Tim, now a university student in Exeter. Unlike his older brother, Tim had not followed his father into physics but instead was studying French and Spanish as his mother had done. Tim managed to interest his father in Formula One racing and even enticed him to rock concerts. Hawking claimed he really enjoyed some of these, but he left one concert for which the tickets had been particularly difficult to obtain after only twenty minutes. This life-long Wagner fan (though his speech synthesizer insisted on pronouncing it 'Werner' or 'Wagoner') was also a discerning enough rock fan to know what he liked and what not – and vote with his wheels.

A Meeting of Theories – No Boundaries and Inflation

In the late 1990s, as the end of the millennium approached, Hawking's reputation among his colleagues and his public celebrity seemed secure, but his and Jim Hartle's no-boundary proposal was still controversial. It predicted a closed universe, the first of Friedmann's models (see Figure 6.1, p. 110). In that model, the universe would eventually collapse to a Big Crunch. The discovery that the expansion of the universe is speeding up, and better-informed estimates of the amount of matter and energy in the universe, were increasingly causing theorists in the late 1990s to doubt the possibility that ours is that kind of universe. It was even looking likely to be an 'open' universe, Friedmann's second model, which expands for ever.

At the same time, inflation theory was predicting that the universe is 'flat', Friedmann's third model, meaning that the amount of matter in it is just right, not too much and not too little, to cause the universe to expand only barely fast enough to avoid collapse. In 1995, Neil Turok, Martin Bucher and Alfred Goldhaber at the State University of New York Stony Brook had written a paper showing that inflation did not necessarily rule out an open universe, expanding for ever – but this was no immediate help for the no-boundary proposal.[23]

With his no-boundary proposal predicting a closed universe, inflation theory a flat or perhaps open one, and observations leaning towards an open one, Hawking began to consider the possibility of bringing these models into agreement. Neil Turok was a good friend, and over cups of tea one day after a Cambridge seminar on open inflation the two began sharing ideas.

The upshot was a model in which a particle of space and time resembling an extremely small, slightly irregular, wrinkled sphere in four dimensions would automatically inflate itself into an infinite, open universe.[24] Because this particle would have lasted for only an instant before undergoing inflation, just a flash in the pan, so to speak, Hawking and Turok named it the 'instanton'. But 'pea' was the name that caught on with the public, because the two theorists announced that though it was unimaginably tinier than a pea, it had the mass of a pea (about one gram). The pea image was also useful because a pea is round, which goes nicely with the rounded 'origin' of the universe, where time was like a fourth dimension of space, in the no-boundary proposal. A pea is not a singularity. Not a point of infinite density. In Neil Turok's words:

> Think of inflation as being the dynamite that produced the big bang. Our instanton is a sort of self-lighting fuse that ignites inflation. To have our instanton, you have to have gravity, matter, space and time. Take any one ingredient away, and our instanton doesn't exist. But if you have an instanton, it will instantly turn into an inflating, infinite universe.[25]

Nothing existed 'outside' the instanton, and nothing

existed 'before' it. In both time and space, it was all there was. However, claims in the media that this theory showed how the universe sprang into being from nothing were far from the mark. It sprang into being from a 'combination of gravity, space, time, and matter packed into a rounded minuscule object'.[26]

It was a good try, bringing together inflation theory, the no-boundary condition and observational evidence, but it was not an immediate hit with Hawking's and Turok's colleagues. One rather embarrassing problem was that many of the possible universes the model predicted had no matter in them at all. That, however, was not too difficult for Hawking to remedy. One could use the anthropic principle and say that, in fact, only one of the possible universes had to support the existence of intelligent life forms.

Criticism came from those who found the no-boundary proposal still too controversial to be used dependably as part of a meaningful theory. Others thought Hawking and Turok were relying too heavily on the anthropic principle. Andrei Linde was very critical. To his way of thinking, a model with universes that included *at best* only about one-thirtieth the density of matter currently observed in our universe was unacceptable. Not even the anthropic principle could save it. Hawking and Turok responded: so far they had been working with only a very simple model. A more realistic model would yield better results.

Because of public interest in any news having to do with Hawking, the media on both sides of the Atlantic picked up the disagreement and treated it as though the theoretical-physics world were staging its own Clash of the Titans. 'Give Peas a Chance!'[27] urged *Astronomy Magazine*, while *Science* showed more restraint with 'Inflation Confronts an Open Universe'.[28] The online news service at Stanford, where Linde was a professor, announced it like a prize fight: 'Hawking, Linde Spar over Birth of the Universe'.[29] Hawking's name headed that title, even though Linde was now a local boy, but perhaps *Stanford Report Online* had decided that no one could be faulted for sticking to alphabetical order.

If Linde had once found Hawking intimidating, that was no longer the case. He called Hawking 'an extremely talented person',[30] 'a very brilliant man',[31] but also commented that Hawking's faith in mathematics bordered on a religion,[32] and that 'Sometimes – this is my interpretation – he trusts mathematics so much that he makes calculations first and interprets them later.'[33] 'You need to make sure that you are applying the mathematics correctly. In this case, my intuition tells me that he has not done so.'[34] In one highly complimentary statement, however, Linde echoed what many others have said about Hawking: 'I consider Stephen my friend and I hope we will remain friends after this is over. A number of times he has come up with surprising conclusions that at first seem like they are wrong. But in several instances he turned out to be right. In other cases, he was wrong. We will just have to wait and see which it is this time.'[35]

The article in the Stanford online news was triggered by a seminar Hawking gave at Stanford in April, at Linde's invitation. The prospect of witnessing these two men 'sparring' in person brought out a mammoth crowd. Hawking also debated with Linde and Alexander Vilenkin the next November in Monterey, California, this time defending his employment of the anthropic principle with a confidence that foreshadowed the much stronger use he would make of it in the future: 'The universe we live in didn't collapse early on or become almost empty. So we have to take account of the anthropic principle: if the universe hadn't been suitable for our existence, we wouldn't be asking why it is the way it is.'[36] When it came to Hawking's and Turok's proposal, the jury would be out a long time.

16

'It seems clear to me'

IN DECEMBER 1999, ON THE EVE OF THE MILLENNIUM, CNN's Larry King came in person to Cambridge to the DAMTP to film an interview with Hawking, who carefully programmed the answers to prearranged questions into his computer to avoid any delay in replying. The interview was broadcast on Christmas Day. When Larry King asked him how he planned to celebrate this special New Year, Hawking replied that he was going to a *Simpsons*-character costume party, dressed as himself. This would not require a costume.[1]

The millennium found Hawking playing in the big leagues. He joined international luminaries – among them Archbishop Desmond Tutu – in signing a 'Charter for the Third Millennium on Disability'. In May 2000 he took issue with Prince Charles for opposing genetically modified food, and in August he recorded a televised tribute to US presidential candidate Al Gore, to be shown at the Democratic National Convention in the United States.

His speaking engagements for audiences numbering in the thousands were continuing to take him worldwide, to South Korea, to Mumbai and Delhi in India, to Granada in Spain. He delivered a lecture at Caltech for Kip Thorne's sixtieth-birthday 'KipFest' in June 2000, and another back home in Cambridge that raised thousands of pounds for the Newnham Croft Primary School to build a new extension. In the summer of 2001 a new documentary, *The Real Stephen Hawking*, aired on BBC4. Hawking called public attention to the need for new technology for the disabled by agreeing to advertise the Quantum Jazzy 1400 Wheelchair. 'It will keep my nurses fit as they try to keep up,' he quipped, and that was no exaggeration.

'Try to keep up' they did. A trip abroad was, and is, like an army on the move: Hawking, his graduate assistant, a medical nurse, and two carers are the minimum. The amount of equipment and luggage that go with them has been prodigious ever since the 1980s, when he began travelling with the computer and apparatus with which he communicates. Joan Godwin, who oversaw the transportation of all this paraphernalia back and forth across the world many times, says it includes the usual suitcases for clothing, the vital suction equipment for keeping his breathing passages open, and an extremely heavy black bag containing everything that might be necessary in case of an emergency. His graduate assistant brings along tools and equipment and spare parts necessary for maintenance and emergency repairs to the wheelchair and computer. Not surprisingly these items sometimes get lost in transit.

Godwin recalls the crisis and emergency shopping trip when the suitcase with all Hawking's clothes got shoved aside by baggage handlers and never made it to the next stop, where he was scheduled to meet Bill Gates. On another occasion, when she wasn't along on the journey, she received a frantic phone call from his staff on the other side of the world. The suction equipment was locked in a car and the keys lost. What should they do?

Since the millennium, Hawking has been flying whenever possible by private jet. On a commercial flight, the wheelchair had to travel inside the plane to avoid a lengthy wait for him on the otherwise deserted plane until the baggage was brought out and the chair brought to him. It had even happened that the chair had its own first class seat. Security checks have been a perennial problem because the chair won't fit through the metal detector. Most of the time no one insists on searching him. It isn't only planes that are a challenge. Complicated special arrangements had to be made for the 'bullet train' in Japan to stop for an extra thirty seconds.

Not long after the millennium, some of the same press that excitedly described all this coming and going and Hawking's public appearances and honours seemed equally eager to report rumours that he was suffering mysterious physical abuse. When the Cambridge police initiated an investigation, Hawking resolutely refused to have any part in it. He dismissed the sometimes frantic concerns of staff and family, and firmly let it be known to police and all others that he wanted no interference in his and his wife's life. Intermittent investigations, rumours, and police interviews of colleagues, staff and family regarding an alleged series of 'unexplained injuries', some of them allegedly life-threatening, went on for five years until, in March 2004, the police dropped the investigation.

P-branes aren't So Dumb

The millennium and the years immediately following found Andrei Linde and his collaborators working to connect string theory with the multiverse of eternal inflation, while Hawking and some of his current and former graduate students were putting their heads together (when Hawking's head wasn't off in some remote part of the world) to bring the no-boundary proposal and a relatively new idea called 'brane' theory into line with one another. They were also continuing to look at black holes in the light of brane theory.

The name 'p-brane' was coined by Peter Townsend, one of Hawking's colleagues in the DAMTP who did fundamental work on these theoretical oddities. The 'p' in p-brane can be any number, the number indicating how many dimensions the 'brane' has. If p = 1, that's a 1-brane. It has one dimension, length. It's a string. If p = 2, that's a 2-brane. It has two dimensions, length and width. It's a sheet or membrane. Continuing in this vein, we might decide that the infamous, deadly 'gelatinous cube' of some adventure games must be a 3-brane – though that isn't part of the theory. Higher numbers for p are possible as well. They are more difficult to picture. The scheme is reminiscent of an idea that had emerged in the fifth century BC with the Pythagoreans and was later picked up by Plato, that the world is created in a progression from point to line to surface to solid. Plato speculated that there might be more dimensions to the progression after that, but these were all that were needed to have the world we know.[2] Modern p-brane theorists are far less restrained when it comes to considering those extra dimensions.

P-branes can absorb and emit particles in the same way black holes do. At least for certain types of black holes, the p-brane model predicts the same emission rate that Hawking's virtual particle-pair model predicts.

The p-branes provide a sort of storage facility for information that falls into a black hole, but they might do better than that: the information *emerges* eventually in the radiation from the p-branes. Considering this possibility, Hawking was thinking again about the implication of the Heisenberg uncertainty principle that every region of space must be full of tiny black holes that appear and disappear as rapidly as the particle pairs in Hawking radiation. These little chaps gobble up particles and information. Of course, they are a hundred trillion times smaller than an atomic nucleus. Nibble might be a better word, which is why, said Hawking, the laws of physics still appear deterministic for everyday intents and purposes. That does not, however, mean the loss of information is any less serious. Could p-branes come to the rescue?

After Hawking's 1981 announcement that information is lost from the universe in black holes, the controversy about the information paradox had continued, though – surprisingly, given the significance of the issue for physics – not at a level that appeared to disturb Hawking or get a rise out of him as his thoughts turned to other matters. Some wondered whether he was simply being stubborn, choosing to ignore interesting arguments that contradicted his own ideas, refusing to move ahead and engage in the discussion. A few feared that he was simply not well enough to respond in a forceful and meaningful way. To him it seemed he had delivered the unassailable verdict, as unfortunate as that verdict was, and none of the arguments he was hearing were significant enough to pull him into the fray.

However, the boss had not been asleep. He knew of Susskind's proposal in 1993; and the p-brane suggestion for solving the information paradox, coming in 1996 from physicists Andrew Strominger and Cumrun Vafa, had definitely caught his attention. Hawking and his colleagues considered that solution to the information loss problem, and some had hopes for it. The giant stirred in his lair a little, but then turned over and remained unconvinced. Hawking stuck to his guns. Information is irretrievably lost in black holes. P-branes – though interesting for other reasons – were not the answer. Nevertheless, when Hawking closed a lecture with the words, 'The future of the universe is not completely determined by the laws of science and its present state as Laplace thought, God still has a few tricks up his sleeve', and the last slide was a drawing showing an elderly, bearded figure with an enigmatic smile, slipping playing cards into the sleeve of his robe, one wondered . . . were those cards 2-branes?

A Prehistory of the 'Nutshell'

The first I saw or heard of *The Universe in a Nutshell* was as a bundle of typewritten pages, sent to me by Ann Harris, Hawking's editor at Bantam in New York in the summer of 2000. There were print-outs of public and scientific lectures and papers, most of them recent, some easy to understand, some full of equations and the language of physics, repetitious of one another in places and occasionally of previous Hawking books – not resembling a coherent book at all. Here were string theory, M-theory, imaginary time, sums-over-histories, the information paradox, the holographic principle, extra dimensions, not to mention summaries of more basic subjects such as quantum mechanics, general relativity, black holes and the Big Bang. It was an awful lot of material and it definitely was not in a nutshell . . . yet. Could it possibly make a book, Ann Harris wanted to know. Stephen Hawking was one of the jewels in her and Bantam's crown. It was unthinkable to send this back to him and say it couldn't be published.

I agreed to take a look at this material and soon found myself completely absorbed. For Ann I produced glosses in the margins: 'This is what he is saying . . .', or 'too technical', or 'understandable', or 'already said this on p. 33', and I created a map showing how the various pieces, and pieces of pieces, could be rearranged and linked in such a way that they would be chapters in a book. Ann Harris took me on board with the brief of 'helping Hawking make the book simpler so that ordinary people can understand it'. The 'ordinary people' was no problem. I was one of them myself.

As it turned out, Stephen Hawking was ahead of both Ann Harris and myself. He had a perfectly good plan of his own for how this book would fit together. The non-linear organization was intentional. There would be separate chapters, each on its own topic, that could be read in any order after mastering a small core of essential material. My task became to help smooth out the un-evenness in the difficulty levels of the chapters, to point out where some of them needed work to bring them to a level accessible to intelligent general readers. Hawking decided to use the opportunity to stress again some of his opinions on controversial issues beyond the realm of his science. After several months communicating by e-mail, we worked together for two weeks in his office in Cambridge. Also on board was a fantastic illustrator, Philip Dunn, from Book Laboratory and Moonrunner Design. At first I worried about the scientific accuracy of some of his drawings, but, in the end, his pictures made *Nutshell* the most innovatively illustrated of all Hawking's books.

Not Silver Street

By the year 2000, there *had* been a dramatic change for the DAMTP. Though the new, ultra-modern complex called the Centre for Mathematical Sciences would not be complete until 2002, the DAMTP had already moved there from the old building in Silver Street when I came to work with Hawking on *The Universe in a Nutshell.* Instead of crossing the Backs and the River Cam to reach the new complex from where I was staying at Clare Hall, I walked in the opposite direction, away from central, ancient Cambridge, through an upscale residential section of town, in the direction of the New Cavendish Labs but not quite so far. The new Centre for Mathematical Sciences was still a work in progress but it already looked like something out of *Star Trek*, except for the 'green' (literally – turf-covered) roof of one section. No longer could I walk straight in. I had to be fetched from Reception by Hawking's personal assistant, whom I hadn't met before, Karen Simes.

Hawking's new office was a vast improvement over the old one in Silver Street. It was spacious, carpeted, modern, light-filled from two walls of windows. A 'corner office'. There was room not only for his desk, computers, bookcases and a blackboard, but also for a plush sofa, chairs and a coffee table for guests, all in muted designer colours. Marilyn Monroe surveyed the scene from a framed, pastel portrait, much more upmarket than the one I remembered hanging in Silver Street. Most of the photographs on the desk and shelves were of William, Lucy's little boy, Hawking's grandchild. The windows overlooked lawns two storeys below and the older, well-to-do residential area surrounding the complex. This view was not to last, I was told, for the Centre was still being built and would soon include another 'pavilion' a few yards outside the windows. Nevertheless, the outlook was a considerable improvement over the old car park and the blank brick wall.

It seemed a serious loss, however, that the only room in the new complex that was vaguely comparable to the common room in Silver Street was a large dining facility a good distance down corridors and across outdoor ramps and bridges from Hawking's office. Its size and remoteness meant that it didn't lend itself as the old shabby common room had to writing equations on tables and having impromptu informal seminars over tea or coffee at 4 p.m. That situation would have to be remedied.

The building was nevertheless impressive in its technical marvels. Window blinds ground up and down without human intervention in response to the intensity of the light. They also went down at night because of complaints from neighbouring homes that the size and brightness of this ultra-modern monstrosity gave the impression that we need no longer wonder whether extraterrestrial life exists . . . it had landed, in force, across the street. The building also 'breathed' occasionally; papers fluttered as air was drawn automatically through vents, doors and windows.

In spite of all this modernity and innovation, the clicking sounds of the little box Hawking held in his hand, the flitting words on the screen, the synthetic voice, were exactly as I remembered them. Some of the nurses were familiar faces.

Though I had been communicating with Hawking by e-mail for several months before coming back to Cambridge, it was a relief to find, in person, that neither the move to the new building nor his continuing world travels and celebrity, nor becoming a grandfather, nor changes in his personal life, had kept him from going right ahead with the work he loved and to which he had devoted so many years of unexpected and triumphant survival. The boss was still awake. Our conversations were not, of course, like ordinary conversation. Using his hand-held device, he would send the cursor on the screen off on a hell-for-leather pursuit of each single word across the half-screens and lines full of words, and finally attempt to capture the word itself, often missing it, and the process would start over again. I knew I must resist the temptation to complete sentences for him, even when I could tell what he was going to say. That would be impolite, and he might very well go right ahead and finish the sentence anyway. So I waited and watched, and silently rooted for the cursor to capture the word. 'Come on! . . . there! . . . got it . . . oh no!' I found myself pumping my fist and stopped. Were there any expletives programmed into his phrase lists? I don't know. After only a few minutes of frustration at the beginning of the two weeks, I calmed down. It was necessary to take his cue, be patient and allow it all to happen his way. He wasn't frustrated – or maybe he was, but he couldn't show it.

Our work together on *The Universe in a Nutshell* consisted in the main of my pointing out paragraphs, sentences and sometimes larger pieces of the manuscript that I thought needed to be stated in simpler language. I had prepared alternative wording, but in all instances, though he listened to my suggestions, he insisted on making the changes in his own words. Sometimes my comment, 'I think that sentence is too difficult, Stephen', would bring forth the vintage Hawking smile, a storm of clicking and flickering words and the reply, 'It seems clear to me.' But he would set to work remedying the problem, laboriously translating the language of theoretical physics into the language of 'ordinary people'. Only occasionally, when the translation still didn't seem simple enough, did I resort to saying, 'I'm sorry, but *I* can't understand that', even if I thought maybe I could. He would reply, 'I'll make it simpler then', and he did.

One of the most interesting suggestions he was making in *The Universe in a Nutshell* was that we may live on a four-dimensional surface within a higher dimensional spacetime. Such a surface had been dubbed a 'brane world'.

If we lived in such a situation, everything in our own four-dimensional brane world, what we normally call 'the universe' – matter and light, for instance – would behave just the way we have found it does in the universe we know, with the exception of gravity. Gravity (thought of as the curved spacetime of general relativity) would spread throughout the *higher* dimensional spacetime, and because this was so, we would find gravity behaving oddly. For one thing, it would grow weaker with distance more rapidly than we have experienced it doing.

Here there was a hitch, for if the gravitational force fell off more rapidly with distance, the planets couldn't orbit as they do. They would fall into the sun or escape to interstellar space. We don't find that happening. However, suppose the extra dimensions didn't spread too far but ended on another brane world fairly near our home brane world – a shadow brane world that we couldn't see because light, as we've said, would be confined to its own brane world and not spread through the space in between brane worlds. It might be only a millimetre away from us but undetectable because that millimetre is measured in some extra spatial dimension. Imagine an analogy in a two-dimensional world: there are insects on a piece of paper, with another piece of paper hovering near it and parallel. The insects are unaware of the other piece of paper because they can't conceive of a third dimension of space. They know only the two dimensions of their piece of paper. If the extra dimensions ended on such a shadow brane world, then for distances *larger* than the separation between the brane worlds, gravity would not be able to spread out freely after all. Just like the other forces of nature, we would find it effectively confined to our home brane world, and it would fall off with distance just as we expect it to – at the right rate for planetary orbits.

However, there would be tell-tale clues. For distances *less* than the separation between the brane worlds, gravity would vary more rapidly, and these variations ought to show up in measurements of the very small gravitational effect between heavy objects placed extremely short distances apart.

There are other interesting implications: a nearby 'shadow' brane world would be invisible to us, because light from that brane world could not spread to ours, but we would feel and observe the gravitational effects of matter on that neighbouring brane. Those effects would be mysterious to us, because they would appear to be produced by sources that we cannot detect at all except through their gravity.

Is that the explanation behind the 'missing mass' and 'dark matter' mystery in astrophysics? For stars, galaxies and galaxy clusters to be situated as they are and to move as they do, there has got to be much more matter in the universe than we can observe in any part of the electromagnetic spectrum. Could it be we are observing the gravitational influence of matter in other brane worlds?

There are other brane world models besides the one involving shadow branes, and speculation about the implications of the models has extended to many subjects which are of enormous interest to Hawking, such as black holes, radiation at the event horizon, black hole evaporation, gravity waves, the relative weakness of gravity in comparison with the other forces of nature, the origin of the universe and its history in imaginary time, inflation theory, the Planck length, and the no-boundary proposal.

How does the no-boundary proposal look when viewed through brane world glasses?

Our home brane world would have a history in imaginary time that was like a four-dimensional sphere, that is, like the surface of the Earth but with two more dimensions. So far – if you have read the previous chapters of this book – that ought to sound familiar. The difference is that in the original no-boundary proposal there was nothing 'inside' the expanding sphere, the 'globe of the world' that Hawking asked us to picture. In the new brane world version that is not the case. Inside the bubble is higher dimensional space, and the volume of that space, as might be expected, increases as the brane world expands.

In the chronological time we experience, our home brane world would expand with an inflationary phase like that described in inflation theory. The most probable scenario would be one in which it expanded for ever at the inflationary rate, never allowing stars and galaxies to form. But we could not exist in such a brane world, and we do obviously exist. So the anthropic principle compels us to find out whether the brane world model offers other less likely but not impossible scenarios. It does. We find that there are imaginary time histories that could correspond to real time behaviour in which the brane world had a phase of accelerating inflationary expansion only at first but then slowed down. After that, galaxies could have formed and intelligent life evolved. That sounds more familiar.

The most mind-boggling suggestion having to do with branes had been inspired by our knowledge of holography. Recall Leonard Susskind's suggestion about the way holography might apply to black holes. In holography, information about what happens in a region of spacetime can be encoded on its boundary. Hawking leaves us with the question of whether we might only *think* we live in a four-dimensional world because we are shadows cast on the brane by what is happening in the interior of the bubble.

The work on *Nutshell* would continue later for several months by e-mail, fine-tuning the editing, but, by and large, those two weeks when I was in Cambridge got the job done. It was exciting work, but it was also tense work. As I exited the car park of the Centre for Mathematical Studies on the final evening, I gave a silent cheer and did, finally, allow myself to pump my fist. We had done it! I had survived. So had Stephen Hawking.

Dinner at Caius

On one cold November evening during those two Cambridge weeks, I rode with Stephen Hawking in his van to Caius for dinner. The van stopped where King's Parade turns into Trinity Street, between the Senate House and Great St Mary's Church and across from Gonville and Caius College. The nurse who had driven us from the Centre for Mathematical Studies set the brakes, left the headlights burning, and came around to the passenger side to release the heavy harnesses that secured Hawking and his wheelchair where the passenger seat would usually be. I, the only back-seat passenger, moved out of her way and waited in the street, since freeing the wheelchair was heavy work and required space to manoeuvre. The ubiquitous Cambridge cyclists dodged with lightning reflexes around me and the van. Soon they were also having to dodge the metal ramp that protruded from the door and allowed Hawking, bundled up against a cold wind, to guide his chair smoothly down on to the pavement.

He made his way at a slow majestic pace through the Gonville and Caius gate and across three courtyards to the doorway that led to the Hall. After all those years, and all his success in the cause of disabled access, there was still no convenient way for Hawking to get to the Senior Common Room and the Hall of his college. There was space on the tiny lift only for him and his nurse. He instructed me and my husband, who had met us at the gate, how to find our way up by another route. We rejoined him as he threaded his way through the kitchens and other rooms that are not on the tourist circuit, although much of Caius is ancient and beautiful. In the richly panelled Senior Common Room a fire was burning, and Fellows of the college greeted Hawking in a manner that indicated they knew him well and had long ago stopped being surprised, shocked or impressed by his disability or his accomplishments. Some of them have equal claims to fame in their own academic fields, though not so much international notoriety. He is to them simply Stephen.

After sherry, everyone proceeded to the Hall and took places at High Table, a step above the level of the long tables crowded with noisy undergraduates – for Caius is a college that still makes a serious commitment to dining in Hall. Graduate students of the college ate somewhat more calmly in the minstrels' gallery, sans minstrels. Amidst the pleasant clink of forks and knives on college china, the rumble of young voices, occasional outbursts of shouts and boisterous laughter, and the more well-modulated voices of the Fellows, we ate and drank a fine wine from the college cellars. Stephen Hawking's nurse wrapped his huge bib around his chest and fed him, while he clicked on his hand-held device and, through his computer, discussed international politics with my husband.

The Hall is hung with portraits of eminent fellows of Caius. Near the centre, prominently displayed, is a modern painting of Hawking. For centuries, men (and, more recently, women) have gone from this college and from those long, rowdy tables to teach, to continue research, to make money, to change the world. We ate, as they had, with this odd amalgam of the new, the antique, the remarkable, the ordinary, the wet-behind-the-ears, the venerable. The occasion was a little like dining at summer camp in a room that just happened to be hundreds of years old and incongruously beautiful. In this generation, Caius has taken in its stride one of the most extraordinarily different men of our time – whom everyone there seems to regard as just another camper.

During the two weeks I spent in Hawking's office in the autumn of 2000, working with him on *The Universe in a Nutshell*, I often would arrive at the DAMTP before he did and wait at the curved modern desk between his office door and the bridge from the lift in this supermodern 'pod'. Each day when the lift doors opened and his chair emerged, it seemed to me there was a tiny but profound alteration in the feel of reality, requiring a slight readjustment on my part. Even after I had learned to expect it, I couldn't escape the impression that a piece of another world – alien to me because of superior intellect and disability, but also for a quality of will that I have never experienced anywhere else – was moving in slow and inexorable progression through our space and time, from the little bridge to the office door, almost running over my toes.

The mechanical voice would say, 'Good morning', or inquire, 'How are you?', or something like that . . . and Hawking's working day would begin.

PART IV
2000–2011

17

'An expanding horizon of possibilities'

IN A STATEMENT IN JANUARY 2000, PART OF AN INTERVIEW for the millennium focusing on his predictions about what the future held in store for the human race, Hawking summed up his thoughts on the issue of genetic engineering. Humans, he said, have had no significant changes in their DNA over ten thousand years. But soon they will no longer have to wait for biological evolution to make changes for them – and they won't wait. In the next thousand years we will probably be able to redesign our DNA completely, increasing our brain size. Whatever bans are placed on genetic engineering in humans, it will surely be allowed on animals and plants for economic reasons, 'and someone is bound to try it on humans unless we have a totalitarian world order. Someone will improve humans somewhere. I am not advocating human genetic engineering. I am just saying it is likely to happen and we should consider how to deal with it.'[1]

A year and a half later, he had changed his mind. In an interview given shortly before September 11, he told the German periodical *Focus* that humans would be well advised to modify their DNA in order to avoid being outdistanced by super-intelligent computers that would eventually rule the world.[2] Are computers really likely to become that intelligent? Hawking had previously commented that computers are 'less complex than the brain of an earthworm, a species not known for its intellectual powers'.[3] But he thought that 'if very complicated chemical molecules can operate in humans to make them intelligent, then equally complicated electronic circuits can also make computers act in an intelligent way'.[4] Intelligent computers will then design even more intelligent and complex computers.[5]

His new position was controversial, but it was largely forgotten in the wake of the 9/11 attacks. The interviews that inevitably followed – for the media felt confident that Stephen Hawking would have wise words to offer on other topics besides physics – gave him the opportunity to introduce another issue that he had been considering. He told an interviewer from the *Guardian*: 'although September 11 was horrible, it did not threaten the survival of the entire human race. The danger is that either by accident or design we may create a virus that destroys us.'[6]

Hawking advised that plans be put in place as soon as possible with the long-range goal of colonizing space to ensure the survival of the human race. This was not an off-the-cuff notion, soon to be forgotten. Already in the millennial interview he had predicted a manned, 'or should I say personned', flight to Mars within the century. But that would be only the first stop. Mars isn't suitable for human habitation. We need either to learn to live in space stations or travel to the next star, and that journey he was sure would not happen within the century. Because we can't travel faster than light (no matter what science fiction claims), the trip would be slow, boring and arduous. Hawking would return to the advisability for humans to colonize space in the children's books he would write with his daughter Lucy a few years later. This need seemed urgent enough for him to try to ingrain it in the minds of children who would be setting the agenda in the future.

Not everyone appreciated pronouncements like these about issues beyond his own field of expertise. Most of Hawking's critics called him not 'wrong' but 'naïve'. Sir Brian Pippard, an eminent twentieth-century physicist, once apologized for himself and his colleagues for being 'inclined to believe that their expertise absolves them from the duty of studying other branches of knowledge before contributing their own penn'orth of wisdom'.[7] Hawking could perhaps be declared guilty, but he knew he had a golden opportunity to reach the public with ideas he strongly believes are important and ought to be heard. He possibly had enough clout to influence public policy.

The TOE Revisited

In his Lucasian inaugural lecture in 1980, Hawking had announced that the most promising candidate to unify the forces and particles was N=8 supergravity. In 1990 he had told me he suspected it might turn out to be superstrings, with his no-boundary proposal answering the question about the boundary conditions of the universe. Now, the close of the millennium had come and gone. Still no end to theoretical physics. Still no Theory of Everything. In April 2002, Hawking told a reporter that 'I still think there's a 50-50 chance that we will find a complete unified theory in the next twenty years'[8] – a much more modest and tentative forecast than he had made in his Lucasian Lecture.

As the months passed, Hawking hedged his bets even more. He was reconsidering one of the main thrusts of his scientific career, beginning to suspect that the fundamental, unified theory – if it exists at all – is at a level that will never be accessible to us. Our understanding will always resemble a patchwork quilt, with different theories holding in different regions, agreeing only in certain overlapping areas. If this is the case, then it would be ill advised to view what seem like inconsistencies among the theories as a sign that they are weak or incorrect. What we would be able to discover about the universe would inevitably be something like a jigsaw puzzle in which it is not so difficult 'to identify and fit together the pieces around the edges' – supergravity and the different string theories – but in which we will never be able to 'have much idea what happens in the middle'.[9] In a lecture for the Paul Dirac Centennial Celebration at Cambridge, in July 2002, he said, 'Some people will be very disappointed if there is not an ultimate theory that can be formulated as a finite number of principles. I used to belong to that camp, but I have changed my mind.'[10]

Hawking asked his audience to recall the Austrian mathematician Kurt Gödel, who in 1931 had shown that mathematics was 'incomplete' because in any mathematical system complex enough to include the addition and multiplication of whole numbers, there are propositions that can be stated – and that we can even *see* are true – but that cannot be proved or disproved mathematically within the system. Hawking thought that it might also be the case in physics that there are things that are true but cannot be proved. Kip Thorne had spoken about the change in Hawking's way of working, from an insistence on rigorous mathematical proof to a quest not for certainty but for 'high probability and rapid movement towards the ultimate goal of understanding the nature of the universe'.[11] Hawking had made large intuitive leaps, expecting others to fill in the gaps he was leaving behind. Was he now about to become even more daring, warning his listeners that things he was sure are true would be impossible to prove? No, even he had to join the human race at the brink of a chasm that no one could cross. Our theories are inconsistent or incomplete, he said, because 'we and our models are both part of the universe we are describing . . . Physical theories are self-referential.'[12]

The new and still elusive candidate he spoke about in his Dirac Centennial lecture was not 'an ultimate theory that can be formulated as a finite number of principles', but it might be the best we could ever do. It was M-theory. A particularly interesting version of that theory incorporated the brane theory that Townsend had suggested. Recall from our discussion of p-branes that when p = 1, that *is* a string. So strings could now be considered members of the larger clan that Townsend had named p-branes. Hawking was not throwing supergravity and string theory, his two earlier favourite candidates for the Theory of Everything, out of the window by any means. The five most promising superstring theories could be grouped in a family of theories that would also include supergravity. The superstring theories and supergravity were the pieces of the 'patchwork quilt', useful for considering different situations, but none of them applied in *all* situations. The unexpected network of relationships that physicists had found among them had led to the suspicion that these theories are all actually different expressions of a deeper, underlying theory – M-theory – which did not, as yet, *have* a single formulation. Hawking had begun to think it never would.

In the M-theory network of mathematical models, spacetime has a total of ten or eleven dimensions. These are usually taken to be nine or ten space dimensions and one time dimension. You may be wondering why no one ever thinks there might be more than one time dimension, so I should tell you that some versions of the theory do allow more time dimensions, as long as the total stays the same.

We, of course, experience only four dimensions. Where are the others? Hawking himself commented in 2001: 'I must say that personally I have been reluctant to believe in extra dimensions. But as I am a positivist, the question "Do extra dimensions really exist?" has no meaning. All one can ask is whether mathematical models with extra dimensions provide a good description of the universe.'[13]

The proposed answer to the question of why we don't see them is that the extras are rolled up very small. Think of a garden hose. We know that a garden hose has thickness to it, but from a distance it looks like a line with length but no other dimension. If the extra dimensions were 'rolled up' like that we would miss noticing them, not only on human scales, but even on atomic or nuclear physics scales.

Could we ever observe them? Suppose one or more of the extra dimensions is not so completely rolled up after all. That suggestion may be testable with a more advanced generation of particle accelerators or by measuring the gravitational force operating over extremely short distances.

Meanwhile, M-theory and extra dimensions had staked out a claim in the future of theoretical physics and cosmology. That future would be the theme of a conference held to celebrate Hawking's sixtieth birthday in 2002.

Sixty or Bust!

Hawking's sixtieth-birthday party almost didn't happen. He and his wheelchair ran into a wall a few days before the event. Hawking brushed this off in the opening of his lecture, 'Sixty Years in a Nutshell': 'It was nearly 59.97 years in a nutshell. I had an argument with a wall a few days after Christmas, and the wall won. But Addenbrooke's Hospital did a very good job of putting me back together again.'[14]

There had been a moment when those planning the birthday party stopped dead and held their breath, but then it was learned that Hawking was working on his own birthday speech in his hospital bed. The preparations continued. No one had to cancel, at the last minute, the Marilyn Monroe impersonator who would fawn over Hawking and croon 'I Want to be Loved By You' . . . or tell all the physics luminaries gathering from around the world that they could come and give their talks, but the man they were celebrating would not be present. The party happened. Hawking thought sixty was well worth celebrating. Many people, he told interviewers, don't welcome turning sixty, but for him it was an accomplishment. He had never expected to live so long.

It was a many-faceted celebration. A serious four-day 'festschrift' conference featured high-level papers presented by giants of theoretical physics and cosmology whose work touched on Hawking's. The public was admitted for a day of popular-level lectures. The real partying took place in the evening, attended by a crowd of as many as 200 guests. There was 'Marilyn', whom Hawking dubbed 'a model of the universe'. There was a choir of former and current graduate students and Hawking's first wife Jane, conducted by her husband, Jonathan Hellyer Jones, and accompanied by the U2 guitarist, Edge. Because Hawking's birthday fell close to the church Feast of St Stephen, they sang 'Good King Wenceslas' with new words written by Bernard Carr. 'Page and Hawking, forth they went' referred to Don Page. At one of the parties, in the Hall of Caius, Martin Rees – by this time Sir Martin Rees, Astronomer Royal – spoke glowingly about his old friend. A party in the Hall of Trinity College featured a sudden burst of colour and music as can-can dancers made a spectacular entrance. Television crews from Channel 4, the BBC, and America's CBS were present, and there was a live webcast to the BBC's website of Hawking's public lecture and the audience's rowdy, spontaneous, painfully off-key singing of 'Happy Birthday' that followed it – one of many renditions of it that week. The BBC later broadcast all the popular lectures as 'The Hawking Lectures'.

Hawking's colleagues took advantage of the opportunity to rib him:

Martin Rees: 'Astronomers are used to large numbers, but few are as huge as the odds I would have offered then [when Hawking was a graduate student at Cambridge] against witnessing this marvellous celebration.'[15]

Roger Penrose: 'I am very glad to note that Stephen has now also officially become an old man, so that he can also get away with saying such outrageous things. Of course Stephen has always done that kind of thing, but he can perhaps feel a little bolder in this even than before.'[16]

Bernard Carr: 'I've often suspected that there must be more than one Stephen Hawking to have made so many important discoveries. I would like to wish all of them a very happy sixtieth birthday!'[17]

Leonard Susskind: 'Stephen, as we all know, is by far the most stubborn and infuriating person in the universe.'[18]

Raphael Bousso: 'It is a pleasure to help celebrate Stephen Hawking's sixtieth birthday (not least because Stephen knows how to party).'[19]

Gary Gibbons, praising 'Stephen's indomitable courage and daring optimism',[20] quoted Robert Browning: 'Ah, but a man's reach should exceed his grasp or what's a heaven for?'

Michael Green recalled the early 1970s in Cambridge, when he first met Stephen, and cosmology was held in such low regard that 'it was considered a sub-branch of astrology and was not discussed at all!'[21]

Neil Turok spoke of Hawking's 'real "lust for life" that keeps him going against all odds'.[22]

Kip Thorne's birthday present was a promise that 'gravitational wave detectors – LIGO, GEO, VIRGO and LISA – will test your Golden-Age black-hole predictions, and they will begin to do so well before your seventieth birthday'.[23]

The papers prepared for the sixtieth-birthday conference gave a splendid summing up of where things stood in theoretical physics and cosmology in 2002, and how they had got there, and, as the conference title suggested, provided a springboard for the future. It brought together the finest minds in the world on the topics that most interested Hawking and impinged on his own work, and also to a remarkable extent brought together the grey eminences of the field with energetic younger people who were going to carry it into the future, many of whom had been Hawking's students. It all went on for a week . . . and why not? This was a birthday party no one for many of the sixty years had expected to attend. Elaine's birthday gift was a thirty-minute flight in a specially designed hot-air balloon. When Hawking had dreamed of such a flight at the time of his tracheotomy in 1985, he had taken it as a symbol of hope. At the age of sixty, it seemed that this hope had been amply fulfilled.

Hawking's colleagues and others who shared his birthday celebration were willing to accept his brushing off his wheelchair accident as something trivial, but it was in truth more serious than that. Wheeling along the uneven old pavement of Malting Lane near his home, accompanied by a nurse, he had lost control and crashed his chair into a wall, turning the chair over and breaking his hip. Neel Shearer, his graduate assistant, shrugged: 'He was late for an appointment and running on Hawking time, as ever.'[24] Hawking's weakened condition precluded the use of a general anaesthetic when the doctors repaired him. With only an epidural anaesthetic, Hawking experienced the whole procedure, something 'like hearing a Black and Decker drill'.[25]

Hawking's sixtieth-birthday year saw the publication of his carefully chosen compilation of excerpts from the writings of Copernicus, Galileo, Kepler, Newton and Einstein. *On the Shoulders of Giants* also featured biographical sketches of the five men and commentary by Hawking.

Unpacking the CMBR

As the new millennium got underway, a new generation of observers and observational instruments were gearing up to test the predictions made by inflation cosmology with unprecedented accuracy.[26] In a continued search for experimental evidence that might, or might not, support inflation predictions, attention focused, not surprisingly, on the CMBR, the afterglow of the Big Bang. George Smoot's discovery had shown that in the extraordinarily evenly dispersed microwave light, the temperature does vary from point to point. In 1998 (results released in 2000), balloon observations had measured the CMBR in detail in certain parts of the sky;[27] and in 2001 the ground-based Degree Angular Scale Interferometer at the South Pole took similar measurements.

Then, in June 2001, NASA launched WMAP, the Wilkinson Microwave Anisotropy Probe.* Its mission: to map the CMBR more precisely than had ever before been possible. It was capable of measuring temperature differences varying by only a millionth of a degree and, because

* WMAP was the result of a partnership between the Goddard Space Flight Center and Princeton University.

it was a satellite rather than a land-based instrument, take these measurements over the entire sky. The expectation was that WMAP would settle once and for all many of the arguments of the last few decades about the basic properties of the universe – its age, its shape, its expansion rate, its composition, its density. Different versions of inflation theory were telling slightly different stories about precisely how inflation happened and making predictions about what pattern of temperature variations we should find in the CMBR if we compare its temperature in different directions.[28] WMAP data were expected to give scientists ways to test these different scenarios.[29]

By February 2003, WMAP was brilliantly living up to its promise. Its data had allowed scientists to nail down precisely after many decades of controversy the age of the universe – 13.7 billion years – as well as the time in the universe's history when patterns in the CMBR froze into place – 380,000 years after the Big Bang. WMAP results showed that space is flat and supported those who were insisting that most of the energy in the universe today is 'dark energy'. WMAP measurements showed that the variations in temperature and density in the CMBR, observed across the sky – the variations that seeded the formation of galaxies – all had roughly the same amplitude regardless of their length, that all forms of energy had the same variation, and that the distribution of variations was random – just as predicted by the standard Big Bang inflationary model.[30]

Nevertheless, important issues remained unresolved by that first February 2003 release of WMAP data. One key piece of evidence was missing: inflation theory makes predictions about what the patterns and characteristics of gravitational waves originating from the Big Bang should be like as they show up in the CMBR. WMAP had not yet found these gravitational wave footprints. Nor was it determined whether the dark energy was due to 'vacuum energy' – the cosmological constant – or 'quintessence'. Interestingly, the observations that were fitting well with inflationary cosmology also fit a cyclical model where the universe expands from a big bang, eventually contracts again to a big crunch, and then reemerges in another big bang, in a cycle that keeps repeating itself – models which Neil Turok and Roger Penrose were favouring.[31]

Slowing down?

Hawking made yet another excursion into popular culture in the spring of 2003 when he agreed to take part in a routine on *Late Night with Conan O'Brien*. Comedian Jim Carrey led off the sketch by discussing cosmology. A mobile phone rang. The caller was Stephen Hawking telling Carrey not to bother: 'Their pea brains cannot possibly grasp the concept.' Having said that, he quickly excused himself. He couldn't stay on the phone because he was watching Carrey's film *Dumb and Dumber*, stunned by the 'pure genius of it'. Hawking's travel schedule that year had him criss-crossing the world, from a month at the Mitchell Institute for Fundamental Physics at Texas A&M to a meeting on cosmic inflation at the University of California-Davis, to Sweden to receive the Oskar Klein Medal from the Royal Swedish Academy of Sciences and to attend a Nobel Symposium on String Theory and Cosmology, back to America for a two-month visit at Caltech and the University of California-Santa Barbara, then over to Case Western Reserve University in Cleveland.

Unease among his physics colleagues that Stephen Hawking was slowing down seemed unfounded, given such a schedule, but they also feared he might be no longer at the height of his intellectual powers. In a poll taken among physicists at the end of the millennium asking who were the most influential physicists, Hawking's name was not anywhere near the top. His use of the little 'clicker' that responded to his faint hand pressure – the only way he could communicate with colleagues – was becoming more and more difficult and slow. One way to get around this frustrating situation was to solicit the help of a research student. This time Hawking chose a young man named Christophe Galfard. The procedure would be for Hawking to consider a problem and suggest possible approaches to Galfard, who would then concentrate on the mathematical details to find out whether Hawking's insights were correct and whether they led anywhere.

Galfard recalls that it took a while for him to get up to steam. Hawking's ideas were coming much more rapidly than he could deal with them. In words that are encouraging to anyone like myself who has read Hawking's papers and laboured to understand them, Galfard has described his chagrin as each sentence seemed to take about six months to decipher. He was lagging behind a half year and catching up was tough going.[32]

In the interest of moving things forward more rapidly, Galfard took a liberty that others had seldom taken – finishing sentences for Hawking when his intention was clear and he was struggling to select the words. In the past Hawking had often gone right ahead and completed the sentence, ignoring this second-guessing even when it was correct, but he allowed Galfard to expedite matters.

Galfard also used Hawking's ability to indicate 'yes' and 'no' with a small movement of his face, rather than wait for him to find the word on his screen. You wonder, seeing video clips of the two of them together, how Galfard avoided getting a bad crick in his neck as he bobbed back and forth between peering straight ahead at the computer screen and dipping forward to see Hawking's face.

Galfard had his work cut out for him. In 2003, a young Argentinian physicist named Juan Martin Maldacena, at the Institute for Advanced Studies in Princeton, finally gave Leonard Susskind's ideas for solving the information paradox a rigorous mathematical treatment that seemed to settle the issue in Susskind's favour.[33] At a conference in Santa Barbara, after-dinner speaker Jeff Harvey, instead of giving the expected talk, introduced a victory song. 'The Maldacena' was set to the music of a Latin dance that was popular in the mid-1990s, the 'Macarena'. Each short verse ended with 'Ehhh! Maldacena!'[34] The audience enthusiastically joined in the singing and dancing, celebrating the rescue of physics from the ogre of the information paradox. Susskind declared the war over. It should have come to an end much earlier, he insisted, but Hawking 'was like one of those unfortunate soldiers who wander in the jungle for years, not knowing that the hostilities have ended'.[35] Although a consensus was indeed growing that Hawking was wrong, Kip Thorne stuck with him. Hawking did not change his mind. Not yet.

He asked Galfard to study Maldacena's paper about the information paradox. Hawking had decided nothing would do but a frontal attack on this piece that was convincing people he was wrong. It was not an easy assignment. After messing with it for a year and a half, Galfard still found it difficult to decide whether information was or was not lost in black holes.

On 1 December 2003, Hawking was rushed to the hospital with pneumonia. On life support for several weeks, while others feared he was probably on his deathbed, he didn't waste the time. He spent it thinking about black holes, intent on finding a fresh approach to the information paradox. Recovery was slow, but after his discharge in the late winter of 2004 he and Galfard began again in earnest discussing the ideas those bedridden months had produced. After many gruellingly long days and late nights, including weekends when the work seemed never to stop,[36] Hawking finally felt he was ready to emerge from the quiet refuge of his office and do battle.

The Dublin Conference

That spring of 2004, Hawking gave a seminar in Cambridge to introduce some of his new ideas in a preliminary, sketchy fashion, and he let it be known that he wished to address his physics colleagues at a major conference. There was just such an event coming up in July in Dublin, the 17th International Conference on General Relativity and Gravitation. Hawking contacted the Chair of the Scientific Committee of the conference, Curt Cutner, requesting a slot on the programme with the words 'I have solved the black hole information problem and I want to talk about it.'[37] It was a big favour to request, for Hawking's paper was a late entry – participants had been asked to submit a title and abstract by 19 March, when he was barely out of the hospital. Also he would not be releasing a preprint of the paper.

Nevertheless, his stature in the field won him an hour-long slot on the conference schedule.

The media and fan frenzy his appearance generated showed that Hawking's superstar celebrity status had not waned. A public relations firm responsible for controlling access to the auditorium charged a reported £4,000, and it earned its fee as reporters and Hawking aficionados stormed the doors. Those fortunate enough to sport press badges soon lined the aisles inside, setting up their cameras and recording paraphernalia.

There was less excitement among Hawking's colleagues, rightfully in the hall as conference participants, as they wondered what to expect. Bursts of camera flashes followed Hawking as he made his smooth, stately, frozen progress along the aisle to the foot of a ramp that led to the stage of the grand concert hall of the Royal Dublin Society. Some believed he was about to deliver a defiant statement that would reiterate what he had been saying for more than twenty years, that information is lost in black holes. Others thought this fading genius would quietly concede the issue. Kip Thorne, who had continued through the years to agree with Hawking about the loss of information in black holes, John Preskill, who had not, Petros Florides, who chaired the conference, and Christophe Galfard were waiting on stage, facing an array of TV cameras ready to record the event. It was not a normal day at a physics conference.

Only Galfard and Thorne knew that Hawking was about to make one of his famous about-faces, but not in the way anyone was expecting. He was not going to concede to Susskind and Maldacena. Yes, the idea he had stubbornly stuck with for more than twenty years was wrong. But Susskind and Maldacena hadn't solved the problem. Hawking would do that himself. He had thought of another way to get round the information paradox.

The session began with an introduction by Petros Florides who quipped that while it is well known that no information can travel faster than the speed of light, this law seemed not to hold when it came to the speed at which the news of an upcoming Hawking appearance travels around the world.

It had become Hawking's custom to introduce his lectures with the question, in his calm, mechanical voice, 'Can you hear me?' Presumably if you couldn't, you wouldn't answer, so there was usually either an agreeable murmur or a cheer in reply. After that trademark beginning, he began by laying out the problem and tracing the history of the information paradox all the way back to the mid-1960s, when it was discovered that all information about a body collapsing to form a black hole was lost from the outside region except for three things: mass, angular momentum and electric charge. John Wheeler called this discovery 'black holes have no hair', and it has been ever since known as the 'no hair theorem'.

None of this presented a problem for information conservation. A classical black hole would last for ever preserving the information inside it, inaccessible, but there. Still part of the universe. The problem had emerged when Hawking discovered that quantum effects would cause a black hole to radiate at a steady rate – the famous Hawking radiation. That radiation carries no information about what made the black hole or what has fallen in. Still no problem until you realize that in this process the black hole would eventually evaporate away and disappear entirely. *Then* what would happen to all that information locked inside? It seemed the only way the information could avoid being lost would be if the Hawking radiation had subtle differences in it that would reflect what had fallen in. No one had found any way for such differences to be produced, though many physicists believed one must exist. Hawking's calculations, however, showed that the radiation was exactly thermal, random and featureless.[38]

In case anyone was wondering whether a baby universe branching off a black hole might solve the information loss problem, Hawking brought them up to date:

> There is no baby universe branching off inside a black hole, as I once thought. The information remains firmly in our universe. I'm sorry to disappoint science fiction fans, but if information is preserved, there is no possibility of using black holes to travel to other universes. If you jump into a black hole, your mass energy will be returned to our universe, but in a mangled form, which contains the information about what you were like, but in an unrecognizable state.[39]

Some in the audience must have pricked up their ears at that. Was Hawking about to say that Hawking radiation could after all be the vehicle of escape and that, as in the description earlier about restoring a burned book, it might at least in principle be possible to recover from that radiation the information hidden in a black hole?

Hawking's new solution to the problem had to do with something else, the possibility that a black hole could have more than one geometry (topology) at the same time. Information would not be trapped because a true event horizon would not form.

Christophe Galfard remembers that Hawking's talk seemed to leave most of his physics audience 'largely bemused'. There were whispers of 'big on claims, short on mathematics . . . not really all that convincing . . . mostly smoke and mirrors'. Kip Thorne commented: 'This looks to me on the face of it to be a lovely argument, but I haven't yet seen all the details.'[40] He said he would have to take some time with Hawking's paper before he could say whether or not Hawking was right. Roger Penrose was not convinced: 'It seems to me that the indication that the information is lost is very powerful, and that is what Stephen originally thought. In Dublin he publicly retracted. In my view he was completely wrong to retract. He should have stuck to his guns.'[41] Hawking, though he expressed intentions of trying to support his idea with a mathematical proof, was convinced enough of his conclusions to concede a bet that he and Kip Thorne had made with John Preskill of Caltech. The bet read as follows:

> Whereas Stephen Hawking and Kip Thorne firmly believe that information swallowed by a black hole is for ever hidden from the outside universe, and can never be revealed even as the black hole evaporates and completely disappears,
>
> And whereas John Preskill firmly believes that a mechanism for the information to be released by the evaporating black hole must and will be found in the correct theory of quantum gravity,

> Therefore Preskill offers, and Hawking/Thorne accept, a wager that:
>
> When an initial pure quantum state undergoes gravitational collapse to form a black hole, the final state at the end of black hole evaporation will always be a pure quantum state.
>
> The loser(s) will reward the winner(s) with an encyclopedia of the winner's choice, from which information can be recovered at will.

The document had been signed by all three men, Stephen Hawking's signature being his thumbprint, and dated 6 February 1997, Pasadena, California.

Hawking ended his talk saying, 'I will give John Preskill the encyclopedia he has requested. John is all-American, so naturally he wants an encyclopedia of baseball. I had great difficulty in finding one over here, so I offered him an encyclopedia of cricket, as an alternative, but John wouldn't be persuaded of the superiority of cricket. Fortunately, my assistant, Andrew Dunn, persuaded the publishers Sportclassic Books to fly a copy of "Total Baseball: The Ultimate Baseball Encyclopedia" to Dublin. I will give John the encyclopedia now. If Kip agrees to concede the bet later, he can pay me back.' Thorne was not convinced that Hawking, or anyone else, had solved the problem of the information paradox. The encyclopedia was brought on stage, and John Preskill held it over his head as though he were holding the men's tennis championship trophy at Wimbledon.

Hawking later commented on the occasion in a lecture he gave at Caltech the following January, in 2005: 'The [information loss] paradox had been argued for thirty years without much progress until I found what I think was its resolution. Information is not lost, but it is not returned in a useful way. It is like burning an encyclopedia. Information is not lost, but it is very hard to read. I gave John Preskill a baseball encyclopedia. Maybe I should have just given him the ashes.'[42]

Hawking had promised to provide his colleagues with a fuller explanation. It would take the form of a paper published in October 2005.

18

'Grandad has wheels'

THE BBC DEBUTED THE TELEVISION FILM *HAWKING* IN April 2004, when Hawking had recently emerged from hospital and was preparing to do battle on the information paradox issue. This was not a full biography but a moving dramatization of two crucial years in Hawking's life when he learned he had ALS, met Jane Wilde, and worked on singularity theorems for his thesis. A 2002 documentary, *Stephen Hawking: Profile*, was re-aired with the film. There were an estimated 4 million viewers.

Hawking helped with the final script, and Benedict Cumberbatch, who played young Stephen Hawking, took the trouble to study carefully the early stages of ALS. Jane, on whose autobiography the film was partially based, watched the video at first in bits and could find little to fault. '[Cumberbatch] was uncanny. He worked so hard in researching the exact progression of motor neurone disease. It brought back that period so very strongly. I think the young lady playing me, however, was much more feisty than ever I was. I was always extremely determined but I was also quite timid. So in that sense many of the things that appear in the film are not quite historically accurate.' However, Jane felt that the film was true to the spirit of those years. 'I can remember vividly the euphoric sense we had about us, that we were doing something exceptional . . . that, despite it all, everything was going to be possible.'[1]

At about the same time the film was aired, the disturbing reports and statements having to do with abuse were finally locked away in the Cambridge police files.

Hawking was not doing any less well in popularity polls. In a 2004 survey of role models that polled five hundred sixteen- to eighteen-year-old English boys, he came in second to rugby star Jonny Wilkinson, edged out again by a sports celebrity, as he had been in Japan. But Hawking was not slowing down as a world traveller and lecturer. His diary in 2005 read: January: Caltech and the University of California-Santa Barbara; February: Washington, DC, and Oxford; March: Spain; June: Hong Kong; October: Germany; November: California again and then Seattle, Washington.

During the January visit to Caltech, Hawking suggested that this time he give a lecture solely for undergraduates. The subject would be his life in physics. The title borrowed the famous *Star Trek* split infinitive: 'To Boldly Go'.

The trip to Washington, DC, in February, with his wife Elaine, was for the purpose of receiving the James Smithson Bicentennial Medal. The Smithsonian Institution arranged, with help from Jim Hartle, to present a retrospective of Hawking's life: 'Stephen Hawking's Alternate Universe'. Hartle introduced his old friend in glowing terms: 'Stephen Hawking's work has been characterized by great mathematical precision and extraordinary physical inventiveness. He is almost always surprising.'[2] Hartle quoted Hawking's words about the career path he had followed: 'You might think I had a grand premeditated design to address the outstanding problems concerning the origin and evolution of the universe. But it was not really like that. I did not have a master plan. Rather I followed my nose and did whatever looked interesting and possible at the time.'[3]

Hawking returned on this occasion to the possibility of extraterrestrial intelligent life, and his lecture was introduced by music that was clearly meant to resemble the sound track of *Star Wars*. 'By intelligent life, I don't mean only the DNA-based humanoid life like you see on *Star Trek*,' he said, 'which all is remarkably like ourselves. The range of possible life forms in the universe is much wider and includes electronic systems like computers.'[4] He criticized the *Star Trek* image of alien civilizations as too static. Even though the suggestion was that their science and technology are more advanced than ours at present, some of the alien beings are pictured as having attained a stable state near perfection, with no further progress or evolution expected. 'I don't believe the *Star Trek* picture. We will never reach a final steady state, the end of development. Instead we shall continue to change at an ever-increasing rate.'[5]

The Oxford visit was to deliver the third lecture in a series honouring his old mentor, Dennis Sciama. In Oviedo, Spain, he helped to celebrate the twenty-fifth anniversary of the Prince of Asturias awards, which are major global awards for scientific, technical, cultural, social and humanitarian work done at an international level.

All this travel and activity would seem enough to exhaust the fittest of men. For Hawking, even after passing his sixtieth birthday, it was apparently exhilarating, and, despite all the moving about, Hawking's work as a popular author continued. October 2005 saw the publication of another version of his first bestseller, this one titled *A Briefer History of Time* and co-authored with physicist Leonard Mlodinow. It was indeed briefer, beautifully illustrated and simpler, and brought the physics up to date. Also that year he began work compiling a collection of historical mathematical works with short biographies of important mathematicians, to be published in 2006. It would be titled *God Created the Integers*.

Here and there, as Hawking lectured and met the media, he went on making provocative comments that had little or nothing to do with cosmology. His low opinion of politicians was increasingly evident. At an anti-war rally in Trafalgar Square in November 2004, he called the USA invasion of Iraq in March 2003 a 'warcrime'.[6] In 2005, when George W. Bush suggested returning astronauts to the moon, Hawking commented that 'sending politicians would be much cheaper, because you don't have to bring them back'.[7] He admonished those who opposed stem cell research: 'The fact that the cells may come from embryos is not an objection because the embryos are going to die anyway. It is morally equivalent to taking a heart transplant from a victim of a car accident.'[8]

In May 2005 he lightened up – *The Simpsons* again. This time, an episode titled 'Don't Fear the Roofer' had the 'Hawking' character announcing that he was now a Springfield resident. He had purchased the pizza parlor. He tried to get his computer to say the company motto but his computer got stuck and kept saying 'pizza pizza'. 'Hawking' had to hit his computer to make it behave, something he cannot do in real life. Later in the story, 'Hawking' saved Homer's sanity (if that is not an oxymoron) by explaining that it was a tear in spacetime and a little black hole, resulting in gravitational lensing, that caused Homer to be the only one able to see one of the other characters in Builder's Barn. Hawking also made a television appearance that year in the docudrama *Alien Planet*, as an expert consultant.

On 22 August 2005, the long-awaited paper that Hawking had been promising would lay out clearly and in detail his solution to the information paradox finally went to *Physical Review*, who published it in their 18 October edition. The paper was only three and a half pages long, with only three equations. Hawking utilized Feynman's sums-over-histories, applying them, as he had before, to the universe. In developing his no-boundary proposal with James Hartle, Hawking had studied different histories the universe could have had, and calculated which were more probable than others. Now he was likewise asking his readers to imagine all alternative histories of universes. Some would have black holes and some would not. Information would be lost in those histories with black holes, but not in those without black holes. His solution depended on the fact that the universe histories where black holes exist would be cancelled out by those where they do not, the upshot being that information wouldn't disappear because there would be no black holes for it to become trapped in in the first place. If you waited long enough, only those histories without black holes would be significant. Information, in the end, would be preserved.

Given Hawking's two explanations, and an earlier firm rejection of the idea that the information would be returned via Hawking radiation, it is a surprise to find that not long after he published this paper, in one of the children's books he wrote with his daughter Lucy – *George's Secret Key to the Universe*[9] – with Christophe Galfard's name also on the title page, he used a completely different solution to the problem, a solution that *did* involve Hawking radiation. It is also intriguing to discover that his top-down approach, which we will discuss a little later, is not easy to reconcile with his 2005 solution. Many of Hawking's colleagues remained unconvinced, wondering why he preferred his solution to Leonard Susskind's and Maldacena's. Perhaps the reason was that, having introduced the problem in the first place, he felt that he should be the one to solve it.

Hawking and Galfard explained the delay of the paper on the grounds of Hawking's increasing difficulty with his hand-held clicker device. In that regard, 2005 was a dispiriting year. The speed at which he could communicate had been slowing down since 2000 until finally his hands were too weak to use the clicker. He exchanged it for a switching device attached to his glasses, developed by Words+ (the Infrared/Sound/Touch (IST) switch). The low-power infrared beam can be controlled by the blink of an eye or by moving a cheek muscle. As of 2011, Hawking controls his with his cheek muscle.

The 2005 travel schedule did not end as planned. His schedule may have read Seattle, but he didn't get there. He was to travel to Seattle from Oakland, California, but while he was being taken off his respirator in the morning shortly before departing from Oakland, something went wrong and Hawking 'basically flat-lined. They had to resuscitate him, and that panicked a few people. But he's been there before.'[10] Hawking stayed in Oakland and delivered the Seattle address by live telecast.

Undaunted by that glitch, Hawking in 2006 scheduled travel to France, Spain, China and Israel. It was not his first time in Israel. He and Jane had been there to accept the Wolf Prize in 1988. But this time he refused to accept the invitation unless he could spend part of his visit with and lecture to Palestinians. That was arranged. In Israel he added a new quip to his long list, commenting that the lack of anonymity when travelling was a down side to celebrity and worse for him than for others: 'It is not enough for me to wear dark sunglasses and a wig. The wheelchair gives me away.'[11]

The prestigious Copley Medal that he received from the Royal Society in November had travelled even further than its recipient. The Royal Society arranged for British astronaut Piers Sellers to fly it into space before presenting it to Hawking.

Hawking's marriage to Elaine ended in divorce in the summer of 2006. Hawking refrained from comment and his personal assistant, Judith Croasdell, brushed off reporters clamouring for a statement with the words 'He is far too busy. This is just a distraction which is really annoying. We don't have any time for any of this . . . We have no interest in any of the gossip that is going on.'[12]

While observations by WMAP continued and theorists looked forward to results that would help them understand inflation theory and settle other questions, they didn't stop in their tracks to wait. There were new models of inflation that went beyond the familiar four dimensions of our universe. In 2000, Hawking had mentioned the possible role of p-branes in inflation. Alan Guth, the founding father of inflation theory, was also studying the possibility of inflation in 'brane world' models. Natalia Shuhmaher and Robert Brandenberger of McGill University in Montreal came up with a model in 2006 in which it was a hot gas of branes that drove inflation. Their model had all the spatial dimensions starting out extremely compact, with the extra dimensions beyond our familiar three tucked together in what is called an 'orbifold'. In the very early universe, the brane gas expanded and its energy density decreased until the three familiar spatial dimensions experienced an inflationary period.[13]

Hawking stayed on in the large, comfortable house that he and his second wife had built together. For the first time since the beginning of his first marriage, he was living on his own, though of course still with nurses in attendance. He had resumed close contact with Jane and their children and grandchildren, relationships that had suffered over the past decade and a half, especially during the unpleasant period of the abuse issues when he had declined to honour their concerns. All seemed to have chosen to put behind them that difficult-to-understand chapter in their lives.

Fat and Lean Years at the Trough

Although string theorist Brian Greene would tell a Cambridge lecture audience in 2011 that string theorists were 'happy as pigs in shit', it had not always been like that. In the years leading up to his remark, beginning in the previous century, the pigsty had not always looked so appealing. As early as 1986, it had become clear to string theorists that the number of different ways the extra dimensions could curl up was distressingly enormous,[14] and this was considered a serious shortcoming of string theory. However, that same year Andrei Linde cheered up his fellow theorists by insisting, in his first paper about eternal inflation, that this multiplicity of types of compactification (curling up) 'should be considered not as a difficulty but as a virtue of these theories, since it increases the probability of the existence of mini-universes in which life of our type may appear.'[15]

A happy moment came in 1997 when Maldacena, then at Harvard, introduced an idea known as ADS–CFT* duality that suggested a link between conventional quantum field theory and a certain type of string theory. Recall that dualities are situations where two very different theories (sometimes seemingly contradictory) both accurately describe the same thing. Maldacena's idea was an unproved conjecture, but it promised eventually to provide a genuine mathematic basis for string theory, and for that reason it had a strong influence on string theory's subsequent development. 'ADS–CFT correspondence' also had significant implications for the relationship between string theory and brane worlds,[16] and Hawking

* ADS stands for anti-de Sitter; CFT for conformal field theory.

thought the idea had something to offer with regard to the information paradox, weighing in on the side of no loss of information.[17]

Then, another downer. In 2000, Joe Polchinski at the University of California-Santa Barbara and Raphael Bousso at the University of California-Berkeley found that the basic equations of string theory had a truly astronomical number of different possible solutions, each representing a different way to describe a universe. It was a while before anyone could figure out whether any of these solutions were stable, but that was settled in 2003. There are (very roughly) 10^{500} of them. The problem this huge number presented was that string theory could never be proved right or wrong. Almost *any* experimental result would be consistent with it. As you will recall from Chapter 2, this is not good news for a theory.

Again it was Andrei Linde who saved the day by pointing out that this was not such terrible news after all. Eternal inflation theory was, in fact, predicting just such a situation.

Let's take a look first at the curling problem . . . or non-problem.

Curling Becomes an Olympic Event

When a new universe emerges, not all the space dimensions predicted by string theory inflate. Some remain invisible but play an important role in deciding what form the new universe takes. Theorists were finding that they are not just curled up willy-nilly. Precisely how they are curled determines the apparent laws of nature in that particular universe.

In the late 1980s and early 1990s, when Hawking had been thinking about wormholes and baby universes, he had speculated that the masses of particles and other fundamental numbers in nature might not be fundamental to the totality of universes, but different for different universes. They might be 'quantum variables' – numbers fixed at random at the moment of creation in each universe. A throw of the dice, with no way to know from a theory how the dice would fall.

In M-theory, it was no longer a dice throw. As Hawking and Mlodinow would put it in *The Grand Design*, 'the exact shape of the internal space determines both the values of physical constants, such as the charge of the electron, and the nature of the interactions between elementary particles'.[18] In other words, the fundamental laws of M-theory permit different laws of nature in different universes, in the way that the Constitution of the USA permits different local laws in the different states. What the local laws of nature *are* in a universe is determined by the way the extra dimensions are curled up.

Linde and his colleagues calculated the different ways the extra dimensions could curl up, each way leading to a unique universe.[19] The number is enormous beyond imagining. Curling had indeed become an Olympic sport. The prosaic word 'multiverse' fails dismally as the name for all this. It's too bad John Wheeler isn't still around to give it a better one.

Top Down

Earlier in the decade, Hawking and Turok had left the pea instanton lying on a shelf and each headed off in a different direction. Turok began to favour 'cyclic' models where the universe expands from a big bang, eventually contracts again to a big crunch, and then reemerges in another big bang, in a cycle that keeps repeating itself.

Hawking was more interested in models of eternal inflation. The fact that there are perhaps infinite possibilities for types of universes in this model makes it difficult, some would argue impossible, to calculate the likelihood of one kind of universe over another. Undeterred by this sticky mathematical problem, Hawking was determined to try, and in doing so he was going to utilize something that was still controversial among some of his colleagues, the anthropic principle.

The controversy over the use of the anthropic principle, which had come to a head when Hawking and Turok introduced their pea instanton theory, had not died away. However, Hawking's old friend Martin J. Rees and Mario Livio had given the anthropic principle and its use added respectability when they wrote in a paper in 2005 that '"anthropic reasoning" not only has a role in valid speculative scientific discourse, but may in fact have predictive power for sorting out allowable cosmological scenarios'. Although 'such arguments raise the blood pressure of many scientists . . . it can indeed be one of many tools in the cosmological toolkit'.[20] Hawking was preparing to treat it as a powerful tool.

In February 2006, he and Thomas Hertog collaborated on a paper reporting on work in which they combined the string landscape of multiple universes with the initial conditions of the no-boundary proposal. Their suggestion was that we picture the early universe as a superposition of all possible ways it could have existed in the string theory landscape – as you might, if you were about to draw a card from an unusually large deck, visualize all the possibilities of your draw at once. Each possibility produces a different future. We are faced with a universe that has many, many possible beginnings and many, many possible histories. As Hawking quipped in a lecture at Caltech that year, 'There will be a history in which the moon is made of blue cheese, but the amplitude for that is low, which is bad news for mice.'[21]

Hawking and Hertog confined themselves to a model with a simple landscape that permitted several different inflationary histories for the universe. Should you place yourself at 'the beginning' and start calculating the probabilities of those universes emerging? No. In Hawking's 'top-down' approach you start with the present, observing the universe as it currently exists, and then work backwards and decide how likely it is that each initial state would allow for the later existence of the populated universe we know. In this way, the present state of the universe 'selects' the past.

Hawking and Hertog studied the observational consequences of the no-boundary initial conditions and came up with a scheme to test their theory. If this is a valid way of working, there will be subtle differences in the CMBR and the gravity wave spectra from what these would be like if standard inflation theory is correct. Future technology, Hawking thought, might be able to discover these subtleties.[22]

Hawking delivered his 2006 lecture at Caltech controlling his computer with his cheek muscle. Titled 'The Origins of the Universe', it included some of the new thinking he and Hertog had been doing. Hawking would describe their top-down approach more fully later in his book *The Grand Design*.

Secret Keys and Cosmic Adventures

When Lucy Hawking's novel *Jaded* had appeared in the spring of 2004, and she'd begun a second one, *The Accidental Marathon*, she'd talked with many interviewers about her books. Lucy was not terribly surprised but somewhat annoyed that questions about her and her writing inevitably drifted into questions about her father.

Lucy's life had taken a difficult turn. Her marriage had been shortlived, and soon after the separation her son William had been diagnosed with autism. 'I actually felt as if my heart was breaking into tiny pieces,' she said.[23] Her mother Jane had insisted she persevere in finding the most effective treatment possible. William, as a result, had done remarkably well. He was very proud of his grandfather, not so much because of his physics accomplishments, but because 'grandad has wheels'. And Hawking returned the compliment. His office was full of William's pictures, and he even included one photo in *The Universe in a Nutshell*.

Lucy's response when the interviewers proved more interested in her father than in her was a practical one. She decided, if you can't beat 'em, join 'em. She would collaborate on a book with him.

In June 2006, Lucy accompanied her father on a trip to Hong Kong and Beijing, where the welcome was even more riotous than usual. When they appeared from their plane, police had to form a body chain to make a human corridor, holding back the crowds, so that they could reach the lift and get away to their hotel. Even with all that effort to protect him, Hawking came near to being pushed over. He was unflappable. He wanted to be photographed with this 'rugby scrum' of excited students, but the front-page newspaper photos captured only him and the rather desperate-looking policemen, until things settled down enough for a picture with a well-behaved group of small school children, unfurling a welcoming banner larger than they were. Two physics undergraduates from the Hong Kong University of Science and Technology, where Hawking was scheduled to lecture, had the honour of presenting him with flowers. Paul Chu Ching-wu, the university's president, commented, 'He's one of the most famous scientists ever. If you could say that Isaac Newton changed the world, then Stephen Hawking has changed the universe.'[24]

Stephen and Lucy Hawking took the opportunity of this well-publicized trip to announce that they would be co-authoring a children's book. *George's Secret Key to the Universe* was the first of their books together. It featured not only the space adventures of its young hero George and his next-door neighbours, the scientist Eric and his daughter Annie, and Cosmos, a supercomputer, but also some of the other issues that concerned Hawking. Confronted with the pollution in Chinese cities, he had expressed his worry that Earth might 'end up like Venus, at 250 degrees centigrade and raining sulphuric acid'.[25]

In the 'George' books, Lucy actually did give people like the interviewers who had annoyed her something of what they had been hoping for. *George's Key to the Universe* and *George's Cosmic Treasure Hunt*,[26] the first two in what we may hope will be a series, are delightful and instructive, and also offer an insight into the personalities and lives of Hawking and his family. The physicist Eric is unmistakably Hawking, imagined without his disability. His passion for physics and insistence on sharing it, his insatiable curiosity, his single-minded devotion to his work, his love of children – all are there. I am told that the wonderful, redoubtable character, George's grandmother Mabel, with her selective deafness, is recognizably a portrait of Hawking's mother Isobel. When George finds himself at a high-level physics conference and dares to raise his hand and ask a question, it is a retelling of a Hawking family story: when Robert, Lucy's older brother, was eight years old, he went with his father to a conference of theoretical physicists. He took a seat in the front row and listened intently, nodding his head, then raised his hand to ask an intelligent question. The description of the thoughtful, serious way the physicists in the book treat George's question is a tribute to Hawking's real-life colleagues. The science in the books is Hawking's science: black holes, Hawking radiation, the information paradox, the search for another planet for humans to colonize.

Lucy told interviewers that there were several reasons for writing the 'George' books. Her own son was ten years old, and she had a nephew as well, Robert's and his wife Katrina's son George. *George's Secret Key to the Universe* was dedicated to him and William. Lucy wanted to create, with her father, a book that would explain for these children some of the work he had done. She had noticed the interest when children and their parents had gathered around her father at William's birthday parties, amazed that he was willing to answer their questions and explain his science to them. She had watched him take the time and trouble to give them good, thoughtful, informative answers and also make them laugh. Hawking pointed out that as children most of us start out filled with curiosity and wonder. Anything is possible. That hasn't changed for him. That's still the way he feels, and he and Lucy want to encourage that feeling in their young readers.

Working with her father on the book gave Lucy a chance to see what her father was like at work in his own field, and that, she said, altered their relationship. Though she doesn't think he has changed much – possibly mellowed – she had never had the opportunity to see this side of him. 'He has the amazing ability to hold enormous amounts of information in his head, but also to pick out relevant details and make brief comments, which can completely transform your way of thinking.'[27] She was awed by how quick and clear-minded he is, and the knack he has for putting things together and having them fit.

In November 2006 Hawking was once again stressing, in a BBC radio interview, that the future of the human race depends on our colonizing another planet, not in our solar system but orbiting another star.[28] He had hopes of going into space himself. A month earlier Hawking had mentioned in an interview that his next goal was to go on a space flight and 'Maybe Richard Branson will help me' – which Branson wasted no time before offering to do. There would be a place for him in 2010 on Branson's sub-orbital space flight venture Virgin Galactic. Paris Hilton and William Shatner would go along. Hawking didn't expect Branson to get him to another habitable planet, but was sure the future could hold such journeys, and the fictional characters in his and Lucy's books followed up on just that suggestion.

Later in 2006, twenty-five thousand people responded in a blog with answers to Hawking's question, 'In a world that is in chaos politically, socially and environmentally, how can the human race sustain another 100 years?' No wonder he was thinking his ideas might actually influence public policy! In his posted follow-up to the blog, he mentioned genetic engineering, this time not as something perhaps undesirable that was going to happen anyway, but with the utopian hope that it might make human beings 'wise and less aggressive'.[29]

On the trip to China with Lucy, Hawking had once again joked about Pope John Paul II's forbidding scientists to study the origin of the universe, and said he was glad the Pope was unaware of the topic of his talk because he 'didn't fancy the thought of being handed over to the Inquisition like Galileo'. The Vatican itself seemed able to overlook comments like that, but Catholic lay leaders had heard the misquotation and flippant remark once too often. In a heated response, Catholic League president Bill Donohue said that Hawking 'should stop distorting the words of the pope': 'There is a monumental difference between saying that there are certain questions that science cannot answer – which is what the pope said – and authoritarian pronouncements warning scientists to back off.'[30] The Pope's statement had, as we have seen, not been an inaccurate description of the state of scientific knowledge at the time he made it, and there was no threat to make anyone fear Galileo's fate. His words were:

> Every scientific hypothesis about the origin of the world . . . leaves unanswered the problem concerning the beginning of the universe. By itself science cannot resolve

> such a question: it requires human knowledge which rises above the physical, the astrophysical, what we call the metaphysical; what is required above all is the knowledge which comes from the revelation of God.[31]

Hawking took Catholic League president Bill Donohue's words to heart. On his visits to the Vatican as a member of the Pontifical Academy, he has not mentioned the incident or Galileo again. Pope John Paul II had acknowledged in a speech in 1992 that the Roman Catholic Church had erred in condemning Galileo, an acknowledgement Hawking had said he hoped would eventually be made, when he visited the Vatican in 1973. It *was* time to bury the hatchet.

Zero Gravity

In April 2007, Hawking undertook an adventure that he hoped would be the first step on the road to a real space flight. This was a flight that offers passengers the experience of weightlessness, zero gravity. No one was certain how his frail body would react. No problem! Hawking took eight turns of weightlessness, four minutes in all, more than anyone had expected, except, perhaps, himself. 'I could have gone on and on!' he said.[32] Four physicians and two nurses, who monitored Hawking's blood pressure, cardiac readings and blood oxygen levels throughout the flight, agreed.

A company called 'Zero Gravity' provides these flights. Here's how it works: The plane follows a parabolic, roller-coaster path. As the plane climbs, passengers feel almost twice the pull of gravity they normally feel on Earth. Near the top of the parabola, they feel they are in free-fall for about twenty-five seconds. The course is then repeated, in Hawking's case eight times.

Hawking had a second motive for taking this flight, and, he hopes, his upcoming space flight – spreading his conviction that colonizing other planets is the only hope for an extended future for us.

> It will be difficult enough to avoid disaster on planet earth in the next 100 years, let alone the next 1000 or million. The human race shouldn't have all its eggs in one basket or on one planet. Getting a portion of the human race permanently off the planet is imperative for our future as a species.[33]

There is, he insisted, a huge future mass market for space-oriented services in tourism. 'We need to engage the entrepreneurial engine that has reduced the cost of everything from airline tickets to personal computers.'[34]

Hawking's third motive for taking the flight was to encourage other disabled people to get out and try things like this. If he can do it, so can others. While that may not be true when it comes to theorizing about the origin of the universe, when it comes to adventures like weightlessness, why not? The answer that most immediately comes to mind is, of course, the price.

Itinerant physicist

2008 was a phenomenally busy year of travel for Hawking, and even more so for his valiant personal assistant, Judith Croasdell, whose task it was to make preparatory journeys to reconnoitre, plan the visits with his hosts, and make all the necessary on-the-spot

Recognition, 1989. (*Right*): the honorary degree procession in Cambridge, with Robert, Jane and Tim (wearing Stephen's hat); (*below*): with Jane at Buckingham Palace for the Companion of Honour investiture.

Celebrity, the early 1990s. (*Above*): Stephen and his mother, Isobel, celebrate his inclusion in the *Guinness Book of World Records*; (*right*): at Steven Spielberg's home with Spielberg and nurse Joan Godwin; (*below*): a scene in his office from the film of *A Brief History of Time*.

In the media, 1990s. (*Above*): starring on *The Simpsons*; (*left*): with an animated Al Gore, Michelle Nichols and Gary Gygax in the TV series *Futurama*; (*below*): winning the *Star Trek* poker game.

(*Above*): Stephen and Elaine on their wedding day in 1995; (*right*): recording the second bet about naked singularities at Caltech with John Preskill and Kip Thorne, 1997.

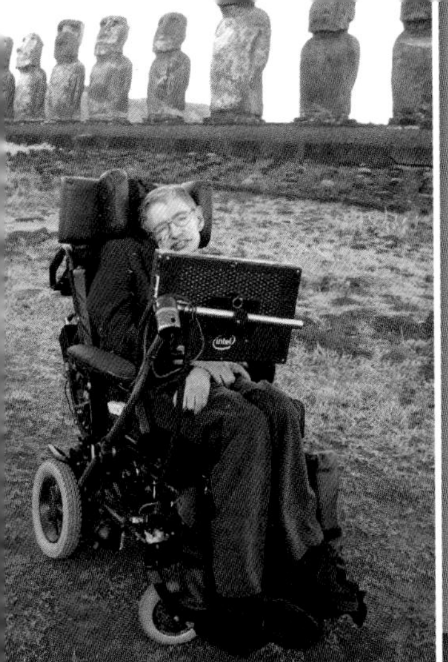

A new millennium. (*Clockwise from top left*): on Easter Island, 2008; in Antarctica, 1997; the zero-gravity flight, 2007; 'Marilyn' shows up at Hawking's sixtieth birthday party.

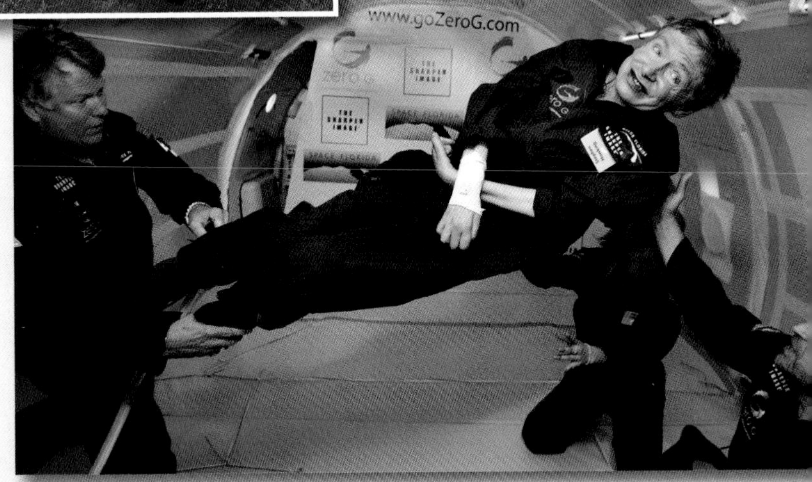

With the great and the good. (*Clockwise from left*): with daughter Lucy, receiving the Presidential Medal of Freedom award from US President Barack Obama; in South Africa with Nelson Mandela; chatting with Queen Elizabeth in the Chelsea Flower Show 'Stephen Hawking Garden for Motor Neurone Disease'; with his three children Tim, Robert and Lucy at the White House.

Cambridge, twenty-first century. (*Left*): the Centre for Mathematical Sciences, new home of the DAMTP; (*below*): Hawking ensconced in his high-tech corner office with the dish of stones visible on the corner of his desk and photos of himself in zero gravity prominently displayed.

(*Left*): unveiling the Time Eater clock at Corpus Christi College, with the clock's inventor John C Taylor; (*below*): the bust by sculptor Ian Walters in the new common room.

Scaling 'the flaming ramparts of the world'. Stephen Hawking and cellist Yo-Yo Ma on stage at the World Science Festival Opening Night Gala at Lincoln Center, New York City, in June 2010.

arrangements. She had been told when she was hired that 'the personal assistant does not travel'. That may have been true for other PAs, but not for Judith.

In January the destination was Chile; the occasion a scientific meeting in Valdivia to celebrate the sixtieth birthday of Chile's most distinguished physicist, the charismatic Claudio Bunster. Bunster had been responsible for Hawking's visit to Chile and Antarctica ten years earlier. From Chile, braving the difficulty of taking Hawking in his wheelchair to a remote part of the world with dauntingly uneven terrain where disabled access was only just beginning to be considered an important addition, his entourage flew with him to Easter Island. For Judith Croasdell, who had lived in the Southern Hemisphere in the Pacific islands herself for many years and been a student of Pacific history, this trip was 'the Holy Grail'.

In May it was South Africa. In Cape Town, Hawking visited the African Institute for Mathematical Sciences (AIMS), an institute for postgraduate study that draws top students from all over Africa and supports the development of mathematics and science across the continent. Hawking's friend and colleague Neil Turok, the founder of AIMS, helped organize that trip. Hawking met Nelson Mandela and launched the Next Einstein Initiative, a programme at AIMS that Hawking said, in his lecture, he hoped would nurture 'An African Einstein'.

In September, hosted by the University of Santiago de Compostela, Hawking touched down in that famous, beautiful Spanish pilgrimage city, to receive the Fonseca Prize, an award that celebrates outstanding individuals in the communication of science to the general public.

Judith Croasdell remembers that 'it was a tough trip, with high expectations and a huge press conference. The press conference was far too long, the press asked far too many questions (over forty questions for Stephen to choose from). Stephen answered about fourteen – which was a lot.' Lucy accompanied him on this journey, to publicize the Spanish edition of *George's Secret Key to the Universe*.

With far-flung travels evidently on his mind that year, in his lecture at Caltech Hawking was a little less pessimistic about the possibility of travelling through a wormhole to another universe. He pointed to a new possibility, black holes in the extra dimensions of space-time. Light wouldn't propagate through the extra dimensions, only through our familiar four, but gravity would affect them and be much stronger than what we experience. That would make it easier for a little black hole to form in the extra dimensions.

In this lecture, while giving background about black holes in general, he described Hawking radiation in a different way. He wasn't reneging on the version involving particle pairs, it was just an alternative way of thinking about it. If a particle is in a very small black hole, you know with fair accuracy where it is. Because of the uncertainty principle, the more certain you are about its position, the less certain you are going to be about its speed. So the smaller the black hole the more uncertainty about a particle's speed. It could even be more than the speed of light, and that would allow the particle to escape from the black hole. In this description, Hawking radiation *does* come from inside the black hole.

Could you possibly fall into a black hole and come out in another universe? He thought that might be possible.

He had not given up on the idea of wormholes. But you couldn't come back again, so he, personally, is not going to try it, in spite of being one of the most intrepid and eager-to-travel people in the world.

19

'I've always gone in a somewhat different direction'

'IT IS ASTONISHING THAT BOLD PREDICTIONS OF EVENTS IN the first moments of the universe now can be confronted with solid measurements,' exclaimed WMAP's principal investigator, Charles Bennett, in March 2008.[1] 'Fifth year results' had shown that the WMAP satellite's data was placing tighter constraints on inflation theories while supporting inflation in general. At the same time, WMAP had discovered something no one had predicted – a mysterious break in the overall random distribution of the temperature variations in the CMBR – a 'cold spot'.[2] All that could be said definitively about it was 'watch this spot', but so far none of the suggested explanations cause a problem for inflation theory.

The race to find experimental and observational evidence to confirm what had been theoretical for a long time was happening not only out in space but also on the ground – and deep under it. With the switch-on of the Large Hadron Collider at CERN, on 11 September 2008, expectations were running high that this long-awaited instrument could at last reveal the Higgs boson.

Hawking v. Higgs

Peter Higgs had proposed the Higgs boson's existence in 1964 and seen it become part of the standard model of cosmological theory. At a press conference in 2008 he reacted heatedly – 'launched an attack' was the way the *Sunday Times* headline put it[3] – to a comment Hawking had made in an interview on the BBC.

The less than cordial relationship between Higgs and Hawking dated back to 1996. Hawking had published a paper in which he had stated that it would be impossible to observe the Higgs particle. By 2000 he had not been proved wrong. When the Large Electron Positron (LEP) experiment at CERN was finally closed down that year without having produced definitive proof of the Higgs particle, Hawking raked in $100 on a bet with colleague Gordon Kane at the University of Michigan. Another Hawking bet against the Higgs particle was still unsettled, awaiting the end of similar experiments at Fermilab near Chicago. The exchange between Hawking and Higgs pushed the usual limits of heated scientific discussions when Higgs, at a dinner in Edinburgh in 2002, said of Hawking that it was 'difficult to engage him in discussion, so he has got away with pronouncements in a way that other people would not. His celebrity status gives him instant credibility that others do not have.' Hawking countered with: 'I would hope one could discuss scientific issues without personal attacks', and Higgs privately made peace by explaining to

Hawking the context of his comments. Hawking said he was not offended, and things settled down, but Hawking never swerved from his opinion that the Higgs particle was beyond the reach of any experiment.

In a press conference shortly before the switch-on date for the LHC in September 2008, Hawking opened old wounds with the comment that he thought 'it will be much more exciting if we don't find the Higgs. That will show something is wrong, and we need to think again. I have a bet of $100 that we won't find the Higgs.'[4] Higgs reacted with disparaging remarks about Hawking's work: 'From a particle physics, quantum theory point of view, you have to put a lot more than just gravity into the theory to have a consistent theory and I don't think Stephen has done that. I am very doubtful about his calculations.'[5] Hawking had argued there might be more interesting outcomes from the LHC, such as the discovery of some of the supersymmetric partners. 'Their existence would be a key confirmation of string theory,' said Hawking, 'and they could make up the mysterious dark matter that holds galaxies together. But whatever the LHC finds, or fails to find, the results will tell us a lot about the structure of the universe.'[6] The *Sunday Times* comment that 'their spat is likely to send shockwaves through the scientific Establishment' was something of an overstatement, but the 79-year-old Higgs could not be blamed too much for being passionately eager for his theory, finally, to be confirmed.

Hawking had other fish to fry in the LHC. He had mentioned in his most recent Caltech lecture that he thought it might be possible to observe microscopic black holes resulting from collisions in the Collider. If so, these black holes should be radiating particles in a pattern we would recognize as Hawking radiation.[7] He might get a Nobel Prize.* He had also mentioned again that fluctuations in the CMBR can be thought of as Hawking radiation from the inflationary period of our universe, now frozen in.

Unfortunately for Higgs and Hawking and a lot of other people, just nine days after the Large Hadron Collider was switched on, CERN had to switch it off again. A faulty electrical connection allowed a helium leak into the tunnel housing the collider and caused the superconductor magnets that steer sub-atomic particles around the collider to malfunction. It took a year to get the LHC back in action.

The Higgs particle, at the time of this writing, continues to be elusive. As of late February 2011, after a short winter maintenance break, researchers were gearing up for yet another try. 'We know that we will either discover the Higgs particle or rule it out, and in either case it will be a big result,' Sergio Bertolucci, CERN's Director for Research and Scientific Computing, said. 'Of course, it's more difficult to sell as a big result if we don't find it, but if the Higgs doesn't exist, there must be something else in its place.'[8]

The Time Eater

That September of 2008 when Higgs and he were renewing their altercation, Hawking had the honour of

* The Nobel Prize is very seldom given for even the most promising theories if there is no experimental or observational evidence to support them.

unveiling a harrowingly beautiful addition to the streetscape in the old centre of Cambridge. Corpus Christi College, which has the most ancient court in Cambridge, became home to this newest phenomenon, a very large mechanical clock on the corner where Bene't Street meets King's Parade.[9] It has no hands but tells the seconds, minutes and hours with what appear to be small flashing blue tear-shaped lights moving in concentric circles around a gleaming clockface that is five feet in diameter. Plated in pure gold, the huge face is designed to look like ripples and troughs radiating out, as if a stone had dropped into a pond of molten metal. The ripples represent the explosion of the Big Bang, sending out pulsating gold.

The shining contraption honours one of history's greatest clockmakers, John Harrison, the eighteenth-century pioneer of longitude, one of whose inventions was a 'grasshopper escapement'. The Corpus clock's maker and donor, John Taylor, was an undergraduate at Corpus in the 1950s and has since been a phenomenally successful inventor. He also has a passion for old clocks. Taylor chose to make his 'grasshopper' a fearsome, crusty giant locust. This beast, at once menacing, beautiful and whimsical, creeps inexorably along the top rim of the clock. It operates by putting its claws into the teeth of the great clockwork escape wheel that rotates around the outer edge of the clock face and, like Harrison's grasshopper, restrains and measures out the speed of rotation. This sinister monster is the 'chronophage' or 'time eater'.[10]

When the Corpus clock strikes the hour it tolls off the numbers not with chimes but with a rattle made by shaking iron chains over a wooden coffin, with a hammer beating the wooden lid, all inside the back of the clock.

It seemed appropriate that Hawking should unveil this awesome device. Nearly everyone associates Stephen Hawking with the Big Bang and time's 'brief history'. He has tamed time by turning it into another space dimension. He has also seemed miraculously to have stretched out his own time – defying, perhaps, that horrific creature on top of the clock.

A quieter celebration

But time was indeed passing, even for Hawking. A year later, on 30 September 2009, obeying a dictum of the University of Cambridge that Lucasian Professors of Mathematics retire at the age of sixty-seven, he relinquished the title he had held for thirty years. His successor would be Michael Green, a distinguished theoretical physicist specializing in string theory.

In contrast to the celebration of Hawking's sixtieth birthday, his stepping down as Lucasian Professor was marked quietly with a champagne reception in the department. His retirement meant little change. His busy schedule, his research, his status in the DAMTP stayed much the same. His title would now be Director of Research for the Cambridge Centre for Theoretical Cosmology. He kept his spacious corner office, and his personal assistant and graduate assistant were not ejected from the cluster of offices around him. In an audio message to BBC *Newsnight*, Hawking reiterated that he was not really retiring, merely changing titles, and added:

> It has been a glorious time to be alive and doing research in theoretical physics. Our picture of the universe has changed a great deal in the last forty years, and I'm happy

> if I have made a small contribution. I want to share my excitement and enthusiasm. There's nothing like the Eureka moment of discovering something that no one knew before. I won't compare it to sex, but it lasts longer.[11]

For the past year, Hawking had been making threatening statements about the possibility of his leaving Cambridge and England – his way of protesting at the proposed draconian cuts in public funding for the kind of basic research he does and the kind of scientific education that he tries to inspire young people to choose. Funds were to be channelled instead into industrial applications of science, science that some thought would bring money into the UK. Hawking had been protesting about such priorities – calling them 'ignorant of the past, and blind to the future' – for over a decade. 'To demand that research projects should all be industrially relevant is ridiculous. How many of the great discoveries of the past that laid the foundations for our modern technology were made through industrially motivated research? The answer is, hardly any.'[12]

If he moved, where would he go? Hawking had enjoyed working as a visitor at the Perimeter Institute for Theoretical Physics, a state-of-the-art research centre in Waterloo, Ontario, where Neil Turok was now Director. There were rumours that in retirement he would take up a post there. However, Hawking did not abandon Cambridge and probably never will. In spite of change of title, funding cuts, and inexorably deteriorating physical condition and ability to communicate his ideas, his goal remained as mind-bogglingly ambitious as ever: 'A complete understanding of the universe, why it is as it is, and why it exists at all.'[13] And how long was that going to take? In an interview on the *Charlie Rose Show*, the previous year, Hawking had been asked that question. He had replied by repeating the words he had used in 1980, in his inaugural lecture as Lucasian Professor: 'by the end of the century'. Then he added with a cagey grin that though his estimate remained the same, there was a lot more of the twenty-first century left than there had been of the twentieth when he first made that prediction.

Hawking's graduate assistant, Sam Blackburn, had set 2009 off with a literal bang by courageously giving him a mini-rocket-launcher for his birthday. With this 'office toy', he could send missiles sailing across the room. In March he had travelled to Los Angeles and met his granddaughter Rose for the first time. He and Lucy had dedicated the second of their 'George' books to her. Rose and her older brother George are children of Robert Hawking and his wife Katrina.

Hawking made another of his lecture stops in Pasadena, California. The frenzy of the occasion was not at all unusual for him. It was happening several times, sometimes even *many* times, every year, and even at Caltech, by then almost home territory.

Space, the Final Frontier

Heralded by the opening fanfare from Richard Strauss's *Also Sprach Zarathustra*, he made his entrance to the convention centre, filled to capacity with 4,500 people. Those who didn't know the music by that name still recognized it as the thundering background music of the film *2001: A Space Odyssey*. Hawking could no longer use his hands to drive his own wheelchair, a sad change, but, hands folded in his lap, he was wheeled down the aisle at considerable speed. The *Blue Danube Waltz* – not quite so huge and impressive, more friendly – replaced Richard Strauss with Johann Strauss as he started up the ramp to the stage. The audience waited. Nothing happened for a little while. A glitch? A way of heightening expectation? Hawking's graduate assistant came out and made some adjustments to Hawking's laptop. Hawking's hands remained immobile in his lap. He was controlling his computer with a movement of his cheek muscle. Soon came the voice and the words everyone was waiting for: 'Can you hear me?' The Caltech crowd cheered. Stephen Hawking was back!

Hawking's lecture was 'Why We Should Go into Space',[14] which he had written the previous year as a fiftieth-birthday present for NASA and delivered in Washington, DC. It was a more adult version of the chapter by that same title in the 'User's Guide to the Universe' in *George's Cosmic Treasure Hunt*, published in 2007. One part of the talk that he had not included in that book had to do with the cost of space travel, which Hawking admitted would not come cheap but would still represent only a small fraction of world GDP, even if the present USA national budget for space exploration were to be increased twenty times. He recommended a goal of a base on the moon by 2020 and a manned landing on Mars by 2025, not only in the interest of space exploration but to reignite public interest in space and science in general. 'A high proportion of space scientists say their interest in science was sparked by watching the moon landings,' he said.

Would we find life out there? Hawking was thinking that even if the probability is small of life appearing on a suitable planet, in a universe as large as ours, life must have appeared somewhere else besides Earth. The distances between the places where it has appeared were likely to be extremely large and the life would almost certainly not all be DNA-based. Another possibility is that meteors may have spread life from planet to planet and even from stellar system to stellar system. If life spread in this way (the process is called panspermia) then it would not be surprising to find other DNA-based life at locations in our own neck of the woods.

One bit of evidence that panspermia may have been the source of life on Earth, he pointed out, is that life here appeared suspiciously quickly after the first moment that it would have been possible. The Earth was formed 4.6 billion years ago and for the first half-billion years was too hot for life to emerge. The earliest evidence of life comes from 3.5 billion years ago. That means life appeared only about a half a billion years after it was first possible. A long time, that might seem, but actually amazingly short.

We haven't of course been visited by aliens (at least, we don't think so – 'Why should they appear only to cranks and weirdos?') and there seem to be no advanced, intelligent beings near us in the galaxy. The SETI project has heard no alien TV quiz shows. There is probably no alien civilization at our stage of development within a few hundred light years of us. 'Issuing an insurance policy against abduction by aliens seems a pretty safe bet.'

Hawking mentioned three possible reasons why we haven't heard from any aliens.

First, the probability of life appearing on a suitable planet may be too low.

Second, even if that probability is high, the probability of this life evolving into intelligent life may be too low. (It isn't clear that intelligence confers long-term survival advantage. Think of bacteria and insects.)

Third, intelligent beings who reach the stage of sending radio signals also have reached the stage of building nuclear bombs or similar weapons of mass destruction, and they may always destroy themselves very soon. Hawking calls that a sick joke, but he has also said that if extra-terrestrial life has not destroyed itself, given the short time-span of life on Earth compared with the age of the universe, it is still unlikely that we would meet an example of alien life at a recognizably human stage. It would either be much more primitive than we are or so advanced that it would regard *us* as impossibly primitive.

Hawking favours the second possibility, the rarity not of life but of intelligent life. 'Some would say it has yet to occur on earth.' Why are we smiling?

Hawking's lecture was long and thoughtful. He answered questions chosen ahead of time from those submitted by students and others in the Caltech community: how close are we to the *Star Trek* world? Don't expect warp drive or replicators. We will have to 'do it the hard way', slower than light speed. To reach distant destinations we'll need more than one generation. The journeys would be so long that the crews would even have time to evolve differently, so that the human race would divide into different species.

At the end of his visit to California, Hawking was not well enough to continue to Phoenix, as had been planned. Lucy appeared there instead and his pre-prepared lecture was broadcast over speakers. Back home in Cambridge, Hawking was in hospital for observation for a brief time, but all this turned out yet again to be only a temporary setback. He was back in fine form well ahead of an August trip to Washington, DC, to receive a Presidential Medal of Freedom from US President Barack Obama. In September, in Switzerland, where he visited CERN and the University of Geneva, his lecture 'The Creation of the Universe' filled one theatre and (by video link) ten other auditoriums.

It was his appearance in Washington to receive his medal that triggered a comment that unexpectedly drew him into the debate raging in the United States while President Obama was struggling to get a health care bill through Congress. One outspoken opponent of all public health care, disparaging the British system, commented that 'If Stephen Hawking had been British, he'd be dead by now!' Hawking responded that he was, of course, British and lived in Cambridge, England, and that 'the National Health Service has taken great care of me for over forty years. I have received excellent medical attention in Britain. I believe in universal health care.'[15] Jane Hawking might not have been so upbeat about the NHS, given her disappointments with them.

In February 2010, the Planetary Society of Pasadena, California, awarded Hawking the Cosmos Award for Outstanding Public Presentation of Science. Previous honorees had been James Cameron, creator of the film *Avatar*, and *NOVA* producer Paula Apsell.[16] With Hawking's health again uncertain, a delegation travelled from California to Cambridge to make the presentation. The society's mission is 'to inspire the people of Earth to explore other worlds, understand our own, and seek life elsewhere'. The press release announcing the presentation event in Cambridge ended with the words: 'Tickets are sold out.'

Hawking unveiled another reminder of the swift passage of time in the spring of 2010. It was an unusual experience, having a garden named for him at the annual Royal Horticultural Society Chelsea Flower Show in London. The 'Stephen Hawking Garden for Motor Neurone Disease: A Brief History of Time' was dedicated not only to him but to all whose lives have been affected by motor neurone disease – patients, families, carers – and it was indeed a garden of mixed emotions. A spiral path representing the Earth's plant history led visitors from an area with some of the most ancient species of plants and ended near the centre of the garden with 'productive Mediterranean type plants that could produce food for us in the future, if climatic conditions allow'. At the centre of the garden was a pool in which water appeared to fall into a dark, hopeless vortex – representing a black hole – the end of time. Nearby, set into a drystone wall, was an antique clock, representing the swiftness with which time disappears for those with motor neurone disease. Queen Elizabeth met Hawking in the garden to admire the design, converse with him, and congratulate him.

Verdict from the Skies

WMAP ended its mission in 2009. A results summary in January 2010 announced that the large-scale temperature fluctuations in the CMBR are slightly more intense than the small-scale ones – a subtle but key prediction of many inflation models – and confirmed that the universe is indeed flat.[17] This second conclusion was supported even more strongly than before by the overall randomness of locations of hot and cold points in the CMBR.[18]

As the WMAP mission was preparing to wind down,* in May 2009 the European Space Agency launched its Planck satellite. Some of its detectors are designed to operate at a temperature of minus 273.05°C, just a tenth of a degree above absolute zero. A formal release of fully prepared CMBR images, analyses and scientific papers was not expected before 2013, but the ESA made a preliminary announcement of some results in January 2011. 'We haven't got to the real treasure yet, the cosmic microwave background itself,'[19] said David Southwood, ESA Director of Science and Robotic Exploration. The project's first goal had been to weed out some foreground sources that hamper studies of the CMBR. Many things may have affected this radiation during the evolution of the universe, 'a whole lot of dirty astrophysics'[20] complicating the picture – irregularities from gravitational lensing, radio sources, black holes, even instrument noise. The Planck scientists had in particular been focusing on the 'anomalous microwave emission', a glow associated with dense dusty regions of the galaxy, and been able to confirm that it comes from dust grains set spinning by collisions with either fast-moving atoms or ultraviolet light. Filtering out this microwave 'fog' from the data would not distort the CMBR. It would leave the CMBR untouched and allow Planck's data to reveal the cosmic microwave background in unprecedented detail.[21]

As observations of the CMBR become more detailed

* WMAP was finally consigned to 'graveyard orbit' in October 2010.

and precise, it becomes more of a challenge for any model to agree with the findings. Success in hitting that target becomes more and more convincing evidence to support a model. Some models are winnowed out. But, so far, the agreement between observations and predictions having to do with the CMBR and the universe's overall shape, large-scale smoothness, and smaller-scale structure look promising for inflationary cosmology.[22] As John Barrow summed it up, 'The growing observational evidence for the distinctive pattern of temperature variations in the microwave background radiation means that we take very seriously the idea that our visible portion of the universe underwent a surge of inflation in its very earliest stages.'[23]

Gravity waves from the moments right after the Big Bang are predicted to have left a distinct footprint in the CMBR,[24] but this footprint was proving elusive. However, there are other potentially better ways to look for gravity waves. Kip Thorne, because of his abiding interest in black holes, has been working with colleagues for some time to develop instruments that can detect and measure more directly gravity waves that originate in black hole events, and in the early universe. One technique is laser interferometry.

The interferometer is a device that splits a laser beam into two beams, perpendicular to one another. Each beam bounces off a mirror that sends it back along its path. The two beams are recombined when they meet. Each of the mirrors has a large mass attached to it, so if a gravity wave passes through the interferometer, stretching and contracting space between the masses (and hence the mirrors), that displaces them slightly and changes the distances the beams

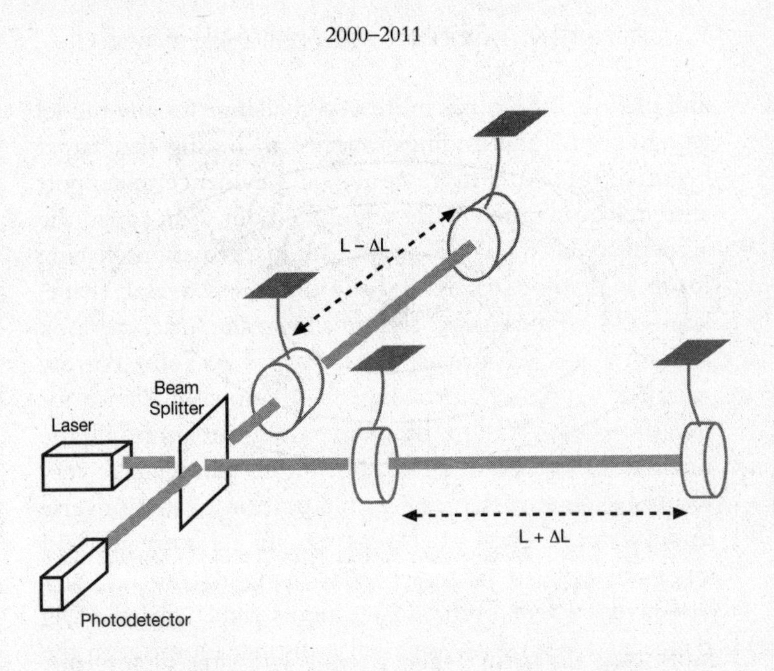

Figure 19.1. Sketch of an Earth-bound gravitational wave interferometer (by courtesy of Kip Thorne)

travel, producing interference patterns in the laser light (see Figure 19.1).

Earth-based gravity wave detectors are already in place in Hanford, Washington (LIGO); Hanover, Germany; and Pisa, Italy; but the grandmother of all such instruments, an astoundingly outsized arrangement, is scheduled to be launched into space in the form of three separate spacecraft, together known as LISA, the Laser Interferometer Space Antenna. Once in place, the three spacecraft will form a triangle with sides 5 million kilometres long. It will take approximately 20 seconds for light to travel between them. (See Figure 19.2.) When gravitational waves, stretching and squeezing space, pass through this

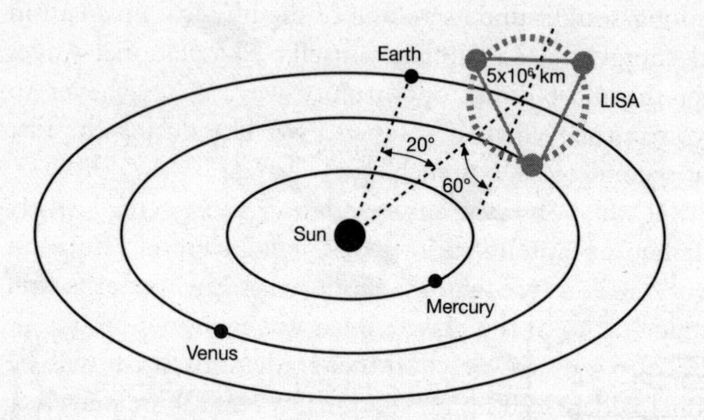

Figure 19.2. LISA, the Laser Interferometer Space Antenna, which ESA and NASA will jointly build, launch and use to monitor low-frequency gravitational waves (by courtesy of Kip Thorne)

enormous 'apparatus', their passage will alter slightly the distance between the spacecraft, and the distance travelled by the light beams between them, causing an interference of the light beams that can be measured with extremely sensitive instruments.[25] LIGO and LISA were two of the instruments that Kip Thorne was talking about when he promised Hawking on his sixtieth birthday that gravitational wave detectors – LIGO, GEO, VIRGO and LISA – would test his 'Golden-Age black-hole predictions' well before his seventieth birthday.[26] They had better get cracking!

Follow-ups to WMAP, Planck and LISA will be the NASA Einstein Inflation Probe, which will focus on the CMBR, and the Big Bang Observer, which will study gravity waves. The two approaches together may lead finally to something no probes or studies, including the highly successful WMAP, have yet been able to give us:

the long-sought understanding of the physical mechanism and energy scales of inflation itself.[27] Gravitational waves offer the most direct opportunity we are likely ever to have to probe what the universe was like during the first split second of its existence.

Will this observational evidence show conclusively whether or not inflation actually did happen? Inflation theory makes predictions about what the patterns and characteristics of the gravitational waves should be like. If they turn out to match those predictions, that will be strong evidence. If no gravitational waves can be detected, that would support another model, the ekpyrotic model of the universe in which inflation does not occur but our universe was created by the exceedingly slow collision of two three-dimensional brane worlds moving in a hidden extra (fourth) dimension of space.

Out on a Limb

When it comes to the bigger picture suggested by eternal inflation, it seems such an idea ought to be impossible to test from our blinkered vantage point within our own universe. What evidence could be lying around within our very limited reach?

Stephen Hawking and his colleagues are far from giving up on the possibility of making relevant predictions that can be set against more precise future observations, perhaps by the Planck satellite. In a paper in September 2010, Hawking, Jim Hartle and Thomas Hertog admitted that there is 'no way the mosaic structure [of universes] can be observed. We don't see the whole universe but only a nearly homogeneous region [that lies within the range of our observations], within our past lightcone'.[28]

However, in spite of the possibility that much larger fluctuations are observable only on scales much more enormous than we can study – on 'super horizon scales'[29] – they thought that the no-boundary wave function could come to their aid in calculating small departures from homogeneity *within* the part of the universe that it is possible for us to observe. The absence or presence of randomness in the spectrum of temperature variations in the CMBR does, they claimed, provide useful information about the bigger picture – and whether there *is* a bigger picture.

Just as the 'quantum wave function' for a particle gives every possible path the particle could follow between two points, the no-boundary wave function represents all the physically possible histories our universe might have had if it began in the way Hartle and Hawking proposed. In a previous paper, in January 2010, they had reported looking at a range of these different universe histories.[30] With an infinite number of possibilities, calculating which are more probable was a questionable undertaking. However, Hartle, Hawking and Hertog – without resorting to renormalization procedures of the sort Richard Feynman had called 'dippy' when he used them himself to handle infinities – felt confident in coming to some conclusions.[31] They divided the universe histories they were studying into two groups.

First, consider universe models in whose histories eternal inflation is *not* likely to have played a part. In other words, these are probably not part of a larger scheme of universes produced by eternal inflation. If we live in that kind of universe, and if it appears to us to be like the universe we know today, and if Hartle, Hawking and

Hertog are correct in thinking they can make valid use of the no-boundary wave function in their calculations, then what do those calculations lead us to expect?

(1) in the CMBR, on scales we can observe, a certain pattern of non-randomness in the spectrum of temperature variations;
(2) beyond our ability to observe, on huge scales, an over-all homogeneity;
(3) only a small amount of inflation in our past.

Observations of the CMBR, though they can't show us item (2), do not seem to back up those predictions.

So, consider another group of universe models. These universes are likely to be part of an eternal inflation picture. If we live in that kind of universe, and, again, if it appears to us to be like the universe we know today – and if Hartle, Hawking and Hertog are correct in thinking they can make valid use of the no-boundary wave function in their calculations – what do these calculations lead us to expect?

(1) in the CMBR, on scales we can observe, a high degree of randomness in the spectrum of temperature variations;
(2) beyond our ability to observe, on huge scales, a significant amount of inhomogeneity;
(3) a longer period of inflation in our past.

That's more like it! Or so it seems, so far. The absence or presence of (and the degree of) randomness that we can observe in the CMBR is a key issue.

Hartle, Hawking and Hertog decided, however, to go further out on a limb: their September 2010 paper reported that they calculate that our universe probably ended its inflationary period at the lowest potential value of the field.[32] They predict, rather precisely, observations in the area we can observe (within our light cone) – not only that there will be a high degree of randomness in the spectrum of temperature variations in the CMBR, but also the degree and manner in which, if eternal inflation is correct, the distribution and spectrum of the variations will depart from complete randomness.* The departure will be extremely small and not easy to detect.

Now, we wait to see whether Planck and other future probes will be able to produce precise enough measurements to test those predictions, as well as show a specific pattern of slight fluctuations in temperature in the CMBR predicted by Hawking and Hartle's no-boundary proposal itself. The Planck satellite may also be able to detect paths of light rays that have been bent in specific ways, indicating that our universe has a geometry predicted by some multiverse and eternal inflation models.[33]

Down-to-Earth Hawking Radiation

While Hawking, Hartle and Hertog had been thinking about what possible evidence might underpin

* Quoting Hartle, Hawking, and Hertog's September 2010 paper directly, to give a feel for what all this sounds like in the language of theoretical physics: '. . . an essentially Gaussian spectrum of microwave fluctuations with a scalar spectral index n8 ~ .97 and a tensor to scalar ratio of about 10%' (James Hartle, S. W. Hawking and Thomas Hertog, 'Eternal Inflation without Metaphysics').

eternal-inflation theory, another group of physicists were working on an experiment that might just possibly create Hawking radiation, not at the border of a black hole or from an event horizon in the early universe but in a laboratory. Daniele Faccio of the University of Insubria, Italy, and his team of researchers reported in a paper accepted by *Physical Review Letters* in late September 2010 that they had succeeded.[34] Their experiment involved firing laser light into a block of glass.

The idea is that as the laser pulse moves through the glass block it changes the speed at which light is capable of travelling there (the 'refractive index' of the glass). Light near the pulse is slowed more and more as the pulse passes through and the refractive index changes. If a pulse (call it pulse A) were sent chasing after a slower, weaker pulse (pulse B), it would gradually catch up and that would reduce the speed of light near pulse B. Pulse B would slow down more and more, eventually so much that it would get stuck. The leading edge of pulse A, acting like the event horizon of a black hole, would have sucked it in.

Recall the discussion of Hawking radiation: pairs of particles continually appear. The two particles in a pair start out together and then move apart. After an interval of time too short to imagine, they come together again and annihilate one another. Near the event horizon of a black hole, before the pair are able to meet again and annihilate, the one with negative energy may cross the event horizon into the black hole. The particle with positive energy might fall into the black hole, too, of course, but it doesn't have to. It's free of the partnership. It can escape as Hawking radiation. To an observer at a distance it appears to come out of the black hole. In fact, it comes from just outside. Meanwhile, its partner has carried negative energy into the black hole.

Faccio and his team watched just such particles – photons in this case – to see whether, as the pulse passed through the glass, its event horizon would sweep in one of a pair, allowing the other to escape as Hawking radiation. They set up a camera, focused it on the block of glass, and then fired 3,600 pulses from the laser. The camera recorded a faint glow in exactly the range of frequencies that Hawking radiation predicts. Carefully ruling out other sources of the glow, the researchers decided that they had in fact observed Hawking radiation.

Might *this* get Hawking a Nobel Prize, which is very seldom given for even the most promising theories if there is no experimental or observational evidence to support them? In November 2010, not long after the announcement of the experiment in Insubria, I asked Hawking whether he thought Faccio and his team had actually discovered Hawking radiation. His reply was enigmatic: 'I will not get the Nobel Prize.'

20

'My name is Stephen Hawking: physicist, cosmologist and something of a dreamer'

HAWKING'S BOOK *THE GRAND DESIGN*, WRITTEN WITH Leonard Mlodinow, appeared in the early autumn of 2010 with a subtitle that sounded rather un-Hawkinglike. *New Answers to the Ultimate Questions of Life*. Nothing here of the wry humour of previous titles – a 'brief' history – the universe in a 'nutshell'. Apparently this book was going to get serious.

The Grand Design pulled together the thinking and work Hawking has been doing for over half a century, to give us a thorough update on the state of the quest for a Theory of Everything. Here are Feynman's sums-over-histories, the anthropic principle, the meaning of 'models' and 'reality', the no-boundary proposal, information loss, the disparagement of modern philosophy (this time on the first page rather than the last), the battle with God. But right from the beginning of the book one dramatic change is clear: the quest for a Theory of Everything has, indeed, fragmented.

Isaac Asimov once wrote that 'of all the stereotypes that have plagued men and women of science, surely one above all has wrought harm. Scientists can be pictured as "evil", "mad", "cold", "self-centred", "absent-minded", even "square" and yet survive easily. Unfortunately they are usually pictured as "right" and that can distort the picture of science past redemption'. Stephen Hawking's startling about-faces, which you have witnessed throughout this book, shatter that stereotype. Hawking has a robustly healthy history of pulling the rug out from under his own assertions. But, as we've seen, what appear to be about-faces have hardly ever actually been steps backwards or reversals. In Hawking's own version of the game Snakes and Ladders, snakes don't take him further from his goal. They just lead him off on more promising pathways. Be that as it may, giving up hope of discovering a fundamental Theory of Everything is an enormous shift – one that Hawking would not have made unless faced with truly unyielding evidence that it was warranted, the only way forward.

Another thing that is clear from the start of *The Grand Design* is that Hawking no longer regards string theory with the suspicion he once did. It's not easy to pin down precisely when he changed his mind on this issue. Most accounts, not his own, have him still rather *anti*-strings well into the 1990s. However, he told me in 1990 (Chapter 13) that he thought superstring theory had become the most promising route towards a Theory of Everything. He was right, but . . . with a twist.

The newest candidate, and perhaps the final claimant

for that title of 'ultimate theory of the universe', is M-theory. As a theory of everything, in the time-honoured description, it is a little disappointing. M-theory is not simple. You can't print it on a T-shirt. It doesn't fulfil the promise of Wheeler's poem. It doesn't measure up to the Pythagorean standard, where beautiful clarity is a guide to truth. Does that mean it might be wrong? Hawking's attitude towards it is not that it is right, or ultimate, but that it is the best we are ever going to do.

M-theory is not a single theory. It is a collection of theories. Hawking describes them as a 'family of theories'. Each member of the family is a good description of observations in some range of physical situations, but none is a good description of observations in *all* physical situations. None can account for 'everything'. The theories may look very different from one another, but all are on an equal footing, and all can be thought of as aspects of the same underlying theory.[1] We don't yet know how to formulate that deeper theory as a single set of equations and arguably never will.

Hawking and Mlodinow compare the situation to a flat map of the Earth. Because the Mercator projection used for such maps makes areas further north and south look larger than they really are in relation to other parts of the world (the distortion becomes more and more pronounced the further north or south on the map you go), and the North and South Poles are not shown at all, the entire Earth ends up much less accurately mapped than it would if we used, instead, a collection of maps, each one covering a limited region, overlapping. Where the maps overlap, they don't conflict; that bit of the landscape looks the same regardless of which of the overlapping maps you're consulting. Each map is reliable and useful for the area it represents. But no single flat map is a good representation of the Earth's surface. Just so, no single theory is a good representation of all observations.[2]

Today, theorists recognize five different string theories and supergravity, a version of which Hawking had high hopes for in 1980, as the family of approximations of the more fundamental theory, M-theory. The six approximations are like the smaller maps in Hawking's map analogy.

While this situation may not meet our most idealistic expectations for a complete understanding of the universe, we needn't sit around too long moaning about our ignorance of the fundamental, comprehensive, underlying theory. There are things we do know about it. There are ten or eleven dimensions of time and space. There are point particles, vibrating strings, two-dimensional membranes, three-dimensional objects, and other objects occupying up to nine (or, in some versions, ten) dimensions of space – in other words, p-branes.

We have already talked about the idea that extra space dimensions beyond the three we experience may be escaping our notice because they are curled up tightly, and that the astounding number of different ways they can curl was at first discouraging to those who were hoping that string theory was going to be the unique theory of everything. Earlier we used the analogy involving a garden hose to help understand the curling up. Hawking and Mlodinow have found a better analogy for it.

They ask us to imagine a two-dimensional plane. It could be, for instance, a piece of paper. It is two-dimensional because two numbers (horizontal and vertical coordinates) are needed to locate any point on it. It may not occur to you that a drinking straw is also two-dimensional. To designate a point on it, you need to show where the point is along its length and also where in its circular dimension. But suppose your straw is very, very thin. You'd hardly feel the need to think about where the point is in its circular dimension. If it were extraordinarily thin, a million-million-million-million-millionth of an inch in diameter, Hawking suggests, you wouldn't think it had a circular dimension at all. That is the way string theorists encourage us to think about the extra dimensions – curled or curved on a scale so small that we don't notice them. They speak of them as being curled up into 'internal space'.

In the early to mid-1990s, theorists were becoming less and less discouraged by the astoundingly many ways dimensions can curl. One change was a new understanding that the different ways of curling up the extra dimensions are nothing but different ways of looking at them from our vantage point in four dimensions. However, as Andrei Linde suggested, the way the extra space dimensions are curled up is crucial. In each universe, it determines the universe's apparent laws of nature. However many solutions there are in M-theory for the ways internal space can be curled, that is how many different types of universe are allowed, all with different laws. The number is too large to comprehend.

Hawking suggests we think of the emergence of these universes by imagining something like Eddington's balloon analogy, the balloon with the ant crawling on it, only this time it isn't a balloon and the ant is missing. In his 2006 lecture at Caltech, he advised his audience to picture the expanding universe as the surface of a bubble.

Imagine, then, the formation of bubbles of steam in boiling water. Many tiny bubbles form and disappear. These are universes that expand only a little and collapse before they get beyond microscopic size. No hope of galaxies, stars, or intelligent life in them. Some, however, start out just as tiny but grow large enough so that they are out of danger of collapsing, at least for a long, long while. These expand at first at an ever-increasing rate, undergoing what we have come to call 'inflation'.

The Grand Design revisits Richard Feynman's idea that a particle travelling from one point to another in quantum physics has no definite position while it is getting to its destination. That has been taken to mean that it takes no path. As we've seen, Feynman realized that it could just as easily be said to take every possible path simultaneously. In this light, consider the possibility of a great many universes, the sort of situation we have in eternal inflation. It doesn't suffice only to say that each universe has a different history. In fact, thinking of sums-over-histories, each universe has *many* possible histories and also many possible states later in its existence. Most of those states are totally unsuitable for the existence of life of any sort. There are only a minuscule few of the universes that would allow creatures like us to exist.

Among all the possible alternative universes, only one is completely uniform and regular. Calculating the probability of this sort of universe, we find that it is very likely indeed. In fact, it is the most probable universe of all, but it isn't our universe. A universe like that, without small irregularities in the early universe that show up now as small variations in the CMBR, could never be a home for us. Ours has to be a universe with some regions slightly more dense than others, so that gravitational attraction can draw matter together to form galaxies, stars, planets and, maybe, us. As Hawking had put it in his 2006 Caltech lecture, 'The [CMBR] map of the microwave sky is the blueprint for all the structure in the universe. We are the product of the fluctuations in the early universe.'[3] Luckily there *are* many universe histories that are only slightly un-uniform and irregular. These are almost as probable as the one that is completely uniform and regular. We don't know how many alternative universes end up producing something like 'us', but we do know it did happen once.

Another familiar concept that is significant in Hawking's thinking about M-theory is the fact that on the quantum level of the universe we cannot observe without interfering, without changing the very thing we are trying to observe. More important and less familiar to most of us, no matter how carefully and completely we observe the present, the part of the past that we cannot observe is, like the future, indefinite. It exists as a range of possibilities, some more probable than others. Putting that together with Feynman's sums-over-histories, Hawking concludes that 'the universe doesn't have just a single history, but every possible history, each with its own probability; and our observations of its current state affect its past and determine the different histories of the universe'.[4] That should not sound entirely unfamiliar. We saw earlier how Hawking and Hartle used sums-over-histories when they were developing their no-boundary proposal. What has occurred in Hawking's thinking is a change of emphasis, the realization that the ability of our observations of the present to decide among those histories has enormously significant implications for our understanding of the universe.

Go back to thinking about Feynman's method of considering all the possible paths a particle might follow from its starting point to its end point. It's not so easy to do with the history of a universe. We don't know point A (the beginning), but we do, in the case of our own universe, know quite a lot about point B, where we are today. Hawking asks us to consider all the histories that satisfy the no-boundary condition (histories that are closed surfaces without boundaries – recall the globe of the Earth) and that end with the universe we know today (point B). There is a vast range of point As, though we can't say they include universe histories starting off in 'every possible way' because we are limiting them to those that satisfy the no-boundary condition. If we were to start our thinking at point A, we'd end up with many possible point Bs, some of which are similar to our universe today, but most of which are not.

Hawking is recommending, instead, what he calls his 'top-down' approach to cosmology, tracing the alternative histories from the top down, backwards from the present time. It is a new view of cosmology, and, for that matter, a new view of cause and effect. The universe doesn't have a unique observer-independent history. We create the history of our universe by being here and observing it. History doesn't create us.

Take, for example, the question of why there are only four un-curled-up dimensions in our universe. In M-theory there is no overall rule that a universe must have four observable dimensions. Top-down cosmology says that there will be a range of possibilities that includes every number of large space dimensions from zero to ten. Three dimensions of space and one of time may not be the most probable situation, but that is the only sort of situation that is of interest to us.

Considering the universe in the old way, from the 'bottom up', there seems to be no discoverable reason why the laws of nature are what they are and not something different, why the universe is fine-tuned for our existence. But we do observe the laws of nature to be what they are, and we are here. Why not start with that? Our presence is hugely significant. Out of the enormous array of possible universes, our presence 'selects' those universes that are compatible with our existence, and makes all the rest of them *almost* irrelevant (though we shall see about that as Hawking continues).

With the no-boundary universe we no longer needed to ask how the universe began. There was no beginning. With M-theory we no longer need to ask why the universe is fine-tuned for our existence. It is our existence that 'chooses' the universe we live in. In effect, we fine-tune it ourselves. The anthropic principle has come to its full strength indeed. As Hawking puts it, 'Although we are puny and insignificant on the scale of the cosmos, this makes us in a sense the lords of creation.'[5]

Now, the question arises: can we test this theory? Hawking writes that there may be measurements capable of differentiating the top-down theory from others, to support it or refute it. Perhaps future satellites can take such measurements. In his 2006 Caltech lecture, Hawking had mentioned the 'new window on the very early universe' that the detection and measurement of gravitational waves would open for us. Unlike light, which was scattered many times by free electrons before freezing out when the universe was 380,000 years old, gravitational waves reach us from the earliest universe without interference from any intervening material.[6]

Hawking extends top-down thinking to the emergence of intelligent life on Earth. He offers an eloquent account of the manner in which our universe, our solar system, and our world are incredibly fine-tuned to allow our existence, far far beyond any reasonable expectation. Nevertheless, in a restatement of the anthropic principle in a simple and unarguable way, he tells us that 'Obviously, when the beings on a planet that supports life examine the world around them, they are bound to find that their environment satisfies the conditions they require to exist.'[7] Just as we, by the fact of our presence, choose our universe, we choose a history of this Earth and our cosmic environment that allows us to exist.

In *The Grand Design*, Hawking seems no longer to entertain doubts that everything is determined. The information paradox, by whatever means it was banished, has ceased to be a bother. He states, unequivocally, that 'The scientific determinism that Laplace formulated is . . . in fact, the basis of all modern science.'[8] *That*, of course, he had never called into question. His earlier suggestion, regarding the implications of information loss, was that all of modern science might be wrong. Such fears apparently have been put to rest, for he goes on to say that scientific determinism is 'a principle that is important throughout this book'.[9] And, later, 'This book is rooted in the concept of scientific determinism.'[10]

Scientific determinism applies to us humans too: 'It seems,' he writes, 'that we are no more than biological machines and that free will is just an illusion . . . Since we cannot solve the equations that determine our behaviour, we use the effective theory that people have free will.'[11] We might wish Hawking had spent a little more time in the book with this issue. Important scientific work has taken place having to do with human free will – some of it supporting his view and some of it not – but Hawking does not discuss it. He has made his own choice. His comment that the world is in a mess because, 'as we all know, decisions are often not rational or are based on a defective analysis of the consequences of the choice', also leaves one wishing for more discussion. The comment comes across as uncharacteristically 'throw-away', compared with Hawking's thoughtful comments about the world situation in his lectures and public statements.

Determinism turns out, however, to be a somewhat complicated concept and not as rigid as we might suppose. As we saw earlier in this book, on the quantum level of the universe we have to accept a somewhat revised version of determinism in which, given the state of a system at any one time, the laws of nature determine the *probabilities* of various different futures and pasts rather than dictating the future and past precisely. As Hawking puts it, 'Nature allows a number of different eventualities, each with a certain likelihood of being realized.'[12] You can test a quantum theory by repeating an experiment many times, noting how frequently different results occur and whether the frequency of their occurrence fits the probabilities the theory predicted.

Hawking mentions again the ideas that he and I discussed in the old DAMTP common room back in 1996, ideas for which some people were criticizing him at that time. His words to me back then were 'We never have a model-independent view of reality. But that doesn't mean there is no model-independent reality. If I didn't think there is, I couldn't go on doing science.' Now, in *The Grand Design*, he writes, in italics for emphasis, '*There is no picture- or theory-independent concept of reality.*' He goes on to say that this is 'a conclusion that will be important in this book'. This statement rephrases the first part of his statement to me, substituting 'concept' for 'view', but not the second part. We are left to wonder whether the rest still holds.

Hawking lists two other ways of thinking about 'reality' that he is rejecting. One is the 'realist' viewpoint of classical science based on the belief that a real, external world exists, a world that can be measured and analysed – that is the same for every observer who studies it. The other is what Hawking calls the 'anti-realist' viewpoint. This viewpoint is so insistent on confining itself to empirical knowledge gleaned through experiment and observation that it has little use for theory and ends up self-destructing with the notion that because anything we learn is filtered through our brains, we can't really count on there *being* such a thing as empirical knowledge.

Hawking believes that his own 'model-dependent realism' makes the argument between realism and anti-realism unnecessary. He insists it is only meaningful to ask whether a model agrees with observation, not whether it is 'real'. If more than one model agrees with observation, you don't have to argue which is more 'real' or 'right'. 'Our perception – and hence the observations upon which our theories are based – is not direct, but rather is shaped by a kind of lens, the interpretive structure of our human brains.'[13] That goes for everyday experience, he says, not just in science. On that level too, whether we are consciously devising models or not, we never have a model-independent view of reality. Nevertheless, our model-dependent views of reality are not worthless. They are the way human beings come to understand and manage their world. Models stand and fall as they continue or cease to match observation and experience.

It isn't difficult to agree with Hawking. Unless I am in a state of denial – which we all probably are in part, sometimes – I do make my learning-progress through life in precisely that way. You and I come from different sets of experiences. Perhaps we might be able to agree to disagree without making claims about who is 'right' and who is 'wrong'. Would Hawking go so far as to apply his philosophy to the more extreme views that divide our world? That, possibly, is where he would invoke something like the second, decidedly Platonic, part of his statement, perhaps to say, 'But that does not mean there are no such things as "right" and "wrong"; if I didn't think there are, I couldn't go on living in any meaningful way.' On the other hand, there are claims that human values are products of our evolutionary history. In this way of thinking, 'right' is what has aided the survival of our species – nothing more profound or fundamental than that. If that is true (and what, after all, does 'truth' come down to in a discussion like this?), then model-independent *morality* is perhaps as illusive as model-independent reality.

Be that as it may, Hawking's discussion of 'reality' helps with something you may have been wondering about since Chapter 2. If no one has actually seen, for example, an electron, how do we know electrons are 'real'?

Though it's true that no one has ever seen an electron,

electrons are a useful 'model' that makes sense of observations of tracks in a cloud chamber or spots of light on a television tube. The model has been applied with enormous success in both fundamental science and engineering. But are electrons real? Though a great many physicists would say, yes, of course they are, that question according to Hawking is meaningless.

'Model-dependent realism', as he calls it, is a useful way of thinking about dualities – those situations in which two different, perhaps mutually exclusive, descriptions are necessary to gain a better understanding than either description alone can provide. Neither theory is 'better' or more 'real' than the other. Recall the most familiar example, wave–particle duality, which emerged in the early twentieth century with the discovery that when light interacts with matter it acts as though it must be particles, while experiments with the way light travels show that it acts as though it were waves.

All of which brings us back to think more knowledgeably about M-theory. As we've said, it appears that no mathematical model is able to describe every aspect of the universe. Each theory in the M-theory family can describe a certain range of phenomena. When these ranges overlap, the theories agree. In this manner they are all parts of the same theory, just as the smaller sections of the map in Hawking's analogy were all parts of the same map. But no single theory in the family is capable of describing all the forces of nature and the particles that we mentioned in Chapter 2, plus the framework of time and space where the universe game plays out. If this seemingly fragmented map is where the great quest must end, so be it, 'it is acceptable within the framework of model-dependent reality'.[14] We have no more fundamental theory that we can claim is *independent* of the models we know.

Hawking and Mlodinow write that all the universes in the multiverse were created out of nothing, arising naturally from physical law, and that they require no creator. They have oversimplified a bit to make their point. In eternal-inflation theory, which Hawking favours, universes don't arise from nothing. They arise from other universes. Somewhere in the past, there may have been a first universe and a first inflation sequence, where it all started off, or the repeating self-replication process may stretch back eternally into the past. Presumably the origin of that first universe (if there *was* a 'first universe') can be explained by the no-boundary-proposal, which leaves us precisely where *A Brief History of Time* stopped, asking those same profound questions that left plenty of room for God.

The Grand Design, however, addresses another puzzle, the fine-tuning mystery. Some who believe in God – *still* not warned off the God-of-the-Gaps theology which clings to instances where something seems unexplainable without God – will no doubt find it distressing that Hawking and Mlodinow have very successfully shown another plausible explanation, using the top-down method and multiverses. If you believe in God only as a necessary explanation, Hawking has once again cut you adrift. More interesting than the media attention Hawking's books gain for God/science issues is the fact that, for careful, thoughtful readers, they do lead to some profound inner debates. Those don't always end the way Hawking might expect.

In their final chapter, Hawking and Mlodinow address

the question of where the physical laws come from, introducing the discussion with the following comment: 'The laws of nature tell us *how* the universe behaves, but they don't answer *why*.' At the end of *A Brief History of Time* Hawking had written that the answer to that question would be to know the mind of God. Now he has broken the question into three parts: 'Why this particular set of laws and not some other?', 'Why is there something rather than nothing?' (the laws being part of the 'something') and 'Why do we exist?'

To help address the first of those questions, Hawking and Mlodinow list the laws that are necessary in a physical universe that looks like our own. They must be a set of laws that have a concept of energy in which the amount of energy is constant, not changing over time. Another requirement is that the laws must dictate that the energy of any isolated body surrounded by empty space will be positive. And there must be a law like gravity. The theory of this gravity must have supersymmetry between the forces of nature and the particles of matter they govern. Adopting the top-down method, the answer to 'Why this particular set of laws and not some other?' can be simply 'Because any other set of laws would make it impossible for us to be here asking that question.' That would be an answer invoking the anthropic principle, but M-theory has a little more to say on the issue than that: because of all the different ways the extra dimensions curl up, with each universe having laws determined by how they curl in that universe, there *will* certainly be a universe around that has these laws.

To help address the third of those questions ('Why do we exist?'), Hawking and Mlodinow introduce us to a computer game known as 'The Game of Life'. It's a fascinating game, invented back in 1970 by John Horton Conway, then a Cambridge mathematician. The layout looks like a chessboard, with some squares 'alive' and some 'dead'. A very simple set of rules dictate 'deaths', 'births' and 'survival' as the game moves from 'generation to generation'. It soon becomes evident that extremely simple rules can play out in very complicated ways. Remember the 'alien who has never experienced our universe' in Chapter 2. Someone coming in on this game after it's been going a while will be in a similar position, able to deduce 'laws' from what's going on, laws that seem to govern the formation and behaviour of elaborate groupings of the live and dead squares – laws that however are not among the simple original laws at all but that arise out of them. The game is a simple example of 'emergent complexity' or 'self-organizing systems'. It helps us comprehend, for example, how the stripes on a zebra or patterns on a flower petal occur from a tissue of cells growing together.

Conway invented this game as an attempt to find out whether in a 'universe' with extremely simple fundamental rules, objects would emerge that were complex enough to replicate themselves. In the game, they do. They could even, in a sense, be thought of as 'intelligent'. The bottom line is that a very simple set of laws is capable of producing complexity similar to that of intelligent life. In Hawking's words, 'It is easy to imagine that slightly more complicated laws would allow complex systems with all the attributes of life.'[15] There is disagreement as to whether such life would be self-aware.

That may seem to answer the question 'Why do we

exist?' Is it a complete answer? In the Game of Life, it doesn't matter what pattern you start with – any 'initial conditions' will give you those same sorts of results – but not just any set of laws, because it is the laws that determine the evolution of the system. Which refers back to the first question: 'Why this particular set of laws and not some other?'

Summing up so far, then, with regard to Hawking's three questions, he and Mlodinow have answered the first ('Why this particular set of laws and not some other?') for our own universe with the idea that we have a particular set of laws because of the way the extra dimensions are curled up. Can they answer it for the overarching laws that govern the entire string theory landscape, the entire multiverse, the laws we do not yet know? They have said that of all the supersymmetric theories of gravity, M-theory is the most general, making it the only candidate for a complete theory of the universe. 'Candidate' it still is, awaiting proof, but Hawking believes it promises to be a model of a multiverse that includes us, because there is no other consistent model.[16]

They have answered the third question ('Why do we exist?') by saying that, in the multitude of possible universes, a universe that allows for our existence is highly probable and, from there, even if there were only a very simple set of laws in place (dictated by the way the extra dimensions curl), it is not difficult to arrive at us (think of the game).

The second of the questions ('Why is there something rather than nothing?') is more fundamental and much more difficult. An answer would have to account for much more than our universe, its laws and us. It would have to account for the very existence of the unknown theory that underlies the M-theory family of theories. Hawking thinks the multiverse ruled by this family of theories 'creates itself', but he does not explain how. Even the often-heard statement that 'nothingness is unstable and tends to decay into something' implies that a certain set of probabilities must already be in place. This question – 'Why is there something rather than nothing?' – Hawking and Mlodinow have left unanswered.

Though many of Hawking's colleagues have high hopes indeed for M-theory, few join him in his enormous optimism about its total explanatory power. There were questions that hung in the air at the close of *A Brief History of Time*, eloquently stated, evoking grand hopes that we might some day solve these mysteries. The attempt actually to answer all of them in *The Grand Design* falls short.

Critics of the book were not enthusiastic, and their lack of enthusiasm seemed based not on disagreement with the authors but on disappointment that this was not a more powerful book. *The Economist* commented that 'whenever the going threatens to get tough, the authors retreat into hand-waving and move briskly on'[17] . . . 'There are actually rather a lot of questions that are more subtle than the authors think.'[18] As for the claim that the ideas presented in the book have passed every experimental test to which they have been put, that is 'misleading' . . . 'It is the bare bones of quantum mechanics that have proved to be consistent with what is presently known of the sub-atomic world. The authors' interpretations and extrapolations of it have not been subjected to any decisive tests, and it is not clear that they ever could be.'[19] Dwight Garner, in

The New York Times, wrote that 'the real news about "The Grand Design" is how disappointingly tinny and inelegant it is. The spare and earnest voice that Mr. Hawking employed with such appeal in "A Brief History of Time" has been replaced here by one that is alternately condescending, as if he were Mr. Rogers explaining rain clouds to toddlers, and impenetrable.' Garner also accused Hawking of 'Godmongering', as the writer Timothy Ferris has called it, when an author who is not religious makes statements about God and religious belief solely to sell books.[20]

There is nothing disappointing or feeble, however, about the 'top-down' approach and Hawking's and Mlodinow's presentation of M-theory. Hawking's statement about their implications for the study of science is one of the best passages in the book. Hawking believes we are:

> at a critical point in the history of science, in which we must alter our conception of goals and of what makes a physical theory acceptable. It appears that the fundamental numbers, and even the form, of the apparent laws of nature are not demanded by logic or physical principle. The parameters are free to take on any values and the laws to take on any form that leads to a self-consistent mathematical theory, and they do take on different values and different forms in different universes. That may not satisfy our human desire to be special or to discover a neat package to contain all the laws of physics, but it does seem to be the way of nature.[21]

What does it mean for banishing belief in God? The

disparagement of belief in *The Grand Design* is much more frequent and assertive than in Hawking's other books. But the more grand the design becomes – and Hawking's is a spectacularly grand design – the more those readers who find Hawking's science convincing *and also* believe in God are bound to find cause for wonder in the elegant complexity of the multiverse vision.

The religious discussion, sometimes heated, erupted as soon as the book reached reviewers and the public. Perusing all that, one is surprised to find how many of the discussants, coming from both viewpoints, clearly have not read Hawking's and Mlodinow's book. Among those who have, it is perhaps not so surprising to find atheists who nevertheless do not feel that Hawking and Mlodinow succeed in banishing the need for a creator, and theists who think he has done a pretty thorough job of it. It seems that one's atheism or theism can remain essentially untouched by Hawking's arguments, perhaps because such choices are very often based on reasons that have nothing to do with science. Among those who disagree with Hawking, the most interesting arguments take two forms.

(1) In spite of the hugely comprehensive explanatory power of Hawking's model – and even if this should some day turn out to be model-*independent* reality – a question that is as old as human thought is still left hanging: why is there something – a grand design – rather than nothing? The 'something' in the M-theory model is far, far grander and more extensive than has ever been proposed before. But why is there anything to have a model of? Granted, a religious answer to that question, 'God', does no better than a scientific one which argues that 'There is a

fundamental mathematical logic that does not allow "nothing"'. 'God' and 'mathematical logic' are both, indeed, 'something', so both prompt a question: who created God? Who set down the mathematical logic?

You might expect Hawking to say that we, the observers, are the answer. Where the buck stops. No need to ask who or what created us. We are here. Our presence 'chooses' that all the rest of it exists. No other argument is possible or needed.

Hawking does not use that argument. Not in his book, nor when he and I spoke in his office in November 2010. I mentioned the question he had posed in *A Brief History of Time*: 'What is it that breathes fire into the equations and makes a universe for them to describe?' Using top-down thinking, might the answer be – us?, I asked. His reply was 'No.'

(2) The second argument relies on the idea of 'model-dependent reality'. Those who present this argument point out that Hawking and Mlodinow have written that each of us has a personal model of the world that fits our life experience and attempts to make sense of it. Our models will in many ways be the same, but not in all ways, because our experiences are different. Hawking's model does not have to include any experience of the presence and power of God. He has evidently had no experience of that and no reports of it from people he considers reliable. Why should he make it part of his model? He has no need of it.

On the other hand, for someone who has had experience of the presence and power of God, Hawking's model is inadequate. Their model must include that experience. (And if you've already decided such experience can't be 'real', then you're already violating the tenets of 'model-dependent reality' and should leave this discussion.) Suppose that in addition to having experience of God you agree with Hawking's science. Perhaps you are a physicist. Then your model will also have to include not only God but all the amazing findings and speculations of twentieth- and twenty-first-century scientists. Are you in trouble?

Luckily, a model that includes both belief in God and in science has not been ruled out by either *A Brief History of Time* or *The Grand Design*. It is possible and not crazy, at least according to many of Hawking's colleagues, and both theists and atheists who have entered the debate. Read *The Grand Design* carefully and with an open mind, and you will probably agree that is so. We have, then, two different models, one with God and one without. According to 'model-dependent reality', it is meaningless even to ask whether one is more 'real' than the other, and inconsistent for Hawking to be so sure that his Godless-universe model represents 'reality'.

You may be thinking that it is not appropriate to apply model-dependent reality to our personal world-views, but Hawking seems to think it is when he states:

> Our brains interpret the input from our sensory organs by making a model of the outside world. We form mental concepts of our home, trees, other people, the electricity that flows from wall sockets, atoms, molecules and other universes. These mental concepts are the only reality we can know. There is no model-independent test of reality.[22]

Undeniably our models will include more than these

physical attributes of the universe. They will, as we've said, include convictions of what is right and what is wrong. In an extreme case: are we required to respect a 'model' created through a lens of hatred, selfishness and prejudice? We know from Hawking's public statements regarding human rights and politics that at least on a practical level he does not extend model-dependent reality to include model-dependent morality.

In My Mind I am Free

When Stephen Hawking and I were first discussing my plans for this book in November 2010, he asked me to be sure to include his latest ideas about eternal inflation and the observations he suggests could help verify it. You read about that in Chapter 19. His second request was that I not fail to mention his new television series that would be shown in the UK early in 2011. It is a three-part documentary called, in Britain, *Stephen Hawking's Universe* (using the same title as an earlier series) and, in the USA, *Into the Universe with Stephen Hawking*.

This time it is not from a lecture hall or his office that Hawking invites us to join him on an adventure through time and space, but from the Hall of his college, Gonville and Caius. The long wooden tables have been pushed back against the walls. Hawking sits in his wheelchair, alone in the splendid wood-panelled room. His portrait can be seen on the wall along with portraits of other college luminaries of the past and present. In the familiar voice, he begins: 'Hello. My name is Stephen Hawking: physicist, cosmologist and something of a dreamer. Although I cannot move, and I have to speak through a computer, in my mind I am free.'[23] That proves to be abundantly correct as we voyage with him far out into the universe, in time and space, to encounter the wonders he knows or has good reason to believe are there, and into his own imagination to find the landscapes and creatures he thinks we *might* find. In this spectacularly filmed triptych, Hawking, and those responsible for the state-of-the-art computer animation and astronomy photography, succeed in evoking a chilling awareness of the enormous distances and numberless galaxies – the sheer, stupefying, inconceivable vastness of space and time.

Hawking doesn't narrate the series himself. Even before his introductory sentence ends, the voice has seamlessly shifted to the voice of Benedict Cumberbatch, the actor who played the young Stephen Hawking in the film *Hawking*. Occasionally Hawking's own voice comes back for a few seconds, and so does the scene in the Hall of Caius, just to remind us who's really telling the tale.

For the first of the trilogy, the animators and Hawking have invented fantastic extraterrestrials, a few of whom he should shoot down immediately with his rocket-launcher if they ever appear in his office – never mind the Prime Directive. Although Hawking insists that the best place to look if we want to enquire about life in our universe is here at home, where there exists the only *known* life, he does lead us far beyond Earth and the solar system and our own galaxy. He tells us that there may be life in the cosmos so strange that we wouldn't recognize it as life. And we hear, with a shudder, that it isn't 'what they're like but what they can *do* that counts': for instance, arrive in swarms of advanced-technology spacecraft and within a few seconds pirate all the energy of the sun by encasing it with mirrors, focusing that energy to create a wormhole.

And yes, when it comes to wormholes, Hawking is back in that game, no matter how discouraging some of his recent statements have been about them. It might indeed be possible for a very advanced civilization to create one. Is all of this highly unlikely? Perhaps, but Hawking ends by reminding us of what we learned earlier in the show about our own improbable emergence here on Earth: 'We only have to look at ourselves to know that extremely unlikely things can and do happen all the time.'[24]

In 'Time Travel', the second of the trilogy, Hawking admits to being 'obsessed by time'. He is particularly curious about 'how our whole cosmic story ends'. This segment is a tour-de-force overview of the possibility of time travel. Tiny wormholes through space and time constantly form and re-form, linking separate spaces and separate times. Could one be captured and enlarged many trillions of times and used as a time machine? Is travel to the past by this or any other means possible? Hawking has printed an invitation, copies of which he hopes might survive several thousand years, laid out champagne and delicious-looking food, and hung a banner, 'Welcome, Reception for Future Time Travellers'. His invitation gives all the necessary information and coordinates required to find him and share this feast. No guests arrive. Perhaps none of the invitations survived long enough, but in a Cambridge college, at least one or two of them probably would. Because no one responds to his invitation, and for other reasons, such as a 'fundamental rule that governs the entire universe' that cause happens before effect, unsolvable paradoxes, and the inevitable radiation feedback that would destroy a wormhole before you could use it, Hawking concludes that we cannot travel to the past. But we can to the future, and we needn't depend on wormholes. Einstein realized that time doesn't flow at the same rate everywhere, and it's a well-established fact that he was right. Matter drags on time, slowing it down, and that means a massive body can serve as a time machine. A spacecraft flying near the event horizon of a super-massive black hole, with skilful enough navigation and sufficient speed not to fall in, would make a noticeable jump forward in time. After about five years (in your personal time) at the black hole, you could find that ten years had passed on Earth. Travelling near the speed of light would work much better, though the same slowing of time that makes this a quicker way to get to the future also makes it impossible to travel *at* or *above* the speed of light. Passengers travelling near light speed (in a wonderfully conceived train that must circumnavigate the Earth seven times a second) could return after only a week measured in their own personal time, to find the world had moved forward by a hundred years.

The third segment of the trilogy is a genuine climax to the series. First, Hawking takes us to the early universe to observe the Big Bang. He admits that it would all have been completely dark, because light didn't exist yet – so you couldn't actually see it. Space didn't exist either. There was no 'outside' this event from which to view it. 'All that there was was inside.'[25] We journey through the inflation era, witness the annihilation of matter and antimatter with the whole future of the universe hanging on there being just a minuscule bit more matter than antimatter, and learn the enormous significance of the force of gravity again and again in creating the universe we know. We see Wheeler's democracy vividly demonstrated as a load of ball-bearings roll through the doors into the Hall of Caius. Animation takes over to show them all perfectly equidistant from one another, forming a gridlock that would never have yielded to make 'our' universe. But remove just a handful of the ball bearings, one here, one there, from the perfect pattern, and gravity gets a foothold – another example, Hawking points out, of how imperfection is absolutely necessary in our universe. Eventually we do get to the universe as we know it, and beyond. Hawking takes us into the future, and he is a truly amazing 'futurist'. His plea is eloquent in support of colonizing other planets, and his vision for Earth harrowing. He is still as concerned as he was in his youth about the hazard of nuclear weapons – 'We are clever enough to design such things, but I'm not sure we are clever enough not to use them'[26] – but some of the other possibilities are even more likely to cause nightmares. He also admits that there are staggering obstacles we will encounter in finding another home in the universe.

What about the question of how our whole cosmic story ends? 'The fate of the universe,' he tells us, 'depends on how dark energy behaves.'[27] Is it increasing? Will it continue to push space apart, driving the expansion? Will all the particles eventually be so far apart that nothing at all can happen? Or might dark energy's strength diminish, allowing gravity to pull everything together again in a Big Crunch? We don't know. That end is a very long way off. Hawking says that on one of his trips to Japan he was requested not to mention the fate of the universe lest it upset the stock market, but he thinks it would be premature to sell up. Hawking's suggestion for survival at that extremely distant time is that we might find out how to travel to another universe. We have about thirty billion years to work out how to do that.

In *Into the Universe with Stephen Hawking*, Hawking chooses not to disparage belief in God directly. After a stunning series of sequences that show many turning points in the history of the universe where it could so easily have gone wrong, and overwhelming us with the sheer grandeur and elegant genius of it all, he says, in his own voice: 'Perhaps science has revealed there is some higher authority at work, setting the laws of nature so that our universe and we can exist. So is there a grand designer, who lined up all the good fortune? In my opinion, not necessarily.'[28] And he goes on to talk about the anthropic principle and the possibility of many different kinds of universes. His goal seems to be to lay out what we know, and what we surmise, and what his opinion is, and to make his audience as excited, awed and curious about the universe as he is. From there, we're on our own. He even contradicts the claim made in the subtitle of *The Grand Design*, that we know the ultimate answers. In the final moments of the documentary, he says, 'Some day we may solve the ultimate mystery . . . discover why the universe exists at all.'[29] *Into the Universe with Stephen Hawking* leaves us not only awed by the universe, but also agape at *ourselves*, us creatures who have actually managed to discover and understand so astonishingly much. But there still remains that question. Errol Morris, director of the film *A Brief History of Time*, said there was a challenge he always set himself when he made a film: 'To extract a situation's truth without violating its mystery.' Hawking has met that challenge.

This magnificent trilogy, with a splendid musical score,

is perhaps close to the dream that was not realized in the film of *A Brief History of Time*, in spite of Morris's skills. It is probably what Hawking originally hoped that film would be like, except that in twenty-five years computer animation has vastly improved, and the dream itself has grown exponentially.

2010–2011

Before I began writing this book, I went to visit Hawking in his office in November 2010 for the first time in several years. The room had changed only a little. The large picture of Marilyn was not on the wall but, inexplicably, lying on the floor. The photographs of Lucy's boy William were still there on the shelves, and there was also a small colour photo of Hawking and Elaine lying among other papers near his computer screen. On the end of his desk nearest the door was a mysterious arrangement of stones in a large flat dish, emitting a faint vapour that seems to have no smell, unusual in that it emerges from the stones in flat, wing-like gauzy clouds that look as though they have a tiny hem at the edge. Judith Croasdell, Hawking's personal assistant, explained that it is a special kind of humidifier, chosen several years ago by Elaine, that makes it easier for Hawking to breathe. The liquid required for it is not ordinary water, and a large supply of bottles takes up a good part of the storage space under the side window. Outside that window is the 'pavilion' that was not yet built in 2000, but it doesn't spoil the view. The office has a tranquil, happy feeling to it.

My conversation with Hawking took place as usual with both of us sitting behind his desk facing his computer screen. Hawking was controlling the cursor on the screen by moving his cheek muscle. A small electronic beep came from the contraption attached to the back of his chair when he made a choice on the screen.

The computer program on the screen looked the same, but he also has the option of using another program. I could not tell how he was making the word choices on that one, and, in fact, it didn't seem to be operating very well for him. His writing speed had slowed down considerably. I was told that if the cheek or eye movement ceases to work there are other possibilities, including direct connections to the brain. He will deal with that if and when it becomes absolutely necessary. Not every bit of communication requires the computer. If he raises his eyebrows, that means yes. Mouth down means no. You can still tell when he is smiling. Hawking recently had surgery for cataracts and probably doesn't even really need to wear glasses now, but still does.

During our conversation the window blinds suddenly went down by themselves as it got darker outside. I had forgotten that such things happen automatically in this ultra-sophisticated building. The nurses changed shift. The one whose shift was ending – a dignified, gentle woman – came over and said goodbye to him without expecting a reply.

Whenever I talk with Stephen Hawking I try to phrase my questions in ways that allow him to answer with a simple yes or no, though he usually goes on to elaborate anyway. That afternoon I was particularly interested in asking him about what seemed a possible change in his views about independent reality (reflected in *The Grand Design*) since we discussed that in 1996. I quoted him the words he had used back then: 'We never have a model-independent view of reality. But that doesn't mean there is no model-independent reality. If I didn't think there is, I couldn't go on doing science.' I asked him whether he would change that now to say, 'Independent reality *is* that there is no independent reality.' His answer was 'I still think there is an underlying reality, it is just that our picture of it is model-dependent.'[30]

A Swipe at Immortality

In an interview in the *Guardian*[31] and a lecture at a 'Google Zeitgeist' meeting in London in the spring of 2011, Hawking bluntly revealed part of his own personal picture of reality. His headlined words were 'There is no heaven or afterlife . . . that is a fairy story for people afraid of the dark.' Hawking was, of course, expressing an opinion regarding something about which no one, including himself, has any scientifically provable knowledge whatsoever either for or against, but he explained his position by stating his own view of the human brain. One school of thought among researchers who study the brain sees the brain as a computer and the 'mind' as nothing more than a product of it, and Hawking apparently had decided to join this club. 'I regard the brain,' he said, 'as a computer which will stop working when its components fail. There is no heaven or afterlife for broken-down computers.' Ergo, no heaven or afterlife for us.

In reply to the interview question 'How should we live?', Hawking said, 'We should seek the greatest value of our action.'

There were, not surprisingly, numerous responses to Hawking's interview. Although some read it as a declaration of atheism, others pointed out that his statement was about belief in human immortality, not about belief in God. Not everyone who believes there is a God also believes in heaven or an afterlife. Other readers noted that it is often possible to transfer the entire intellectual content of a computer to a new computer or even a 'memory stick' as an old computer dies, and asked facetiously whether this might represent a sort of transmigration of the soul.

The *Guardian* printed one thoughtful response in full, even though it was longer than the original interview article itself.[32] The respondent, Michael Wenham, like Hawking, has ALS. 'For someone "facing the prospect of an early death",' Wenham wrote, 'with probably an unpleasant prelude, the idea of extinction holds no more fear than sleep. It really is insulting to accuse me of believing there might be life after death because I'm afraid of the dark.' Wenham called Hawking's statement 'both sad and misinformed. Openness to the theoretical possibility of there being 11 dimensions and fundamental particles "as yet undiscovered" shows an intellectual humility strangely at odds with writing off the possibility of other dimensions of existence.'

Wenham ended his response: 'I can't prove it of course, but on good grounds I'd stake my life on it, that beyond death will be another great adventure; but first I have to finish this one.'

Perhaps Stephen Hawking is up for another bet?

Going On

Hawking currently has two graduate students, and he still holds court in the common room at teatime, for there *is* now a common room only a short distance from

Hawking's office door, around the curve in the corridor, past the lift. A sign over the entrance reads 'Potter Room', but officially it is the 'Centre for Theoretical Cosmology' and is used not only at teatime but for meetings, lectures and conferences. The room is large and pleasant, with low tables, chairs and a counter for serving food and beverages in one corner, dimly lit as the common room in Silver Street was for most of the day. Large black chalk-boards – something the old common room lacked – cover a good part of two walls. I have never seen them without equations scribbled on them. In one corner of the room a bust of Hawking – the Director of Research for the Centre – stands on a plinth. It is a splendid likeness by the sculptor Ian Walters.

Hawking still lives in the large house that he built for himself and Elaine. He still goes often to concerts and the opera, especially Wagner – *Tannhäuser* planned for the next week at Covent Garden, when I visited him in November – though he hasn't been to Bayreuth recently. He still travels, when possible by private jet. In January 2011, he was once again at Caltech. When he attended a play in Los Angeles, *33 Variations*, in which Jane Fonda portrayed a musicologist in the early stages of ALS, Fonda seemed, in news reports, as thrilled to meet him as many fans would be to meet Jane Fonda. In March 2011, I had to rush to get a set of questions to Hawking before he headed back across the Atlantic to a conference at 'Cook's Branch' near Houston, Texas, a rural conference centre in a nature reserve where physicists from around the world gather yearly for meetings, eager to see one another, eager to sink their teeth into theoretical questions that leave the rest of us gaping, eager to rough it a bit, sleeping in bungalows with lazily turning ceiling fans.

In Cambridge, and when he visits Seattle or Arizona, where Lucy spends part of her time, Hawking's family – now including three grandchildren (Lucy's William, and Robert and Katrina's two children), and Jane and Jane's husband Jonathan – are comfortable with him. There is the new closeness to Lucy that developed when they wrote their books together. In an interview in April 2011, Hawking was asked, if time travel were possible, what moment in his own past he would revisit, what was the best moment of all. His answer was: 'I would go back to 1967, and the birth of my first child, Robert. My three children have brought me great joy.'[33] His mother Isobel, at the time of writing, is still alive, well into her nineties, and still orders him about occasionally. She has said, frankly:

> Not all the things Stephen says probably are to be taken as gospel truth. He's a searcher, he is looking for things. And if sometimes he may talk nonsense, well, don't we all? The point is, people must think, they must go on thinking, they must try to extend the boundaries of knowledge; yet they don't sometimes even know where to start. You don't know where the boundaries are, do you?[34]

John Wheeler called those boundaries, those frontiers not just of science but of human knowledge, 'the flaming ramparts of the world'. And, yes, we do know where those are. Not only somewhere out there in the distance. They fill our world.

As Hawking has said about his own adventures on the ramparts:

> With hindsight it might appear that there had been a grand and premeditated design to address the outstanding problems concerning the origin and evolution of the universe. But it was not really like that. I did not have a master plan; rather I followed my nose and did whatever looked interesting and possible at the time.[35]

Stephen Hawking returned to Cambridge from Texas and Arizona – where he had been visiting Lucy and William – in mid-April 2011, on the day I finished writing this book. Joan Godwin went over to cook him something to eat. His office was all set for him, the stones were emitting their vapour. 'The Boss' was back, ready to go on with his adventures as long as his health and his ability to communicate hold out . . . a child who has never grown up . . . still asking how and why questions . . . occasionally finding an answer that satisfies him . . . for a while.

Glossary

anthropic principle The idea that the answer to the question 'Why is the universe, as we find it, so suitable for our existence?' is that if it were different, we could not be here to ask the question.

antimatter Matter consisting of antiparticles.

antiparticle For every type of particle there exists an antiparticle with opposite properties, such as the sign of its electrical charge (for example, the electron has negative electrical charge; the antielectron, or positron, has positive electrical charge), and other qualities that we haven't dealt with in this book. However, the antiparticles of photons and gravitons are the same as the particles.

arbitrary element Something which isn't predicted in a theory but must be learned from observation. For example, an alien who had never seen our universe couldn't take any theory we have so far and use it to work out what the masses and charges are of the elementary particles. These are arbitrary elements in the theories.

Big Bang singularity A singularity at the beginning of the universe.

Big Bang theory The theory which says that the universe began in a state of enormous density and pressure and exploded outwards and expanded until it is as we see it today.

black hole A region of spacetime shaped like a sphere (or a slightly bulged-out sphere in the case of a rotating black hole) which cannot be seen by distant observers because gravity there is so strong that no light (or anything else) can escape from it. Black holes may form from the collapse of massive stars. This was the 'classical' definition of a black hole. Hawking showed that a black hole does radiate energy and may not be entirely 'black'. (See also **primordial black hole**.)

boson Particle with spin expressed in whole numbers. The messenger particles of the forces (gluons, W^+, W^-, Z°, photons and gravitons) are bosons.

boundary conditions What the universe was like at the instant of beginning, before any time whatsoever had passed. Also what it is like at any other 'edge' of the universe – the end of the universe, for example, or the centre of a black hole.

classical physics Physics that doesn't take quantum mechanics into account.

conservation of energy The law of science that says that energy (or its equivalent in mass) cannot be either created or destroyed.

cosmological arrow of time The direction of time in which the universe is expanding.

cosmological constant Albert Einstein introduced a 'cosmological constant', to counteract gravity, into his theory of general relativity. Without it, the theory predicted that the universe ought to be either expanding or collapsing, neither of which Einstein believed to be true. He later called it 'the greatest blunder of my life'. We now use the term to mean the energy density of the vacuum.

cosmology The study of the very large and of the universe as a whole.

dark energy The mysterious energy that makes up about 73 per cent of the universe and is thought to be responsible for the current acceleration of the universe's expansion.

determinism The idea that the future is completely predictable from the present, completely determined by the present.

Einstein's general theory of relativity (1915) The theory of gravity in which gravity is explained as a curvature in four-dimensional spacetime caused by the presence of mass or energy. It provides a set of equations that determine how much curvature is generated by any given distribution of mass or energy. It is a theory that we use to describe gravity at the level of the very large.

Einstein's special theory of relativity (1905) Einstein's new view of space and time. The theory is based on the idea that the laws of science should be the same for all freely moving observers, no matter what their speed. The speed of light remains unchanged, no matter what the velocity of the observer measuring it is.

electromagnetic force One of the four fundamental forces of nature. It causes electrons to orbit the nuclei of atoms. At our level it shows up as light and as all other electromagnetic radiation, such as radio waves, microwaves, X-rays and gamma rays. The messenger particle (boson) of the electromagnetic force is the photon.

electromagnetic interaction The interaction in which an electron emits a photon and another electron absorbs it.

electromagnetic radiation All forms of radiation that make up the electromagnetic spectrum, such as radio waves, microwaves, visible light, X-rays and gamma rays.

All electromagnetic radiation is made up of photons.

electroweak theory A theory developed in the 1960s by Abdus Salam at Imperial College, London, and Steven Weinberg and Sheldon Glashow at Harvard which unified the electromagnetic force and the weak force.

elementary particle A particle that we believe is not made up of anything smaller and that cannot be divided.

entropy The measurement of the amount of disorder in a system. The second law of thermodynamics states that entropy always increases, never decreases. The universe as a whole, or any isolated system, can never become more orderly.

escape velocity The speed necessary to escape the gravity of a massive body such as the Earth and escape to elsewhere in space. Escape velocity for the Earth is about 7 miles (11 kilometres) per second. Escape velocity for a black hole is slightly greater than the speed of light.

event A point in spacetime, specified by its position in time and space, as on a spacetime diagram.

event horizon The boundary of a black hole; the radius where escape velocity becomes greater than the speed of light. It is marked by hovering photons, which (moving at the speed of light) cannot escape and also cannot be drawn into the black hole. Light emitted inside it is drawn down into the black hole. To calculate the radius at which the event horizon forms, multiply the solar mass of the black hole (the same as for the star that collapsed to form it, unless that star lost mass earlier in the collapse) by 2 for miles or 3 for kilometres. Thus, a 10-solar-mass black hole has its event horizon at a radius of 20 miles or 30 kilometres. You can see that if the mass changes, the radius where the event horizon is will also change, and the black hole will change in size.

fermion For the purposes of this book you need to know that particles of ordinary matter (the particles in an atom, such as electrons, neutrons and protons) belong to a class of particles called fermions, and, like all fermions, they exchange messenger particles. A more technical definition of a fermion is a particle with half-integer spin which obeys the Pauli exclusion principle. We have not dealt with the exclusion principle in this book.

forces of nature The four basic ways that particles can interact with one another. They are, in order from strongest to weakest, the strong force, the weak force, the electromagnetic force and the gravitational force.

fractal A geometric pattern in which parts of the pattern repeat when viewed at any scale.

frequency For a photon, the rate at which the electromagnetic field associated with the photon changes. For the purposes of this book, all you need to know is that the higher the frequency, the greater the energy of the photon.

gamma rays Electromagnetic radiation of very short wavelengths.

gluon The messenger particle which carries the strong force from one quark to another and causes the quarks to hold together in protons and neutrons in the nucleus of the atom. Gluons also interact with one another.

grandfather paradox The idea that someone could travel back in time and prevent their grandparents from giving birth to their parents, thus preventing their own birth.

gravitational force One of the four fundamental forces of nature, and also the weakest. Gravity usually attracts (but not during inflation), and can work over extremely long distances.

gravitational radius Photons cannot escape from a black

hole to the outside universe from within this radius. You can think of it in the same way as the event horizon, though the two terms are used differently. To calculate roughly what this radius will be, multiply the solar mass of the black hole by 2 for miles and 3 for kilometres. Thus a 10-solar-mass black hole will have a radius of 20 miles or 30 kilometres.

graviton The messenger particle which carries the gravitational force among all particles in the universe, including gravitons themselves. None has ever been directly observed.

gravity *See* **gravitational force**

Hawking radiation Radiation produced by a black hole when quantum effects are taken into account. You can think of it as a type of virtual particle pair production near the event horizon of a black hole in which one of the two falls into the hole, allowing the other to escape into space.

Heisenberg uncertainty principle In quantum mechanics, it is impossible to measure precisely, at the same time, the position and the momentum of a particle. Likewise it is impossible to measure precisely, at the same time, the value of a field and the way the field is changing over time.

helium The second lightest of the chemical elements. The nucleus of a helium atom contains two protons and either one or two neutrons. There are two electrons orbiting the nucleus.

homogeneous Having the same quality and appearance at all places.

hydrogen The lightest of the chemical elements. The nucleus of ordinary hydrogen consists of just one proton. There is a single electron orbiting the nucleus. Hydrogen is fused into helium in the cores of stars.

imaginary numbers Numbers that when squared give a negative result. Thus the square of imaginary two is minus four. The square root of minus nine is imaginary three.

imaginary time Time measured using imaginary numbers.

inflationary-universe model Model in which the early universe went through a short period of extremely rapid expansion.

initial conditions The boundary conditions at the beginning of the universe, before any time whatsoever had passed.

isotropic Looking the same in all directions.

microwave radiation Electromagnetic radiation that has wavelengths longer than those of visible light and shorter than radio waves. The particles of microwave radiation, as of all radiation in the electromagnetic spectrum, are photons. A background of microwave radiation that we detect in the universe is evidence used to support the idea of the Big Bang model.

N=8 supergravity A theory that attempts to unify all the particles, both bosons and fermions, in a supersymmetric family, and to unify the forces. This was the theory Hawking spoke of in his 1980 Lucasian Lecture and which he thought might turn out to be the Theory of Everything.

naked singularity A singularity that is not hidden inside an event horizon.

neutron One of the particles that make up the nucleus of an atom. Neutrons have no electrical charge. Every neutron is made up of three smaller particles called quarks.

neutron star The final stage of a star that is too massive to form a white dwarf star but not massive enough to collapse to a black hole.

Newton's theory of gravity Each body in the universe is

attracted towards every other by a force that is stronger the more massive the bodies are and the closer they are to one another. Stated more precisely: bodies attract each other with a force that is proportional to the product of their masses and inversely proportional to the square of the distance between them.

no-boundary proposal The idea that the universe is finite but has no boundary (in imaginary time).

nucleus The central part of an atom, made up of protons and neutrons (which in turn are made up of quarks). The nucleus is held together by the strong force.

optical telescope A telescope that produces images of stars and galaxies in the part of the electromagnetic spectrum that is visible to human eyes.

particle pairs Pairs of particles that are being created everywhere in the vacuum and all the time. They are usually thought to be virtual particles, are extremely short-lived, and cannot be detected except indirectly by observing their effect on other particles. In a fraction of a second the two particles in a pair must find each other again and annihilate one another.

photon The messenger particle of the electromagnetic force. At our level, photons show up as visible light and as all the other radiation in the electromagnetic spectrum, such as radio waves, microwaves, X-rays and gamma rays. Photons have zero mass and move at the speed of light.

Planck length Thought to be the smallest meaningful length. It is 10^{-33} centimetres.

positron Antiparticle of the electron. It has positive electric charge.

primordial black hole Tiny black hole created not by the collapse of a star but the pressing together of matter in the very early universe. According to Hawking the most interesting ones are about the size of the nucleus of an atom, with a mass of about a billion tons.

proton One of the particles that make up the nucleus of the atom. Protons have a positive electric charge. Every proton is made up of three smaller particles called quarks.

psychological arrow of time Our everyday experience of the way time passes, from past to future.

pulsar A neutron star that rotates very rapidly and sends out regular pulses of radio waves.

quantum fluctuations The constant appearance and disappearance of virtual particles that occur in what we think of as empty space (the vacuum).

quantum gravity The scientific theory that successfully unites general relativity and quantum mechanics. At present we do not have such a theory.

quantum mechanics or quantum theory The theory developed in the 1920s that we use to describe the very small, generally things the size of the atom and smaller. According to the theory, light, X-rays and any other waves can only be emitted or absorbed in certain 'packages' called 'quanta'. For instance, light occurs in quanta known as photons, and it can't be divided up into smaller 'packages' than one photon. You can't have half a photon, for example, or one and three-quarter photons. In quantum theory, energy is said to be 'quantized'. The theory includes the uncertainty principle.

quantum wormhole A wormhole of an unimaginably small size. (*See also* **wormhole**.)

quarks The fundamental particles (meaning they can't be divided into anything smaller) which, banded together in groups of three, make up protons and neutrons. Quarks also band together in groups of two (one quark and one antiquark) to form particles called mesons.

radio waves Electromagnetic waves with longer wavelengths than those of visible light. The particles of radio waves, as of all radiation in the electromagnetic spectrum, are photons.

radioactivity The spontaneous breakdown of one type of atomic nucleus into another.

radius The shortest distance from the centre of a circle or sphere to the circumference or surface.

renormalization A process that is used to remove infinities from a theory. It involves putting in other infinities and allowing the infinities to cancel one another out.

second law of thermodynamics Entropy, the amount of disorder, in an isolated system can only increase, never decrease. If two systems join, the entropy of the combined system is as great as or greater than the entropy of the two systems added together.

singularity A point in spacetime at which spacetime curvature becomes infinite, a point of infinite density. Some theories predict that we will find a singularity at the centre of a black hole or at the beginning or end of the universe.

solar mass Mass equalling the mass of our sun.

spacetime The combination of the three dimensions of space and one dimension of time.

spacetime curvature Einstein's theory of general relativity explains the force of gravity as the way the distribution of mass or energy in spacetime causes something that resembles the warping, denting and dimpling in an elastic surface by heavy pellets of different weights and sizes lying on it.

strong nuclear force The strongest of the four fundamental forces of nature. It holds the quarks together, in neutrons and protons for instance, and is responsible for the way protons and neutrons hold together in the nucleus. The messenger particle (boson) of the strong force is the gluon.

supernova An enormous explosion of a star in which all but the inner core is blown off into space. The material blown off in a supernova forms the raw material for new stars and for planets.

superstring theory The theory that explains the fundamental objects in the universe not as pointlike particles but as tiny strings or loops of string. It is a leading candidate for unifying all the particles and forces.

Theory of Everything Sometimes called the TOE, this is the nickname for the theory that explains the universe and everything that happens in it.

thermodynamic arrow of time Entropy (disorder) increases over time.

uncertainty principle A particle cannot have both a definite position and a definite momentum at the same time. The more precisely you measure the one, the less accurate your measurement of the other will be. Similarly you cannot measure precisely the value of a field and its rate of change over time. There are other pairs of quantities that present the same problem. The uncertainty principle was discovered by the German physicist Werner Heisenberg and is more properly called the Heisenberg uncertainty principle.

unified theory A theory that explains all four forces as one 'superforce' showing up in different ways and that also unites both fermions and bosons in a single family.

vacuum energy The energy that is there in what we think of as empty space.

velocity The speed at which something is moving away

from some fixed place, and the direction in which it is moving.

virtual particle In quantum mechanics a particle that can never be directly detected, but whose existence we know about because we can measure its effect on other particles.

W^+, W^-, Z° The messenger particles (bosons) of the weak force.

wave function In quantum theory, a wave function describes all the possible paths a particle could follow between two points. If the value of the wave function is high for a particular path, then the particle is more likely to be found on that path.

wave function of the universe Hartle and Hawking treat the universe like a quantum particle. The Hartle–Hawking wave function of the universe represents all the physically possible histories our universe might have. If the value of the wave function is high for a history, then that is a more likely history.

wavelength The distance from one wave crest to the next.

weak nuclear force One of the four fundamental forces of nature. The messenger particles (bosons) of the weak force are the W^+, W^-, and the Z°. The weak force is responsible for radioactivity, such as a type called beta radioactivity in the nuclei of atoms.

wormhole A hole or tunnel in spacetime, which may end in another universe or another part (or time) of our own universe.

References

PART I: 1942–1975

2 ‘Our goal is nothing less than a complete description of the universe we live in’

1 Richard Feynman, *QED: The Strange Theory of Light and Matter* (Princeton: Princeton University Press, 1985), p. 4.

2 Stephen W. Hawking, *A Brief History of Time: From the Big Bang to Black Holes* (New York: Bantam Books, 1988), p. 9.

3 Ibid.

4 *Professor Hawking’s Universe*, BBC broadcast, 1983.

5 Hawking, *Brief History of Time*, p. 174.

6 John A. Wheeler, unpublished poem.

7 Feynman, p. 128.

8 Stephen W. Hawking, ‘Is the End in Sight for Theoretical Physics?’, inaugural lecture as Lucasian Professor of Mathematics, April 1980.

9 Stephen W. Hawking, ‘Is Everything Determined?’, unpublished, 1990.

10 Bryan Appleyard, ‘Master of the Universe: Will Stephen Hawking Live to Find the Secret?’, *Sunday Times*, 19 June 1988.

11 Murray Gell-Mann, lecture.

3 'Equal to anything!'

1 Stephen Hawking (ed., prepared by Gene Stone), *A Brief History of Time: A Reader's Companion*, New York and London: Bantam Books, 1992, p. 24.
2 Except where otherwise noted, all quotations in Chapter 3 come from two unpublished articles by Stephen Hawking, 'A Short History' and 'My Experience with Motor Neurone Disease'.
3 Hawking interview with Larry King.
4 Kristine Larsen, *Stephen Hawking: A Biography*, Amherst, NY: Prometheus Books, 2007, p. 22.
5 Nigel Hawkes, 'Hawking's Blockbuster Sets a Timely Record', *Sunday Times*, May 1988, p. 8.
6 Hawking, *Reader's Companion*, p. 4.
7 Ibid.
8 Ibid., p. 9.
9 Ibid., p. 10.
10 Ibid., p. 13.
11 Ibid.
12 Ibid., p. 12.
13 Larsen, p. 22.
14 All quotations from Isobel Hawking, in Hawking, *Reader's Companion*, pp. 7, 8.
15 Hawking, *Reader's Companion*, p. 12.
16 Jane Hawking, *Music to Move the Stars: A Life with Stephen Hawking*, London: Pan Books, 2000, p. 9.
17 Information about Majorca visit is from Larsen, p. 24.
18 Hawking, *Reader's Companion*, p. 23.
19 Ibid., p. 13.
20 Stephen Hawking, *Black Holes and Baby Universes and Other Essays*, London: Bantam Books, 1994, p. 3.
21 Melissa McDaniel, *Stephen Hawking: Revolutionary Physicist*, New York: Chelsea House Publications, 1994, p. 28.
22 Hawking, *Black Holes and Baby Universes*, p. 3.
23 Judy Bachrach, 'A Beautiful Mind, an Ugly Possibility', *Vanity Fair*, June 2004, p. 145.

24 Larsen, pp. 25–6.
25 Michael Harwood, 'The Universe and Dr. Hawking', *The New York Times Magazine*, 23 January 1983, p. 57.
26 Hawking, *Black Holes and Baby Universes*, p. 46.
27 Hawking, *Reader's Companion*, p. 38.
28 Ibid., p. 36.
29 Ibid., p. 42.
30 Harwood, 'Universe and Dr. Hawking', p. 57.
31 Hawking, *Reader's Companion*, p. 38.
32 Ibid., p. 39.
33 Harwood, p. 57.
34 Ibid.
35 Gregg J. Donaldson, 'The Man behind the Scientist', *Tapping Technology*, May 1999, www.mdtap.org/tt/1999.05/1-art.html.
36 Larsen, p. 34.
37 Jane Hawking, *Music to Move the Stars*, p. 11.
38 Jane Hawking, *Travelling to Infinity*, London: Alma Books, 2008, p. 15.

4 'The realization that I had an incurable disease . . .'

1 Stephen Hawking and Roger Penrose, *The Nature of Space and Time*, Princeton and Oxford: Princeton University Press, 1996, p. 75.
2 Larsen, p. 39.
3 Denis W. Sciama, *The Unity of the Universe*, Garden City, NJ: Doubleday and Company, 1961, p. vii.
4 Stephen Hawking, 'Sixty Years in a Nutshell', in G. W. Gibbons, E. P. S. Shellard and S. J. Rankin (eds.), *The Future of Theoretical Physics and Cosmology: Celebrating Stephen Hawking's Contributions to Physics,* Cambridge: Cambridge University Press, 2003, (Stephen Hawking 60th Birthday Workshop and Symposium, January 2002), p. 106.
5 Hawking, *Reader's Companion*, p. 50.
6 For information about Stephen and Jane Hawking's courtship I have relied on Jane Hawking, *Music to Move the Stars* and

Travelling to Infinity, and on Hawking, 'Short History'.
7 Jane Hawking, *Music to Move the Stars*, p. 17.
8 Ibid.
9 Ibid., p. 23.
10 Ibid., p. 25.
11 Ibid., p. 26.
12 Ibid., p. 29.
13 Ibid.
14 Hawking, *Brief History of Time*, p. 49.
15 Hawking, *Reader's Companion*, p. 53.
16 Jane Hawking, *Music to Move the Stars*, p. 43.
17 Veash, Nicole Tuesday, 'Ex-Wife's Kiss-and-Tell Paints Hawking as Tyrant', *Indian Express, Bombay*, 3 August 1999, p. 1.
18 Jane Hawking, *Travelling to Infinity*, p. 43.
19 Ibid., p. 44.
20 Ibid.
21 *Master of the Universe: Stephen Hawking*, BBC, broadcast 1989.
22 Appleyard.
23 Jane Hawking, personal interview with the author, Cambridge, April 1991.
24 Larsen, p. 45.
25 Ibid., pp. 45–6.
26 Jane Hawking, *Travelling to Infinity*, p. 56.
27 ABC, *20/20*, broadcast 1989.
28 Jane Hawking, *Music to Move the Stars*, p. 68.

5 'The big question was, was there a beginning or not?'

1 Hawking, 'Short History', p. 5.
2 Jane Hawking, *Music to Move the Stars* (rev. edn, 2004), p. 80.
3 Larsen, p. 52.
4 Hawking, Jane, *Music to Move the Stars* (rev. edn), p. 91.
5 BBC, *Horizon*, 'The Hawking Paradox', 2005.
6 Jane Hawking, *Music to Move the Stars* (rev. edn), pp. 113–14.
7 ABC, *20/20.*
8 Bob Sipchen, 'The Sky No Limit in the Career of Stephen

Hawking', *West Australian*, 16 June 1990.
9 Appleyard.
10 John Boslough, *Beyond the Black Hole: Stephen Hawking's Universe*, Glasgow: Fontana/Collins, 1984, p. 107.
11 Hawking, 'Short History', p. 34.
12 BBC, *Horizon*: 'The Hawking Paradox'.
13 Bryce S. DeWitt, 'Quantum Gravity', *Scientific American*, vol. 249, no. 6 (December 1983), p. 114.
14 S. W. Hawking, 'Black Holes in General Relativity', *Communications in Mathematical Physics* 25 (1972), pp. 152–66.

6 'There is a singularity in our past'

1 Larsen, p. 54.
2 Hawking, 'Sixty Years in a Nutshell', p. 111.
3 Hawking, *Brief History of Time*, p. 42.
4 Stephen W. Hawking, Ph.D. thesis, University of Cambridge, March 1966.
5 S. W. Hawking and R. Penrose, 'The Singularities of Gravitational Collapse and Cosmology', *Proceedings of the Royal Society of London* A314 (1970), pp. 529–48.
6 BBC, *Horizon*, 'The Hawking Paradox'.
7 Hawking, *Brief History of Time*, p. 103.
8 Larsen, p. 57.
9 S. W. Hawking, 'Gravitational Radiation from Colliding Black Holes', *Physics Review Letters* 26 (1971), pp. 1344–6.
10 Jacob D. Bekenstein, 'Black Hole Thermodynamics', *Physics Today*, January 1980, pp. 24–6.
11 Kip Thorne, *Black Holes and Time Warps*, New York: W. W. Norton and Company, 1994, p. 427.
12 J. M Bardeen, B. Carter and S. W. Hawking, 'The Four Laws of Black Hole Mechanics', *Communications in Mathematical Physics* 31 (1973), p. 162.
13 Hawking, *Brief History of Time*, p. vi.
14 Ibid., p. 105.
15 Stephen Hawking, personal interview with author, Cambridge, December 1989.

16 Hawking, *Brief History of Time*, p. 108.
17 Dennis Overbye, 'The Wizard of Space and Time', *Omni*, February 1979, p. 106.
18 Hawking, *Reader's Companion*, pp. 93–4.
19 Boslough, p. 70.
20 Stephen Hawking, 'Black Hole Explosions?', *Nature* 248 (1974), pp. 30–31.
21 J. G. Taylor and P. C. W. Davies, paper in *Nature* 248 (1974).
22 Hawking, *Reader's Companion*, p. 108.
23 Boslough, p. 70.
24 Bernard Carr, 'Primordial Black Holes', in Gibbons, Shellard and Rankin (eds.), p. 236.
25 Hawking, *Reader's Companion*, p. 110.
26 Stephen Hawking, *Hawking on the Big Bang and Black Holes*, Singapore: World Scientific, 1993, p. 3.
27 J. B Hartle and S. W. Hawking, 'Path-Integral Derivation of Black Hole Radiance', *Physical Review* D13 (1976), pp. 2188–203.

PART II: 1970–1990

7 'These people must think . . .'

1 Thorne, *Black Holes and Time Warps*, p. 420.
2 Gerald Jonas, 'A Brief History', *The New Yorker*, 18 April 1988, p. 31.
3 E-mail correspondence: Jane Hawking to Alex Gallenzi, 24 February 2012.
4 Ellen Walton, 'Brief History of Hard Times' (interview with Jane Hawking), *Guardian*, 9 August 1989.
5 Ibid.
6 Ibid.
7 Jane Hawking, personal interview with author, Cambridge, April 1991.
8 Ibid.
9 Harwood, p. 58.
10 Jane Hawking, *Music to Move the Stars*, p. 88.
11 Ibid., p. 178.

12 *Master of the Universe*, BBC.
13 ABC, *20/20*.
14 Harwood, p. 53.
15 *Master of the Universe*, BBC.
16 'Hawking Gets Personal', *Time*, 27 September 1993, p. 80.
17 *Master of the Universe*, BBC.
18 Ibid.
19 Ibid.
20 Ibid.
21 Robert Matthews, 'Stephen Hawking Fears Prejudice against Fundamental Research Threatens the Future of Science in Britain', *CAM: The University of Cambridge Alumni Magazine*, Michaelmas Term, 1995, p. 12.
22 Harwood, p. 58.
23 Information about the lead-up to the Hawkings' year in California, including the choice and preparation of the West Road house and the decision to have graduate or post-graduate students live in to help Stephen, is available in S.W. Hawking, 'In His Own Words: On Disability', *Abilities*, Fall 2010, p. 35; Jane Hawking, *Music* pp. 233–6, 242–5, 273, 275; S.W. Hawking, *Black Holes and Baby Universes*, pp. 24, 25, 168; S.W. Hawking, *Readers Companion*, p. 113; S.W. Hawking, 'My Experience with Motor Neurone Disease', p. 3; Larsen, p. 69; White and Gribbin, pp. 166, 169–71; and the author's experience of student/faculty housing in Cambridge.
24 Except where otherwise indicated, the information in these paragraphs about the Hawking family's experiences in Pasadena, California, comes from Jane Hawking, *Music to Move the Stars*, pp. 249ff.
25 Jane Hawking, *Travelling to Infinity*, p. 222.
26 D. N. Page and S. W. Hawking, 'Gamma Rays from Primordial Black Holes', *Astrophysical Journal* 206 (1976), pp. 1–7.
27 Hartle and Hawking, 'Path-Integral Derivation', p. 2188.
28 Faye Flam, 'Plugging a Cosmic Information Leak', *Science* 259 (1993), p. 1824.

29 Jane Hawking, *Travelling to Infinity*, p. 232.
30 Ibid.

8 'Scientists usually assume . . .'

1 Information about the electric wheelchair is in Boslough, p. 23; Larsen, p. 74; Jane Hawking, *Music*, pp. 284–5. Information about S.H.'s journey to his office comes from the author's knowledge of Cambridge and the routes through King's College; photographs of him making the journey; the *Horizon* 1983 broadcast; *Professor Hawking's Universe*; White and Gribbin, pp. 172–3; and Jane Hawking, *Music*, p. 85.
2 Information about the hiring of Judy Fella and her great value to Hawking and Jane Hawking is to be found in Larsen, pp. 74, 75, 88; and in more detail in Jane Hawking, *Music* pp. 281, 382–3. See also the extraordinary note of thanks to Fella in G.W. Gibbons, Stephen W. Hawking, S.T.C. Siklos, 'Superspace and Supergravity: Proceedings of the Nuffield Workshop, Nuffield Foundation, 1981 and 1985'. For Jane's taking up her thesis again see Larsen p. 74, and Jane Hawking, *Music* pp. 287 ff, where she discusses her work in detail. For Jane's beginning her voice training, see *Music*, p. 292 and Larsen pp. 74–5.
3 Nigel Farndale, 'A Brief History of the Future', *Sydney Morning Herald*, 7 January 2000.
4 See Larsen, pp. 76, 77; and Lisa Sewards, 'A Brief History of Our Time Together', *Daily Telegraph*, 27 April 2002. Jane Hawking has written in detail about Jonathan Hellyer Jones's increasingly important and supportive role in the family in *Travelling to Infinity*, pp. 280–5.
5 Jane Hawking, *Travelling to Infinity*, p. 285.
6 Sewards, quoted in Larsen, p. 96.
7 Information in this paragraph can be found in Larsen, p. 96, Sewards; Jane Hawking, *Travelling to Infinity*, pp. 284–5, and comes also from e-mail correspondence between author and Jane Hawking, June and July 2011.
8 Larsen, p. 78; Jane Hawking, *Music* pp. 357–9.

9 S. W. Hawking and W. Israel (eds.), *General Relativity*, Cambridge: Cambridge University Press, 1979, p. xvi.
10 Hawking, *Reader's Companion*, pp. 151–2.
11 See Larsen p. 81; Jane Hawking, *Music*, p. 377. The change to limited home nursing is also mentioned in S.W. Hawking 'My Experience with Motor Neurone Disease', p. 3.
12 Jane Hawking, *Music to Move the Stars*, pp. 410–12.
13 Ellen Walton, 'A Brief History of Hard Times', the *Guardian*, 9 August 1989.
14 *Master of the Universe*, BBC.
15 Dennis Overbye, 'Cracking the Cosmic Code with a Little Help from Dr. Hawking', *The New York Times*, 11 December 2001.
16 Kip Thorne in Hawking, *Reader's Companion*, p. 120.
17 Story comes from Leonard Susskind, *The Black Hole War*, New York, Boston and London: Back Bay Books, 2008, pp. 20–21.
18 Ibid., pp. 17–18.
19 Andrei Linde, 'Inflationary Theory versus the Ekpyrotic/Cyclic Scenario', in Gibbons, Shellard and Rankin (eds.), p. 801.
20 BBC, *Horizon*, 'The Hawking Paradox'.
21 Tim Folger, 'The Ultimate Vanishing Act', *Discover*, October 1993, p. 100.
22 This example is paraphrased, with some modifications, from ibid.
23 BBC, *Horizon*, 'The Hawking Paradox'.
24 'Out of a Black Hole', lecture at Caltech, 9 April 2008.
25 Ibid.
26 'Out of a Black Hole', lecture.
27 BBC, *Horizon*, 'The Hawking Paradox'.
28 Leonard Susskind, *Black Hole War*, p. 340.

9 'The odds against a universe . . .'

1 Boslough, p. 100.
2 Ibid.
3 Ibid., p. 101.
4 Ibid., p. 105.

5 Hawking, *Brief History of Time*, p. 133.
6 S. W. Hawking and G. F. R. Ellis, 'The Cosmic Black-Body Radiation and the Existence of Singularities in our Universe', *Astrophysical Journal* 152 (1968), pp. 25–36.
7 Hawking, *Brief History of Time*, pp. 71–2.
8 Ibid., pp. 132–3.
9 Except where otherwise noted, the paragraphs about Andrei Linde and his inflation theory are based on Linde, 'Inflationary Theory', in Gibbons, Shellard and Rankin (eds.), pp. 801–2.
10 Ibid., p. 802.
11 Ibid.
12 Ibid.
13 Ibid.
14 Hawking talks about this in *Brief History of Time*, p. 131.
15 A. D. Linde, 'A New Inflationary Universe Scenario: A Possible Solution of the Horizon, Flatness, Homogeneity, Isotropy, and Primordial Monopole Problems', *Physics Letters* B108 (1982), pp. 389–93.
16 S. W. Hawking and I. G. Moss, 'Supercooled Phase Transitions in the Very Early Universe', *Physics Letters* B110 (1982), p. 35.
17 S. W. Hawking, 'The Development of Irregularities in a Single Bubble Inflationary Universe', *Physics Letters* B115 (1982), pp. 295–7.

10 'In all my travels . . .'

1 Stephen W. Hawking, 'The Edge of Spacetime', in Paul C. W. Davies (ed.), *The New Physics*, Cambridge: Cambridge University Press, 1989, p. 67.
2 Ibid.
3 Ibid., p. 68.
4 Ibid.
5 Ibid.
6 Jerry Adler, Gerald Lubenow and Maggie Malone, 'Reading God's Mind', *Newsweek*, 13 June 1988, p. 59.
7 Hawking, 'Short History', unpublished, p. 6.

8 *Master of the Universe*, BBC.
9 Don N. Page, 'Hawking's Timely Story', *Nature*, 332, 21 April 1988, p. 743.
10 Hawking, *Brief History of Time*, p. 174.
11 Hawking, *Reader's Companion*, p. 122.
12 Ibid.
13 Andrei Linde, in an e-mail to the author, 21 March 2011.
14 John Barrow, *The Book of Universes*, London: The Bodley Head, 2011, p. 202.
15 Ibid., p. 205.

11 'It's turtles all the way down'

1 Hawkes, p. 8.
2 Information about the travels that summer and the phone call to Switzerland comes from Jane Hawking, *Travelling to Infinity*, pp. 350, 357–9.
3 Private e-mail correspondence, Jane Hawking to Alex Gallenzi, 24 February 2012.
4 Walton.
5 Ibid.
6 Ibid.
7 Hawking, *Reader's Companion*, p. 155.
8 Matthews, p. 12.
9 Robert Crampton, 'Intelligence Test', *The Times Magazine*, 8 April 1995, p. 27.
10 Hawking, 'My Experience with Motor Neurone Disease', unpublished.
11 Jane Hawking, *Music to Move the Stars* (rev. edn), p. 443.
12 Bachrach, p. 149.
13 Hawking, *Reader's Companion*, p. 161.
14 Hawking, *Brief History of Time*, p. viii.
15 Hawking, *A Reader's Companion*, p. 154.
16 Stephen W. Hawking, 'A Brief History of *A Brief History*', *Popular Science*, August 1989, p. 70.
17 S. W. Hawking and W. Israel (eds.), *300 Years of Gravitation*, Cambridge: Cambridge University Press, 1987.

18 Hawkes.
19 Hawking, 'Brief History of *A Brief History*', p. 72.
20 Hawking, *Reader's Companion*, p. viii.
21 Hawking, 'Brief History of *A Brief History*', p. 72.
22 Matthews, p. 12.
23 ABC, *20/20*.

12 'The field of baby universes is in its infancy'

1 Walton.
2 ABC, *20/20*.
3 Ibid.
4 Ibid. and *Master of the Universe*, BBC.
5 ABC, *20/20*.
6 Quoted in Larsen, p. 82.
7 Hawkes, p. 8.
8 Walton.
9 ABC, *20/20*.
10 Crampton, p. 28.
11 Stephen Hawking personal interview with author, Cambridge, December 1989.
12 Ibid.
13 Ibid.
14 David H. Freedman, 'Maker of Worlds,' *Discover*, July 1990, p. 49.
15 M. Mitchell Waldrop, 'The Quantum Wave Function of the Universe', *Science*, 242, 2 December 1988, p. 1248.
16 Stephen Hawking, personal interview with author, Cambridge, June 1990.
17 Stephen W. Hawking, 'Black Holes and Their Children, Baby Universes', unpublished, p. 7.
18 Kip Thorne, 'Warping Spacetime', in Gibbons, Shellard and Rankin (eds.), pp. 102–3.
19 Information and quotations in this paragraph are from ibid.
20 Hawking, *Black Holes and Baby Universes*, p. 154.
21 Thorne, 'Warping Spacetime', p. 103.

PART III: 1990–2000

13 'Is the end in sight for theoretical physics?'

1 ABC, *20/20* broadcast.
2 Bob Sipchen, 'The Sky No Limit in the Career of Stephen Hawking,' *The West Australian*, Perth, 16 June 1990.
3 *Master of the Universe*, BBC.
4 Appleyard.
5 Sipchen.
6 Hawking, *Brief History of Time*, p. 174.
7 M. Mitchell Waldrop, 'The Quantum Wave Function of the Universe,' *Science*, 242, 2 December 1988, p. 1250.
8 Hawking, *Brief History of Time*, p. 175.
9 *Master of the Universe*, BBC.
10 Biographical information about Errol Morris comes from Philip Gourevitch, 'Interviewing the Universe', *The New York Times Magazine*, 9 August 1992, http://www.errolmorris.com/content/profile/bhot_ gourevitch.html.
11 David Stevens, IMDb Mini Biography of Errol Morris, http:// www.imdb.com/name/nm0001554/bio.
12 Gourevitch.
13 Ibid.
14 Hawking, *Reader's Companion*, pp. viii–ix.
15 Gourevitch.
16 Bachrach, p. 149.
17 Gordon Freedman, 'Afterword', in Hawking, *Reader's Companion*, p. 182.
18 Gourevitch.
19 David Ansen, 'Off the Beaten Track', *Newsweek*, 21 September 1992, p. 50B.
20 Richard Schickel, 'The Thrust of His Thought', *Time*, 31 August 1992, pp. 66, 69.
21 The three quotations from Errol Morris are from Gourevitch.

14 'Between film roles I enjoy solving physics problems'

1 Andrei Linde, 'The Self-Reproducing Inflationary Universe', *Scientific American*, November 1994, p. 48.

2 Barrow, p. 231.
3 Linde, 'Inflationary Theory' in Gibbons, Shellard and Rankin (eds.), p. 811.
4 David Gross, 'String Theory', in Gibbons, Shellard and Rankin (eds.), p. 465.
5 Information about *The Voyage* comes from Edward Rothstein, 'Glass on Columbus, Hip on a Grand Scale', *International Herald Tribune*, 15 October 1992; and Katrine Ames, 'Santa Maria and Spaceships', *Newsweek*, 2 November 1992.
6 The information in the paragraphs about Hawking's appearance on *Star Trek* comes from 'Trek Stop', *People Magazine*, 28 June 1993, pp. 81–2.
7 Ibid., p. 81.
8 Ibid., p. 82.
9 Ibid.
10 Ibid., p. 81.
11 Stephen Hawking, *The Universe in a Nutshell*, New York and London: Bantam Books, 2001, p. 157.
12 Information in this paragraph about Hawking's appearance before an audience of disabled teenagers comes from Michael D. Lemonick, 'Hawking Gets Personal', *Time*, 27 September 1993, p. 80.
13 Ibid.
14 Crampton, p. 28.
15 Sharon Begley and Jennifer Foote, 'Why Past is Past', *Newsweek*, 4 January 1993, p. 50.
16 The information in these paragraphs about computer viruses is from Fred Tasker, 'Deep Thinkers Abuzz over Idea of Computer Virus as Life', *Richmond Times–Dispatch*, 10 August 1994, p. 4; and Mike Snider, 'Are Computer Viruses a Form of Life?' *U.S.A. Today*, 3 August 1964, p. 1. The quotations from Hawking are from Tasker.
17 G. W. Gibbons and S. W. Hawking (eds.), *Euclidean Quantum Gravity*, Singapore: World Scientific Publishing Company, 1993.
18 Hawking, *Hawking on the Big Bang and Black Holes*, Singapore:

World Scientific Publishing Company, 1993.

19 S. W. Hawking, 'The No-Boundary Proposal and the Arrow of Time', in J. J. Halliwell, J. Perez-Mercader and W. H. Zurek (eds.), *Physical Origins of Time Asymmetry*, Cambridge: Cambridge University Press, 1992, p. 268.

20 Begley and Foote, p. 50.

21 Hawking, *Brief History of Time*, p. 148.

22 Ibid., p. 149.

23 Don N. Page, 'Will Entropy Decrease If the Universe Recollapses', *Physical Review* D32 (1985), pp. 2496–9.

24 Hawking, *Reader's Companion*, p. 166.

25 S. W. Hawking, 'The Arrow of Time in Cosmology', *Physical Review* D32 (1985), p. 2495.

26 Leonard Susskind, 'Twenty Years of Debate with Stephen', in Gibbons, Shellard and Rankin (eds.), p. 330.

27 Quoted in Dugald Murdoch, *Niels Bohr's Philosophy of Physics*, Cambridge: Cambridge University Press, 1987, p. 52.

28 Susskind, 'Twenty Years of Debate', p. 334.

29 Susskind, *Black Hole War*, p. 287.

15 'I think we have a good chance . . .'

1 Crampton, pp. 27–8.

2 Both quotations from ibid.

3 The information in these paragraphs about the concert in Aspen is from Richard Jerome, Vickie Bane and Terry Smith, 'Of a Mind to Marry: Physicist Stephen Hawking Pops the Most Cosmic Question of All to His Nurse', *People Magazine*, 7 August 1995, p. 45–6.

4 Jerome, Bane and Smith, p. 45.

5 From the Associated Press, reported in *The New York Times*, 16 September 1995, p. L-20.

6 Jerome, Bane, and Smith, p. 45.

7 Bachrach, p. 144.

8 Lemonick, p. 80.

9 Larsen, p. 123; Jane Hawking, *Music*, pp. 584–6. See also Adams 'Brief History of a First Wife'. 'My Life with Stephen

Hawking', a lecture given at the Australian National University at Canberra, 10 March 2011, in which Jane Hawking discusses her reasons for writing her book and reads from it, can be viewed on line at http://www.abc.net.au/tv/bigideas/sotrie/201.

10 Hawking *Black Holes and Baby Universes*, p. 44.

11 Hawking and Penrose, *The Nature of Space and Time*, p. 4.

12 Stephen Hawking, conversation with author, spring of 1996.

13 Hawking and Penrose, *Nature of Space and Time*, p. 4.

14 See Kitty Ferguson, 'Devouring the Future: A Profile of Stephen Hawking', *Astronomy Magazine*, December 1998.

15 Grice; Jane Hawking, *Music* 588; Larsen, p. 151; correspondence Jane Hawking to Alexandro Gallenzi, 17 January 2012, and 24 February 2012.

16 Information on these wagers and naked singularities comes from Malcolm W. Browne, 'A Bet on a Cosmic Scale, and a Concession, Sort Of', *The New York Times*, 12 February 1997, p. A-22.

17 Stephen Hawking and Roger Penrose, 'Afterword to the 2010 Edition: The Debate Continues', *The Nature of Space and Time*, Princeton and London: Princeton University Press, 2010, p. 139.

18 Ibid., p. 140.

19 Stephen Hawking, 'Remarks by Stephen Hawking', White House Millennium Council 2000, http://clinton4.nara.gov/Initiative/Millennium/shawking.html.

20 Robin McKie, 'Master of the Universe', *Observer*, 21 October 2001.

21 Martin Durrani, 'Hawking Slams "Stupid, Worthless" Play', *Physics World*, August 2000, p. 8.

22 Elizabeth Grice, 'Dad's Important, But We Matter, Too,' *Telegraph*, 13 April 2004, http://www.telegraph.co.uk/arts/main.jhtml?xml+/arts/ 2004/04/13.bohawk13.xml.

23 M. Bucher, A. S. Goldhaber, and N. Turok, 'Open Universe from Inflation', *Physical Review* D52 (1995), pp. 3314-37.

24 S. W. Hawking and N. Turok, 'Open Inflation without False

Vacua', *Physics Letters* B425 (1998), pp. 25–32.
25 Neil Turok, quoted in 'All Things Came from a Pea', +*Plus Magazine . . . Living Mathematics*, University of Cambridge Centre for Mathematical Sciences Millennium Maths Project, 23 November 2007, http://web.uvic.ca/%7Ejtwong/Hawking-Turok.htm.
26 'All Things Came from a Pea'.
27 Tom Yulsman, 'Give Peas a Chance', *Astronomy Magazine*, September 1999, pp. 38–9.
28 Andrew Watson, 'Inflation Confronts an Open Universe', *Science* 279 (1998), p. 1455.
29 David Salisbury, 'Hawking, Linde Spar Over Birth of the Universe', *Stanford Report Online*, 19 April 1998, http://news-service.stanford. edu/news/1998/april29/hawking.html.
30 Yulsman, p. 39.
31 Salisbury.
32 Yulsman, p. 38.
33 Ibid., p. 39.
34 Salisbury.
35 Ibid.
36 S. W. Hawking, 'A Debate on Open Inflation', in David O. Caldwell (ed.), *COSMO-98: Second International Workshop on Particle Physics and the Early Universe*, College Park, Md: American Institute of Physics, 1999, p. 21.

16 'It seems clear to me'

1 Stephen Hawking, interview with Larry King.
2 Kitty Ferguson, *The Music of Pythagoras*, New York: Walker Publishing, 2008, pp. 107 and 136.

PART IV: 2000–2011

17 'An expanding horizon of possibilities'

1 Nigel Farndale, 'A Brief History of the Future', *The Hindu Magazine*, 15 January 2000, p. 1.
2 Nick Paton Walsh, 'Alter Our DNA or Robots Will Take Over, Warns Hawking', *Observer*, 2 September 2001.

3 Farndale, 'Brief History of the Future', *Sydney Morning Herald*, 7 January 2000, p. 2.
4 Ibid.
5 Ibid.
6 'Space Colonies Needed for Human Survival', *Guardian*, 16 October 2001, p. 3.
7 Brian Pippard, 'The Invincible Ignorance of Science', *The Great Ideas Today, 1990*, Encyclopaedia Britannica, Inc., p. 325.
8 Gregory Benford, 'Leaping the Abyss', *Reason Online*, April 2002, http//reason.com/0204/fe.gb.leaping.shtml
9 Hawking, *Universe in a Nutshell*, p. 57.
10 Stephen Hawking, 'Gödel and the End of Physics', lecture for Dirac Centennial Celebration, 20 July 2002.
11 Kip Thorne in Hawking, *Reader's Companion*, p. 120.
12 Hawking, 'Gödel and the End of Physics'.
13 Hawking, *Universe in a Nutshell*, p. 54.
14 Hawking, 'Sixty Years in a Nutshell', p. 106.
15 Martin Rees, 'Our Complex Cosmos and its Future', in Gibbons, Shellard and Rankin (eds.), p. 17.
16 Roger Penrose, 'The Problem of Spacetime Singularities: Implications for Quantum Gravity?', in Gibbons, Shellard and Rankin (eds.), p. 51.
17 Bernard Carr, who was Hawking's graduate assistant in the 1970s, 'Primordial Black Holes', in Gibbons, Shellard and Rankin (eds.), p. 236.
18 Susskind, 'Twenty Years of Debate', p. 330.
19 Raphael Bousso, 'Adventures in de Sitter Space', in Gibbons, Shellard and Rankin (eds.), p. 539.
20 Gary Gibbons, 'Euclidean Quantum Gravity: The View from 2002', in Gibbons, Shellard and Rankin (eds.), p. 370.
21 Michael Green, 'A Brief Description of String Theory', in Gibbons, Shellard and Rankin (eds.), p. 473.
22 Neil Turok, 'The Ekpyrotic Universe and Its Cyclic Extension', in Gibbons, Shellard and Rankin (eds.), p. 781.

23 Thorne, 'Warping Spacetime', in Gibbons, Shellard and Rankin (eds.), p. 74.
24 Natalie Clarke, 'Professor Hawking in assault probe', *Daily Mail*, January 2004, http://www.dailymail.co.uk/news/article-206323/Professor-Hawking-assault-probe.html#ixzz1GNPJUR1q.
25 'Hawking Extols Joy of Discovery', *BBC News*, 11 January 2002.
26 Alan H. Guth and David I. Kaiser, 'Inflationary Cosmology: Exploring the Universe from the Smallest to the Largest Scales', *Science*, vol. 307, no. 5711 (11 February 2005), pp. 884–90.
27 Paul Preuse, 'Strong Evidence for Flat Universe reported by BOOMERANG Project', *Berkeley Lab Research News*, 26 April 2000, http://www.lbl.gov/Science Articles/boomerang-flat.html, BOOMERANG stands for 'Balloon Observations of Millimetric Extragalactic Radiation and Geophysics'.
28 Barrow, p. 206.
29 NASA/WMAP Science Team, National Aeronautics and Space Administration, 'First Year Results on the Oldest Light in the Universe', 11 February 2003, http://wmap.gsfc.nasa.gov/news/PressRelease_03-064.html.
30 Sarah L. Bridle, Ofer Lahav, Jeremiah P. Ostriker and Paul J. Steinhardt, 'Precision Cosmology? Not Just Yet . . .', 10 March 2003, http://arxiv.org/pdf/astro-ph/0303180.
31 Ibid.
32 BBC, *Horizon*, 'The Hawking Paradox'.
33 'This Week's Finds in Mathematical Physics' (Week 207), 25 July 2004: 'John Baez's Stuff', math.ucr.edu/home/baez/.
34 Susskind, *Black Hole War*, p. 420.
35 Ibid., p. 419.
36 BBC, *Horizon*, 'The Hawking Paradox'.
37 Jenny Hogan, 'Hawking Cracks Black Hole Paradox', *New Scientist*, 14 July 2004.
38 Stephen Hawking, 'Out of a Black Hole', Caltech lecture, 9 April 2008.

39 Stephen W. Hawking, paper at the 17th International Conference on General Relativity and Gravitation, Dublin, July 2004.
40 David Whitehouse, 'Black Holes Turned Inside Out', *BBC News*, 22 July 2004.
41 Tim Folger, 'Return of the Invisible Man', *Discover Magazine*, July/August 2009, p. 48.
42 Stephen Hawking, 'To Boldly Go', lecture for undergraduates at Caltech, 14 January 2005.

18 'Grandad has wheels'

1 Tim Adams, 'Brief History of a First Wife', *Observer*, 4 April 2004, http://observer.guardian.co.uk/review/story/0,1185067,00.html.
2 'Stephen Hawking's Alternate Universe', video at the Smithsonian Institute, Washington, DC, 14 February 2005.
3 Ibid.
4 Stephen W. Hawking, lecture at the Smithsonian Institute, Washington, DC, 14 February 2005.
5 Ibid.
6 Associated Press, 'Scientist Stephen Hawking Decries Iraq War', *USA Today*, 3 November 2004.
7 Alan Boyle, 'The Show Goes On for Stephen Hawking', 15 June 2006, http://www.msnbc.msn.com/id/10086479 (no longer accessible).
8 Steve Connor and Stephen Castle, 'Hawking Criticizes EU States Trying to Ban Stem Cell Research', *Independent*, 24 July 2006.
9 Lucy and Stephen Hawking, *George's Secret Key to the Universe*, London: Doubleday, 2007.
10 Alan Boyle, 'The Show Goes On for Stephen Hawking', MSNBC, 15 November 2005, http://www.msnbc.msn.com/id/10086479.
11 'Hawking Humor', *Israel Today*, 28 January 2007, http://www.israeltoday.co.il.
12 'Stephen Hawking to divorce second wife', *Mail Online*, last

updated 19 October 2006, http://www.dailymail.co.uk/news/article-411349/ Stephen-Hawking-divorce-second-wife.html#ixzz1GNl5y1yx.

13 Natalia Shuhmaher and Robert Brandenberger, 'Brane Gas-Driven Bulk Expansion as a Precursor State to Brane Inflation', *Physics Review Letters* 96 (2006), 161301.

14 See W. Lerche, D. Lust and A. N. Schellekens, 'Chiral Four-Dimensional Heterotic Strings from Selfdual Lattices', *Nuclear Physics* B287 (1987), p. 477.

15 A. D. Linde, 'Eternally Existing Self-Reproducing Chaotic Inflationary Universe', *Physics Letters* B175 (1986), p. 395.

16 Juan Martin Maldacena, 'The Large N Limit of Superconformal Field Theories and Supergravity', November 1997, http://inspirebeta.net/ record/451647.

17 Hawking and Penrose, 'Afterword to the 2010 Edition', *Nature of Space and Time*, p. 142.

18 Stephen Hawking and Leonard Mlodinow, *The Grand Design*, London: Transworld, 2010, p. 118.

19 Tim Folger, 'Our Universe is Perfectly Tailored for Life, *Discover Magazine*, December 2008, http://discovermagazine.com/2008/dec/10-sciences-alternative-to-an-intelligent-creator/article_view?searchterm=Andrei%20Linde&b_start:int=1.

20 Mario Livio and Martin J. Rees, 'Anthropic Reasoning', *Science*, vol. 309, no. 5737 (12 August 2005), pp. 1022–3.

21 Stephen Hawking, 'The Origins of the Universe', lecture at Caltech, 4 April 2006.

22 S. W. Hawking and Thomas Hertog, 'Populating the Landscape: A Top Down Approach', February 2006, http://inspirebeta.net/record/710178; and Amanda Gefter, 'Mr. Hawking's Flexiverse', *New Scientist* 189 (2006), no. 2548, pp. 28–32.

23 Emine Saner, 'Lucy Hawking's Fears', *Evening Standard* (London), 14 April 2004, http://www.thisislondon.co.uk/showbiz/article-10226902-lucy-hawkings-fears.do (accessed June 2011).

24 Steve Cray, 'Rock-Star Welcome for Top Scientist', *South China Morning Post*, 13 June 2006, City section, p. 1.
25 Alexa Olesen, 'Stephen Hawking: Earth Could Become Like Venus', 22 June 2006, http://www.livescience.com/environment/ap_060622_ hawking_climate.html.
26 Lucy and Stephen Hawking, *George's Secret Key to the Universe*, London: Doubleday, 2007; and Lucy and Stephen Hawking, *George's Cosmic Treasure Hunt*, London: Doubleday, 2009.
27 Q. and A.: Stephen Hawking and daughter Lucy, *TODAY, Al's Book Club*, 1 November 2007, http://today.msnbc.msn.com/id/21550559/ns/ today-books/ (accessed June 2011).
28 Harry MacAdam, 'Search is Vital, Says Hawking', *Sun*, 28 December 2006.
29 Yahoo Searchblog, 1 August 2006, http://www. ysearch-blog.com/archives/999336.html.
30 'Hawking Misrepresents Pope John Paul II', *Catalyst* 31 (2006), no. 6, http://www.catholicleague.org/catalyst/2006_catalyst/07806.htm#broward.
31 'Hawking Misrepresents Pope John Paul II', Catholic League for Religious and Civil Rights, http://www. catholicleague.org/catalyst.php?year= 2006&month=July-August&read=2078.
32 Alan Boyle, 'Hawking Goes Zero-G: "Space, here I come" ', *Space on msnbc.com*, http://www.msnbc.msn.com/id/18334489/ns/technology_ and_science-space/.
33 Ibid.
34 Ibid.

19 'I've always gone in a somewhat different direction'

1 Information about 2008 results release and quotation come from NASA/WMAP Science Team, National Aeronautics and Space Administration, 'Fifth Year Results on the Oldest Light in the Universe', 7 March 2008.
2 M. Cruz, E. Martinez-Gonzalez, P. Vielva, J. M. Diego, M. Hobson, N. Turok, 'The CMB Cold Spot: Texture, Cluster

or Void?', April 2008, http://inspirebeta.net/record/783713.

3 Mike Wade, 'Peter Higgs Launches Attack against Nobel Rival Stephen Hawking', *Sunday Times*, 11 September 2008, http://www.timesonline.co.uk/tol/news/science/article4727894.ece.

4 'Hawking Bets CERN Mega-Machine Won't Find "God's Particle"', 9 September 2008, http://afp.google.com/article/ALeqM5jaOONGqv- xW-JhBOWgiNCVi6Rsmw.

5 Mike Wade, 'Peter Higgs Launches Attack Against Nobel Rival Stephen Hawking', *Sunday Times*, 11 September 2008.

6 'Hawking Bets CERN Mega-Machine Won't Find "God's Particle"'.

7 Stephen Hawking, 'Out of a Black Hole', lecture at Caltech, 9 April 2008.

8 Ian Sample, 'Large Hadron Collider Warms Up for Final Drive to Catch a Higgs Boson', *Guardian*, 26 February 2011, http://www.guardian.co.uk/ science/2011/feb/28/large-hadron-collider-higgs-boson.

9 Information about the Corpus clock comes from Christopher de Hamel, *The Corpus Clock*, Isle of Man, Fromanteel, 2008, and from the author's own observations of the clock.

10 Roger Highfield, 'Stephen Hawking to Unveil Strange New Way to Tell the Time', *Telegraph*, 14 September 2008.

11 BBC *Newsnight*, 1 October 2009, http://news.bbc.co.uk/1/hi/programmes/newsnight/8285100.stm.

12 Matthews, p. 12.

13 Quoted in Folger, 'Return of the Invisible Man', p. 44.

14 Quotations and information in these paragraphs are from Stephen Hawking, 'Why We Should Go into Space', lecture at Caltech, 2009, video copyright, Caltech Digital Media Services (Information Management Systems and Services).

15 Claudia Dreifus, 'Life and the Cosmos, Word by Painstaking Word: A Conversation with Stephen Hawking', *The New York Times,* 9 May 2011, Science Section, p. 1.

16 'Stephen Hawking to Accept Cosmos Award in Cambridge, England', The Planetary Society, press release,

24 February 2010, http://www.planetary.org/about/press/releases/ 2010/0224.

17 NASA/WMAP Science Team, National Aeronautics and Space Administration, 'WMAP Produces New Results', 26 January 2010, http://wmap.gsfc.nasa.gov/news/.

18 Adrian Cho, 'A Recipe for the Cosmos', *Science*, vol. 330, no. 6011 (17 December 2010), p. 1615.

19 'Planck's New View of the Cosmic Theatre', http://www.esa.int/SPECIALS/Planck/SEMK4D3SNIG_0.html.

20 Ibid.

21 Ibid.

22 Alan H. Guth and David I. Kaiser, 'Inflationary Cosmology: Exploring the Universe from the Smallest to the Largest Scales', *Science*, vol. 307, no. 5711 (11 February 2005), pp. 884–90.

23 Barrow, p. 212.

24 Lawrence M. Krauss, Scott Dodelson and Stephan Meyer, 'Primordial Gravitational Waves and Cosmology', *Science*, vol. 328, no. 5981 (21 May 2010), pp. 989–92.

25 'Catching Waves with Kip Thorne', *+Plus Magazine . . . Living Mathematics*, 23 November 2007.

26 Thorne, 'Warping Spacetime', p. 74.

27 James Bock, et al., 'Study of the Experimental Probe of Inflationary Cosmology Intermediate Mission for NASA's Einstein Inflation Probe', Cornell University Library website http://arXiv:0906.1188v1 [astro-ph.CO].

28 James Hartle, S. W. Hawking and T. Hertog, 'Eternal Inflation Without Metaphysics', http://arxiv.org/find/all/1/all:+AND+inflation+AND+hawking+eternal/0/1/0/all/0/1, September 2010.

29 Ibid.

30 James Hartle, S. W. Hawking and Thomas Hertog, 'The No-Boundary Measure in the Regime of Eternal Inflation', *Physical Review* D82 (2010), 063510.

31 Ibid.

32 Hartle, Hawking and Hertog, 'Eternal Inflation Without Metaphysics'.
33 Folger, 'Our Universe is Perfectly Tailored for Life'.
34 Information about Daniele Faccio and his team comes from 'Dr. Hawking's Bright Idea', *The Economist*, 2 October 2010, pp. 93–4.

20 'My name is Stephen Hawking . . .'

1 Hawking and Mlodinow, *Grand Design*, p. 8.
2 Ibid.
3 Stephen Hawking, 'The Origins of the Universe', lecture at Caltech, 4 April 2006.
4 Hawking and Mlodinow, *Grand Design*, p. 72.
5 Ibid., p. 9.
6 Hawking, 'The Origins of the Universe', lecture.
7 Hawking and Mlodinow, *Grand Design*, p. 153.
8 Ibid., p. 30.
9 Ibid.
10 Ibid., p. 34.
11 Ibid., p. 32–3.
12 Ibid., p. 72.
13 Ibid., p. 46.
14 Ibid., p. 58.
15 Ibid., p. 178.
16 Ibid., p. 181.
17 'Understanding the Universe: Order of Creation', *The Economist*, 11 September 2010, p. 85.
18 Ibid.
19 Ibid.
20 Dwight Garner, 'Many Kinds of Universes, and None Require God', *The New York Times*, 7 September 2010.
21 Hawking and Mlodinow, *Grand Design*, p. 144.
22 Ibid., p. 172.
23 *Into the Universe with Stephen Hawking*, Discovery Channel, broadcast 2011.
24 Ibid.

25 Ibid.

26 Ibid.

27 Ibid.

28 Ibid.

29 Ibid.

30 Stephen Hawking, conversation with author, November 2010.

31 Ian Sample, ' "There is no heaven or afterlife . . . that is a fairy story for people afraid of the dark" ', *Guardian*, 16 May 2011, p. 3.

32 Michael Wenham, 'I'd stake my life that Stephen Hawking is wrong about heaven', *Guardian*, 17 May 2011, www.guardian.co.uk/commentisfree/belief/2011/may/17/stephen-hawking-heaven?intcmp=239.

33 Dreifus, Science Section, p. 1.

34 Hawking, *Reader's Companion*, p. 174.

35 Introduction to Hawking, *Hawking on the Big Bang and Black Holes*, p. 1.

Suggested Further Reading

All of Stephen Hawking's non-academic books (listed in Bibliography below). *Black Holes and Baby Universes* and *A Brief History of Time: A Reader's Companion* include particularly interesting accounts (autobiographical and from relatives and close acquaintances) of his childhood and university years and his life outside of science.

Jane Hawking's two memoirs, *Music to Move the Stars* and *Travelling to Infinity: My Life with Stephen.* These provide the most in-depth account ever likely to be written of their life together.

The three children's books on which Lucy Hawking and Stephen Hawking have collaborated. Though fictional, these include superb explanations at a non-academic level of the science that has most interested Stephen Hawking: *George's Secret Key to the Universe*, *George's Cosmic Treasure Hunt*, and *George and the Big Bang.*

Kristine Larsen's excellent, succinct biography, *Stephen Hawking: A Biography*, written by a scientist who works in Stephen Hawking's field and writes well for non-scientists.

Bibliography

Adams, Tim, 'Brief History of a First Wife', *Observer*, 4 April 2004, http://observer.guardian.co.uk/review/story/0,1185067,00.html

Adler, Jerry, Gerald Lubenow and Maggie Malone, 'Reading God's Mind', *Newsweek*, 13 June 1988, p. 59

'All Things Came from a Pea', *+Plus Magazine . . . Living Mathematics*, University of Cambridge Centre for Mathematical Sciences Millennium Maths Project, 23 November 2007, http://web.uvic.ca/%7Ejtwong/ Hawking-Turok.ht

Ames, Katrine, 'Santa Maria and Spaceships', *Newsweek*, 2 November 1992

Ansen, David, 'Off the Beaten Track', *Newsweek*, 21 September 1992, p. 50B

Appleyard, Bryan, 'Master of the Universe: Will Stephen Hawking Live to Find the Secret?', *Sunday Times*, 19 June 1988

Associated Press, 'Scientist Stephen Hawking Decries Iraq War', *USA Today*, 3 November 2004

Bachrach, Judy, 'A Beautiful Mind, an Ugly Possibility', *Vanity Fair*, June 2004

Bardeen, J. M., B. Carter and S. W. Hawking, 'The Four Laws of Black Hole Mechanics', *Communications in Mathematical Physics* 31 (1973), p. 162

Barrow, John, *The Book of Universes*, London: The Bodley Head, 2011

BBC, *Horizon*, 'The Hawking Paradox', 2005

BBC, *Newsnight*, 1 October 2009, http://news.bbc.co.uk/1/hi/programmes/newsnight/8285100.stm

Begley, Sharon and Jennifer Foote, 'Why Past is Past', *Newsweek*, 4 January 1993, p. 50

Bekenstein, Jacob D, 'Black Hole Thermodynamics', *Physics Today*, January 1980, pp. 24–6

Benford, Gregory, 'Leaping the Abyss', *Reason Online*, April 2002, http//reason.com/0204/fe.gb.leaping.shtml

Bock, James, et al., 'Study of the Experimental Probe of Inflationary Cosmology Intermediate Mission for NASA's Einstein Inflation Probe', Cornell University Library website: arXiv:0906.1188v1 [astro-ph.CO]

Boslough, John, *Beyond the Black Hole: Stephen Hawking's Universe*, Glasgow: Fontana/Collins, 1984

Bousso, Raphael, 'Adventures in de Sitter Space', in G. W. Gibbons, E. P. S. Shellard and S. J. Rankin (eds.), *The Future of Theoretical Physics and Cosmology: Celebrating Stephen Hawking's Contributions to Physics*, Cambridge: Cambridge University Press, 2003 (Stephen Hawking 60th Birthday Workshop and Symposium, January 2002), p. 539

Boyle, Alan, 'Hawking Goes Zero-G: "Space, here I come"', *Space on msnbc.com*, http://www.msnbc.msn.com/id/18334489/ns/technology_ and_science-space/

—, 'The Show Goes On for Stephen Hawking', 15 June 2006, http://www.msnbc.msn.com/id/10086479 (no longer accessible)

Bridle, Sarah L., Ofer Lahav, Jeremiah P. Ostriker, Paul J. Steinhardt, 'Precision Cosmology? Not Just Yet . . .', 10 March 2003, http://arxiv.org/pdf/astro-ph/0303180

Browne, Malcolm W., 'A Bet on a Cosmic Scale, and a Concession, Sort Of', *The New York Times*, 12 February 1997, p. A-22

Bucher, M., A. S. Goldhaber and N. Turok, 'Open Universe from Inflation', *Physical Review* D52 (1995), pp. 3314–37

Carr, Bernard, 'Primordial Black Holes', in G. W. Gibbons, E. P. S. Shellard and S. J. Rankin (eds.), *The Future of Theoretical Physics and Cosmology: Celebrating Stephen Hawking's Contributions to Physics*, Cambridge: Cambridge University Press, 2003 (Stephen Hawking 60th Birthday Workshop and Symposium, January 2002), p. 236

'Catching Waves with Kip Thorne', *+Plus Magazine . . . Living Mathematics*, University of Cambridge Centre for Mathematical Sciences Millennium Maths Project, 23 November 2007, http://plus.maths.org/content/catching-waves-kip-thorne

Cho, Adrian, 'A Recipe for the Cosmos', *Science*, vol. 330, no. 6011 (17 December 2010), p. 1615

Clarke, Natalie, 'Professor Hawking in Assault Probe', *Daily Mail*, January 2004, p. 3

Connor, Steve and Stephen Castle, 'Hawking Criticizes EU States

Trying to Ban Stem Cell Research', *Independent*, 24 July 2006, p. 14

Crampton, Robert, 'Intelligence Test', *The Times Magazine*, 8 April 1995, p. 27

Cray, Steve, 'Rock-Star Welcome for Top Scientist', *South China Morning Post*, 13 June 2006, City section, p. 1

Cruz, M., E. Martinez-Gonzalez, P. Vielva, J. M. Diego, M. Hobson and N. Turok, 'The CMB Cold Spot: Texture, Cluster or Void?', April 2008, http://inspirebeta.net/record/783713

de Hamel, Christopher, *The Corpus Clock*, Isle of Man: Fromanteel, 2008

DeWitt, Bryce S., 'Quantum Gravity', *Scientific American* 249 (6) (December 1983), p. 114

Donaldson, Gregg J., 'The Man behind the Scientist', *Tapping Technology*, May 1999, http://www.mdtap.org/tt/1999.05/1-art.html

'Dr. Hawking's Bright Idea', *The Economist*, 2 October 2010, pp. 93–4

Dreifus, Claudia, 'Life and the Cosmos, Word by Painstaking Word: A Conversation with Stephen Hawking', *The New York Times*, 9 May 2011, Science Section, p. 1

Durrani, Martin, 'Hawking Slams "Stupid, Worthless" Play', *Physics World*, August 2000, p. 8

Farndale, Nigel, 'A Brief History of the Future', *Sydney Morning Herald*, 7 January 2000

—, 'A Brief History of the Future', *The Hindu Magazine*, 15 January 2000, p. 1

Ferguson, Kitty, 'Devouring the Future: A Profile of Stephen Hawking', *Astronomy Magazine*, December 1998

—, *The Music of Pythagoras*, New York: Walker Publishing, 2008. Published in Great Britain as *Pythagoras: His Lives and the Legacy of a Rational Universe*, London: Icon, 2010

Feynman, Richard, *QED: The Strange Theory of Light and Matter*, Princeton: Princeton University Press, 1985

Flam, Faye, 'Plugging a Cosmic Information Leak', *Science* 259 (1993), p. 1824

Folger, Tim, 'Our Universe is Perfectly Tailored for Life', *Discover Magazine*, December 2008, http://discovermagazine.com/2008/dec/10-sciences-alternative-to-an-intelligent-creator/article_view?searchterm=Andrei%20Linde&b_start:int=1

—, 'Return of the Invisible Man', *Discover Magazine*, July/August 2009, p. 44

—, 'The Ultimate Vanishing Act', *Discover Magazine*, October 1993, p. 100

Freedman, David H., 'Maker of Worlds', *Discover Magazine*, July 1990, p. 49

Garner, Dwight, 'Many Kinds of Universes, and None Require God', *The New York Times*, 7 September 2010

Gefter, Amanda, 'Mr. Hawking's Flexiverse', *New Scientist* 189, no. 2548 (2006), pp. 28–32

Gell-Mann, Murray, lecture

Gibbons, G. W. and S. W. Hawking (eds.), *Euclidean Quantum Gravity*, Singapore: World Scientific Publishing Company, 1993

Gibbons, Gary, 'Euclidean Quantum Gravity: The View from 2002', in G. W. Gibbons, E. P. S. Shellard and S. J. Rankin (eds.), *The Future of Theoretical Physics and Cosmology: Celebrating Stephen Hawking's Contributions to Physics*, Cambridge: Cambridge University Press, 2003 (Stephen Hawking 60th Birthday Workshop and Symposium, January 2002), p. 370

Gourevitch, Philip, 'Interviewing the Universe', *The New York Times Magazine*, 9 August 1992, http://www.errolmorris.com/content/profile/bhot_gourevitch.html

Green, Michael, 'A Brief Description of String Theory', in G. W. Gibbons, E. P. S. Shellard and S. J. Rankin (eds.), *The Future of Theoretical Physics and Cosmology: Celebrating Stephen Hawking's Contributions to Physics*, Cambridge: Cambridge University Press, 2003 (Stephen Hawking 60th Birthday Workshop and Symposium, January 2002), p. 473

Grice, Elizabeth, 'Dad's Important, But We Matter, Too', *Telegraph*, 13 April 2004, http://www.telegraph.co.uk/arts/main.jhtml?xml+/arts/2004/04/ 13.bohawk13.xml

Gross, David, 'String Theory', in G. W. Gibbons, E. P. S. Shellard and S. J. Rankin (eds.), *The Future of Theoretical Physics and Cosmology: Celebrating Stephen Hawking's Contributions to Physics*, Cambridge: Cambridge University Press, 2003 (Stephen Hawking 60th Birthday Workshop and Symposium, January 2002), p. 465

Guth, Alan H. and David I. Kaiser, 'Inflationary Cosmology: Exploring the Universe from the Smallest to the Largest Scales', *Science*, vol. 307, no. 5711 (11 February 2005), pp. 884–90

Hartle, J. B. and S. W. Hawking, 'Path-Integral Derivation of Black Hole Radiance', *Physical Review* D13 (1976), pp. 2188–203

Hartle, James, S. W. Hawking and Thomas Hertog, 'The No-Boundary Measure in the Regime of Eternal Inflation', *Physics Review* D82 (1 January 2010), 063510

—, 'Eternal Inflation without Metaphysics', http://arxiv.org/find/all/1/all:+AND+inflation+AND+hawking+eternal/0/1/0/all/0/1, September 2010

Harwood, Michael, 'The Universe and Dr. Hawking', *The New York Times Magazine*, 23 January 1983

Hawkes, Nigel, 'Hawking's Blockbuster Sets a Timely Record', *Sunday Times*, May 1988

Hawking, Jane, personal interview with author, Cambridge, April 1991

—, *Music to Move the Stars: A Life with Stephen Hawking*, London: Pan Books, 2000

—, *Music to Move the Stars: A Life with Stephen Hawking*, updated edition, 2004

—, *Travelling to Infinity: My Life with Stephen*, London: Alma Books, 2008. This book is a much expanded version of *Music to Move the Stars*, 1999

Hawking, Lucy and Stephen, *George's Cosmic Treasure Hunt*, London: Doubleday, 2007

—, *George's Secret Key to the Universe*, London: Doubleday, 2007

Hawking, S. W./Stephen/Stephen W.

— 'The Arrow of Time in Cosmology', *Physical Review* D32 (1985), p. 2495

— 'Black Hole Explosions?' *Nature*, 248 (1974), pp. 30–1
— *Black Holes and Baby Universes and Other Essays*, London: Bantam Press, 1993
— 'Black Holes in General Relativity', *Communications in Mathematical Physics* 25 (1972), pp. 152–66
— 'A Brief History of *A Brief History*', *Popular Science*, August 1989, p. 70
— (ed., prepared by Gene Stone), *A Brief History of Time: A Reader's Companion*, New York and London: Bantam Books, 1992
— *A Brief History of Time: From the Big Bang to Black Holes*, London and New York: Bantam Books, 1988
— and G. F. R. Ellis, 'The Cosmic Black-Body Radiation and the Existence of Singularities in Our Universe', *Astrophysical Journal* 152 (1968), pp. 25–36
— 'A Debate on Open Inflation', in David O. Caldwell (ed.), *COSMO-98: Second International Workshop on Particle Physics and the Early Universe*, College Park, Md: American Institute of Physics, 1999, p. 21.
— 'The Development of Irregularities in a Single Bubble Inflationary Universe', *Physics Letters* B115 (1982), pp. 295–7
— 'The Edge of Spacetime', in Paul C. W. Davies (ed.), *The New Physics*, Cambridge: Cambridge University Press, 1989, p. 67
— and W. Israel (eds.), *General Relativity*, Cambridge: Cambridge University Press, 1979
— *The Grand Design*, with Leonard Mlodinow, London and New York: Bantam Books, 2010
— 'Gravitational Radiation from Colliding Black Holes', *Physics Review Letters* 26 (1971), pp. 1344–6
— *Hawking on the Big Bang and Black Holes*, Singapore: World Scientific Publishing Company, 1993
— and Roger Penrose, *The Nature of Space and Time*, Princeton and Oxford: Princeton University Press, 1996, 2010
— 'The No-Boundary Proposal and the Arrow of Time', in J. J. Halliwell, J. Perez-Mercader and W. H. Zurek (eds.), *Physical*

Origins of Time Asymmetry, Cambridge: Cambridge University Press, 1992, p. 268
— and N. Turok, 'Open Inflation without False Vacua', *Physics Letters* B425 (1998), pp. 25–32
— Ph.D. thesis, University of Cambridge, March 1966
— and Thomas Hertog, 'Populating the Landscape: A Top Down Approach', February 2006, http://inspirebeta.net/record/710178
— and R. Penrose, 'The Singularities of Gravitational Collapse and Cosmology', *Proceedings of the Royal Society of London* A314 (1970), pp. 529–48
— 'Sixty Years in a Nutshell', in G. W. Gibbons, E. P. S. Shellard and S. J. Rankin (eds.), *The Future of Theoretical Physics and Cosmology: Celebrating Stephen Hawking's Contributions to Physics*, Cambridge: Cambridge University Press, 2003 (Stephen Hawking 60th Birthday Workshop and Symposium, January 2002), p. 106
— and I. G. Moss, 'Supercooled Phase Transitions in the Very Early Universe', *Physics Letters* B110 (1982), p. 35
— and W. Israel (eds.), *300 Years of Gravitation*, Cambridge: Cambridge University Press, 1987
— *The Universe in a Nutshell*, New York and London: Bantam Books, 2001

Lectures and papers

'Gödel and the End of Physics', lecture for Dirac Centennial Celebration, 20 July 2002
'Is the End in Sight for Theoretical Physics', inaugural lecture as Lucasian Professor of Mathematics, Apri1 1980
Lecture at the Smithsonian Institute, Washington, DC, 14 February 2005
'The Origins of the Universe', lecture at Caltech, 4 April 2006
'Out of a Black Hole', lecture at Caltech, 9 April 2008
Paper at the 17th International Conference on General Relativity and Gravitation, Dublin, July 2004
'Remarks by Stephen Hawking', White House Millennium

Council 2000, http://clinton4.nara.gov/Initiative/Millennium/shawking.html
'To Boldly Go', lecture for undergraduates at Caltech, 14 January 2005
'Why We Should Go into Space', lecture at Caltech, 2009, video copyright, Cal Tech Digital Media Services (Information Management Systems and Services)

Unpublished
'Black Holes and Their Children, Baby Universes'
'Is Everything Determined?', 1990
'Is the End in Sight for Theoretical Physics?'
'My Experience with Motor Neurone Disease'
'A Short History'

Personal interviews with author
Cambridge, December 1989; June 1990; November 2010
Conversation with author, spring 1996, October 2000

TV interviews
ABC, *20/20*, broadcast 1989
Interview with Larry King, *Larry King Live Weekend*, Cable News Network, 25 December 1999
'Hawking Bets CERN Mega-Machine Won't Find "God's Particle"', 9 September 2008, http://afp.google.com/article/ALeqM5jaOONGqv-xW-JhBOWgiNCVi6Rsmw
'Hawking Extols Joy of Discovery', *BBC News*, 11 January 2002
'Hawking Gets Personal', *Time*, 27 September 1993, p. 80
'Hawking Humor', *Israel Today*, 28 January 2007, http://www.israeltoday.co.il
'Hawking Misrepresents Pope John Paul II', *Catalyst* 31, no. 6 (2006) http://www.catholicleague.org/catalyst/2006_catalyst/07806.htm#broward
'Hawking Misrepresents Pope John Paul II', Catholic League for Religious and Civil Rights, http://www.catholicleague.org/catalyst.php?year= 2006&month=July-August&read=2078

Highfield, Roger, 'Stephen Hawking to Unveil Strange New Way to Tell the Time', *Telegraph*, 14 September 2008

Hogan, Jenny, 'Hawking Cracks Black Hole Paradox', *New Scientist*, 14 July 2004

Into the Universe with Stephen Hawking, Discovery Channel, 2011 (title in Great Britain is *Stephen Hawking's Universe*, repeating the title of an earlier television special)

Jerome, Richard, Vickie Bane and Terry Smith, 'Of a Mind to Marry: Physicist Stephen Hawking Pops the Most Cosmic Question of All to His Nurse', *People Magazine*, 7 August 1995, p. 45

Jonas, Gerald, 'A Brief History', *The New Yorker*, 18 April 1988, p. 31

Krauss, Lawrence M., Scott Dodelson and Stephan Meyer, 'Primordial Gravitational Waves and Cosmology', *Science*, vol. 328, no. 5981 (21 May 2010), pp. 989–92

Larsen, Kristine, *Stephen Hawking: A Biography*, Amherst, NY: Prometheus Books, 2007

Lemonick, Michael D., 'Hawking Gets Personal', *Time*, 27 September 1993, p. 80

Lerche, W., D. Lust and A. N. Schellekens, 'Chiral Four-Dimensional Heterotic Strings from Selfdual Lattices', *Nuclear Physics* B287 (1987), p. 477

Linde, A. D., 'Eternally Existing Self-Reproducing Chaotic Inflationary Universe', *Physics Letters*, B175 (1986), p. 395

—, 'A New Inflationary Universe Scenario: A Possible Solution of the Horizon, Flatness, Homogeneity, Isotropy, and Primordial Monopole Problems', *Physics Letters* B108 (1982), pp. 389–93

Linde, Andrei, 'Inflationary Theory versus the Ekpyrotic/Cyclic Scenario', in G. W. Gibbons, E. P. S. Shellard and S. J. Rankin (eds.), *The Future of Theoretical Physics and Cosmology: Celebrating Stephen Hawking's Contributions to Physics*, Cambridge: Cambridge University Press, 2003 (Stephen Hawking 60th Birthday Workshop and Symposium, January 2002), pp. 801–2

—, 'The Self-Reproducing Inflationary Universe', *Scientific American*, November 1994, pp. 48–55.

Livio, Mario and Martin J. Rees, 'Anthropic Reasoning', *Science*, vol. 309, no. 5737 (12 August 2005), pp. 1022–3

MacAdam, Harry, 'Search is Vital, Says Hawking', *Sun*, 28 December 2006

Maldacena, Juan Martin, 'The Large N Limit of Superconformal Field Theories and Supergravity', November 1997, http://inspirebeta.net/record/451647

Master of the Universe: Stephen Hawking, BBC broadcast, 1989

Matthews, Robert, 'Stephen Hawking Fears Prejudice against Fundamental Research Threatens the Future of Science in Britain', *CAM: The University of Cambridge Alumni Magazine*, Michaelmas Term, 1995, p. 12

McDaniel, Melissa, *Stephen Hawking: Revolutionary Physicist*, New York: Chelsea House Publications, 1994

McKie, Robin, 'Master of the Universe', *Observer*, 21 October 2001

Murdoch, Dugald, *Niels Bohr's Philosophy of Physics*, Cambridge: Cambridge University Press, 1987

NASA/WMAP Science Team, National Aeronautics and Space Administration, 'First Year Results on the Oldest Light in the Universe', 11 February 2003, http://wmap.gsfc.nasa.gov/news/PressRelease_03-064.html

—, 'Fifth Year Results on the Oldest Light in the Universe', 7 March 2008

—, 'WMAP Produces New Results', 26 January 2010, http://wmap. gsfc.nasa.gov/news/

'No End of Universe Creation Theories', in *+Plus Magazine . . . Living Mathematics*, University of Cambridge Centre for Mathematical Sciences Millennium Maths Project, 23 November 2007, http://web.uvic.ca/ %7Ejtwong/Hartle-Hawking.html

Olesen, Alexa, 'Stephen Hawking: Earth Could Become Like Venus', 22 June 2006, http://www.livescience.com/environment/ap_060622_hawking_climate.html

Overbye, Dennis, 'Cracking the Cosmic Code with a Little Help from Dr. Hawking', *The New York Times*, 11 December 2001

—, 'The Wizard of Space and Time', *Omni*, February 1979, p. 106

Page, D. N. and S. W. Hawking, 'Gamma Rays from Primordial Black Holes', *Astrophysical Journal* 206 (1976), pp. 1–7.

Page, Don N., 'Hawking's Timely Story', *Nature* 332, 21 April 1988, p. 743

—, 'Will Entropy Decrease If the Universe Recollapses', *Physical Review* D32 (1985), pp. 2496–9

Paton Walsh, Nick, 'Alter Our DNA or Robots Will Take Over, Warns Hawking', *Observer*, 2 September 2001

Penrose, Roger, 'The Problem of Spacetime Singularities: Implications for Quantum Gravity?', in G. W. Gibbons, E. P. S. Shellard and S. J. Rankin (eds.), *The Future of Theoretical Physics and Cosmology: Celebrating Stephen Hawking's Contributions to Physics*, Cambridge: Cambridge University Press, 2003 (Stephen Hawking 60th Birthday Workshop and Symposium, January 2002), p. 51

Pippard, Brian, 'The Invincible Ignorance of Science', *The Great Ideas Today, 1990*, Encyclopaedia Britannica, Inc.

'Planck's New View of the Cosmic Theatre', http://www.esa.int/SPECIALS/Planck/SEMK4D3SNIG_0.html

Preuse, Paul, 'Strong Evidence for Flat Universe reported by BOOMERANG Project', *Berkeley Lab Research News*, 26 April 2000, http://www.lbl.gov/ Science-Articles/boomerang-flat.html

Professor Hawking's Universe, BBC broadcast, 1983

Rees, Martin, 'Our Complex Cosmos and Its Future', in G. W. Gibbons, E. P. S. Shellard and S. J. Rankin (eds.), *The Future of Theoretical Physics and Cosmology: Celebrating Stephen Hawking's Contributions to Physics*, Cambridge: Cambridge University Press, 2003 (Stephen Hawking 60th Birthday Workshop and Symposium, January 2002), p. 17

Rothstein, Edward, 'Glass on Columbus, Hip on a Grand Scale', *International Herald Tribune*, 15 October 1992

Salisbury, David, 'Hawking, Linde Spar Over Birth of the Universe', *Stanford Report Online,* 19 April 1998, http://news-service.stanford.edu/news/ 1998/april29/hawking.html

Sample, Ian, 'Large Hadron Collider Warms Up for Final Drive to Catch a Higgs Boson', *Guardian*, 26 February 2011, http://www.guardian.co.uk/science/2011/feb/28/large-hadron-collider-higgs-boson

—, '"There is no heaven or afterlife . . . that is a fairy story for people afraid of the dark"', *Guardian*, 16 May 2011, p. 3, http://www.guardian.co.uk/science/2011/may/15/stephen-hawking-interview-there-is-no-heaven?INTCMP=SRCH

Saner, Emine, 'Lucy Hawking's Fears', *Evening Standard* (London), 14 April 2004, http://www.thisislondon.co.uk/showbiz/article-10226902-lucy-hawkings-fears.do (accessed June 2011)

Schickel, Richard, 'The Thrust of His Thought', *Time*, 31 August 1992, pp. 66, 69

Sciama, Denis W., *The Unity of the Universe*, Garden City, NJ: Doubleday and Company, 1961

Shuhmaher, Natalia and Robert Brandenberger, 'Brane Gas-Driven Bulk Expansion as a Precursor State to Brane Inflation', *Physical Review Letters* 96 (2006), 161301

Sipchen, Bob, 'The Sky No Limit in the Career of Stephen Hawking', *West Australian*, 16 June 1990

Snider, Mike, 'Are computer viruses form of life?' *U.S.A. Today*, 3 August 1964, p. 1

'Space Colonies Needed for Human Survival', *Guardian*, 16 October 2001, p. 3.

'Stephen Hawking to Accept Cosmos Award in Cambridge, England', The Planetary Society, press release, 24 February 2010, http://www.planetary.org/about/press/releases/2010/0224

'Stephen Hawking to Divorce Second Wife', *Mail Online*, last updated 19 October 2006

'Stephen Hawking's Alternate Universe', video at the Smithsonian Institute, Washington, DC, 14 February 2005

Stevens, David, IMDb Mini Biography of Errol Morris, http://www. imdb.com/name/nm0001554/bio

Susskind, Leonard, *The Black Hole War: My Battle with Stephen*

Hawking to Make the World Safe for Quantum Mechanics, New York, Boston and London: Back Bay Books, 2008

—, 'Twenty Years of Debate with Stephen', in G. W. Gibbons, E. P. S. Shellard and S. J. Rankin (eds.), *The Future of Theoretical Physics and Cosmology: Celebrating Stephen Hawking's Contribution to Physics*, Cambridge: Cambridge University Press, 2003 (Stephen Hawking 60th Birthday Workshop and Symposium, January 2002), p. 330

Tasker, Fred, 'Deep Thinkers Abuzz Over Idea of Computer Virus as Life', *Richmond Times–Dispatch*, 10 August 1994, p. 4

Taylor, J. G. and P. C. W. Davies, paper in *Nature* 248 (1974)

'This Week's Finds in Mathematical Physics' (Week 207), 25 July 2003, website 'John Baez's Stuff', math.ucr.edu/home/baez

Thorne, Kip, *Black Holes and Time Warps*, New York: W. W. Norton and Company, 1994

—, 'Warping Spacetime', in G. W. Gibbons, E. P. S. Shellard and S. J. Rankin (eds.), *The Future of Theoretical Physics and Cosmology: Celebrating Stephen Hawking's Contributions to Physics*, Cambridge: Cambridge University Press, 2003 (Stephen Hawking 60th Birthday Workshop and Symposium, January 2002), pp. 74–103

'Trek Stop', *People Magazine*, 28 June 1993, pp. 81–2

Turok, Neil, 'The Ekpyrotic Universe and Its Cyclic Extension', in G. W. Gibbons, E. P. S. Shellard and S. J. Rankin (eds.), *The Future of Theoretical Physics and Cosmology: Celebrating Stephen Hawking's Contributions to Physics*, Cambridge: Cambridge University Press, 2003 (Stephen Hawking 60th Birthday Workshop and Symposium, January 2002), p. 781

'Understanding the Universe: Order of Creation', *The Economist*, 11 September 2010, p. 85

Veash, Nicole Tuesday, 'Ex-Wife's Kiss-and-Tell Paints Hawking as Tyrant', *Indian Express*, Bombay, 3 August 1999

Wade, Mike, 'Peter Higgs Launches Attack against Nobel Rival Stephen Hawking', *Sunday Times*, 11 September 2008

Waldrop, M. Mitchell, 'The Quantum Wave Function of the Universe', *Science*, vol. 242 (2 December 1988), p. 1248

Walton, Ellen, 'Brief History of Hard Times' (interview with Jane Hawking), *Guardian*, 9 August 1989

Watson, Andrew, 'Inflation Confronts an Open Universe', *Science* 279 (1998), p. 1455

Wenham, Michael, 'I'd stake my life that Stephen Hawking is wrong about heaven', *Guardian*, 17 May 2011, www.guardian.co.uk/commentisfree/belief/2011/may/17/stephen-hawking-heaven?intcmp=239

Wheeler, John A., unpublished poem

Whitehouse, David, 'Black Holes Turned Inside Out', *BBC News*, 22 July 2004

Yahoo Searchblog, 1 August 2006, http://www.ysearchblog.com/archives/999336.html

Yulsman, Tom, 'Give Peas a Chance', *Astronomy Magazine*, September 1999, pp. 38–9

Picture Acknowledgements

Every effort has been made to contact copyright holders. Those who have not been acknowledged are invited to get in touch with the publishers.

Photos not credited have kindly been supplied by Stephen Hawking.

Illustration on p. 6: portrait of Stephen Hawking by Oliver Wallington, 2010. Photo Joan Godwin.

Colour inserts

Credits read clockwise, starting in the top left-hand corner.

First section

Pages 2/3: house in Little St Mary's Lane, Cambridge: Kitty Ferguson.

Pages 4/5: house on West Road: Yale Ferguson; Hawking family and sandpit: Homer Sykes/Time Life Pictures/Getty Images; all other photos on this spread: © Ian Berry/Magnum.

Pages 6/7: SH and Chris Hull: David Montgomery/Getty

Images; SH and Don Page: Ian Berry/Magnum Photos; Stephen Hawking fan club: Steve Kagan/Time Life Pictures/Getty Images; SH in Berkeley: James A Sugar/National Geographic Stock; SH with Roger Penrose and Kip Thorne: photograph taken and supplied by Stephen Wall Morris of eventphotographer.ie; SH and Elaine Mason: Miriam Berkeley; SH in his office: © Ian Berry/Magnum; SH with students, Cambridge: © Ian Berry/Magnum.

Page 8: main picture: © Ian Berry/Magnum; inset: AIP Emilio Segre Visual Archives, Wolf Foundation.

Second section

Page 1: SH receives his honorary degree: Michael Manni, courtesy of Christine Manni.

Pages 2/3: SH and his mother: Tim Rooke/Rex Features; SH with animated clipping of characters from *The Simpsons*: AP Photo/Sherwin Crasto; SH's guest appearance on *Star Trek: the Next Generation*: Julie Markes/AP/Press Association Images; SH as a character in Futurama: 20thC Fox/Everett/Rex Features; SH and Stephen Spielberg: Joan Godwin; still from *A Brief History of Time*: Triton/The Kobal Collection.

Pages 4/5: SH and Elaine Mason, wedding day: Justin Williams/PA Archive; SH on Easter Island: Judith Croasdell; SH in zero gravity: Jim Campbell, Aero-News Network; SH at his sixtieth birthday party: Anna Zytkow; SH in Antarctica: Tom Kendall; John Preskill, Kip Thorne and SH: courtesy of the Archives, California Institute of Technology.

Pages 6/7: SH meets Obama: the White House; Cambridge Centre for Mathematical Sciences: Yale Ferguson; SH in his office and view of his outer door: both Frank Zauritz/Laif/Camera Press, London; portrait bust: photo

Kitty Ferguson; SH and the Grasshopper Clock: Jeremy Pembrey/Camera Press; SH with Queen Elizabeth at the Chelsea Flower Show: Matt Dunham/WPA Pool/Getty Images; SH at the White House: Joan Godwin; SH with Nelson Mandela: Denis Farrell/Press Association Images.

Page 8: SH and Yo-Yo Ma at the World Science Festival, New York, 2 June 2010: Wire Images.

Index

INDEX

INDEX

Tycho and Kepler

Kitty Ferguson

The extraordinary, unlikely tale of Tycho Brahe and Johannes Kepler and their enormous contribution to astronomy and understanding of the cosmos is one of the strangest stories in the history of science.

Kepler was a poor, devoutly religious teacher with a genius for mathematics. Brahe was an arrogant, extravagant aristocrat who possessed the finest astronomical instruments and observations of the time, before the telescope. Both espoused theories that seem off-the-wall to modern minds, but their fateful meeting in Prague in 1600 was to change the future of science.

Set in one of the most turbulent and colourful eras in European history, when medieval was giving way to modern, *Tycho and Kepler* is a double biography of these two remarkable men.

'Kitty Ferguson has written a book that has romance . . . it has love, sword fighting, murder, deception, betrayal, trust gone wrong, incredible riches, amazing poverty, reaching for the stars and abject failure, all wrapped in this historical book . . . and it's all one hundred per cent true, the most fascinating read about two incredibly interesting people!'
RICHARD NEWSOME, BOOK CRITIC, *612 ABC BRISBANE RADIO*

'In *Tycho and Kepler*, we are given the sense of science as a quintessentially human activity, conducted not by disembodied spirits squirreled away in ivory towers but by living, breathing, and distinctly idiosyncratic subjects'
LOS ANGELES TIMES (BEST BOOKS OF 2003)

'By putting together Brahe as Smaug the dragon sitting on a fabulous golden hoard with Kepler as Bilbo Baggins who wrests the treasure from him, and expounding the science with conscientious clarity, Kitty Ferguson has written an absorbing non-fiction fable that simultaneously stimulates our imagination and satisfies our scientific curiosity'
THE TIMES

Prisons of Light

Kitty Ferguson

What is a black hole? Could we survive a visit to one? Perhaps even venture inside? What would we find? Have we yet discovered any real black holes?

And what do black holes teach us about what physicist John Archibald Wheeler called "the deep, happy, mysteries of the universe"?

These are just a few of the tantalizing questions examined in this jargon-free review of one of the most fascinating topics in modern science. In search of the answers, we trace a star from its birth to its death throes, take a fabulous hypothetical journey to the border of a black hole and beyond, spend time with some of the world's leading theoretical physicists and observational astronomers scanning the cosmos for evidence of real black holes, and take a whimsical look at some of the wild ideas black holes have inspired.

'An astute blend of entertainment and enlightenment, the sort of book that might have come from George Gamow as part of his series *Mr Tompkins in Wonderland*. Ferguson's grip on her material is firm, her style crisp, lucid and lively'
WERNER ISRAEL, *PHYSICS WORLD*

'Kitty Ferguson marshals her argument well, taking a bite at a time and giving the reader the opportunity to digest before cramming in the next. Her text is carefully thought out, vivid and accurate. I'm glad I read it'
ROGER O'BRIEN, *JOURNAL OF THE BRITISH ASTRONOMICAL SOCIETY*

'The reader will be amply rewarded not only with knowledge, but also with the humor, fantasy, poetry and awe Ferguson brings to the subject'
PUBLISHERS WEEKLY

Measuring the Universe

Kitty Ferguson

Suppose you and I still wondered whether all of the pinpoints of light in the night sky are the same distance from us. Suppose none of our contemporaries could tell us whether the Sun orbits the Earth, or vice versa, or even how large the Earth is. Suppose no one had guessed there are mathematical laws underlying the motions of the heavens. How would – how did – anyone begin to discover these numbers and these relationships without leaving the Earth? What made anyone even think it was possible to find out 'how far', without going there?

In *Measuring the Universe* we join our ancestors and contemporary scientists as they tease this information out of a sky full of stars. Some of the questions have turned out to be loaded, and a great deal besides mathematics and astronomy has gone into answering them. Politics, religion, philosophy and personal ambition: all have played roles in this drama.

There are poignant personal stories, of people like Copernicus, Kepler, Newton, Herschel, and Hubble. Today scientists are attempting to determine the distance to objects near the borders of the observable universe, far beyond anything that can be seen with the naked eye in the night sky, and to measure time back to its origin. The numbers are too enormous to comprehend.

Nevertheless, generations of curious people have figured them out, one resourceful step at a time. Progress has owed as much to raw ingenuity as to technology, and frontier inventiveness is still not out of date.

'It is one of the great stories of science, and Ferguson tells it well'
SUNDAY TELEGRAPH

'Ferguson offers lucid accounts of the reasoning behind important leaps of insight, but it's the little details that delight'
DISCOVER

'Occasionally, an author comes along who dares to describe how science works, who dares to find its underbelly and remind us that the romance and pleasure of cosmic discoveries lies not necessarily in experimental results but in the journey of measurements that led to them. Such an author is Kitty Ferguson, a musician turned science writer, who is distinguished as one who can explain complex things – from the life and times of cosmic objects like black holes to the life and times of cosmic physicists such as Stephen Hawking'
NEIL DEGRASSE TYSON, DIRECTOR OF THE HAYDEN PLANETARIUM AT THE AMERICAN MUSEUM OF NATURAL HISTORY IN NEW YORK CITY

A Brief History of Time
From the Big Bang to Black Holes

Stephen Hawking
Introduction by Carl Sagan

The internationally bestselling phenomenon

'He can explain the complexities of cosmological physics with an engaging combination of clarity and wit . . . His is a brain of extraordinary power'
OBSERVER

'This book marries a child's wonder to a genius's intellect. We journey into Hawking's universe, while marvelling at his mind'
SUNDAY TIMES

Was there a beginning of time? Could time run backwards? Is the universe infinite or does it have boundaries? These are just some of the questions considered in the internationally acclaimed masterpiece by one of the world's greatest thinkers. It begins by reviewing the great theories of the cosmos from Newton to Einstein, before delving into the secrets which still lie at the heart of space and time – from the Big Bang to black holes, via spiral galaxies and string theory. To this day *A Brief History of Time* remains a staple of the scientific canon, and its succinct and clear language continues to introduce millions to the universe and its wonders.

'The most brilliant British scientist of his generation'
NEW STATESMAN